HANDBOOK
OF
DIGITAL IC
APPLICATIONS

HANDBOOK
OF
DIGITAL IC
APPLICATIONS

DAVID L. HEISERMAN

Research and Development Consultant

PRENTICE-HALL, INC., *Englewood Cliffs, New Jersey* 07632

Library of Congress Cataloging in Publication Data

HEISERMAN, DAVID L   1940—
    Handbook of digital IC applications.

    Includes index.
    1. Digital electronics. 2. Digital integrated
circuits. I. Title.
TK7868.D5H37   621.381'73   79–4415
ISBN 0–13–372698–3

Editorial/production supervision and interior
        design by Gary Samartino
Cover design by A Good Thing, Inc.
Manufacturing buyer: Gordon Osbourne

Printed in the United States of America

10 9 8 7 6 5 4 3 2 1

PRENTICE-HALL INTERNATIONAL, INC., *London*
PRENTICE-HALL OF AUSTRALIA PTY. LIMITED, *Sydney*
PRENTICE-HALL OF CANADA, LTD., *Toronto*
PRENTICE-HALL OF INDIA PRIVATE LIMITED, *New Delhi*
PRENTICE-HALL OF JAPAN, INC., *Tokyo*
PRENTICE-HALL OF SOUTHEAST ASIA PTE. LTD., *Singapore*
WHITEHALL BOOKS LIMITED, *Wellington, New Zealand*

This book is dedicated to my dear wife,
Judy,
whose confidence and understanding
made the project possible.

CONTENTS

6 Basic Flip-Flop Circuits *98*

7 Monostable and Astable Multivibrators *132*

PREFACE

It is no longer very difficult to appreciate the importance and power of digital electronics in the industrial and consumer marketplaces. Digital electronics has not always enjoyed this high degree of popularity, however, being largely restricted to applications in computer technology through the 1940s, '50s and the first half of the 1960s.

The advent of modern automated industrial machinery, automated communications systems and a host of revolutionary, consumer-oriented digital devices (namely pocket calculators, digital clocks and wristwatches, and TV games) have given digital technology a place in the sun that can be shared with very few other major disciplines of modern electronics. And there is every reason to suspect that digital electronics will become even more powerful and important in the future.

It is safe to say that at least 80% of the graduates from today's electronics engineering and technical schools will be placed in jobs calling for a knowledge of digital principles, devices and systems. A sound education in digital electronics is thus vital to a sussessful career in electronics, both now and in the forseeable future.

This book presents digital electronics from three distinct points of view: basic principles, available devices and systems that implement the principles and devices. Although the overall field of digital technology is rapidly and continually evolving and becoming more sophisticated, the fundamental principles remain nearly unchanged.

In regard to available digital devices, this book uses the popular TTL and

CMOS technologies as models for implementing the basic principles. To be sure, these devices will one day be replaced with other circuits having more versatile and superior operating characteristics; but even so, a thorough appreciation of how TTL and CMOS devices operate will make any forthcoming transition to another type of device easier and more meaningful.

Thus, this book combines the unchanging elements of digital fundamentals with the TTL and CMOS devices that are currently used for putting the principles to work. There is more to the story, however. The discussions in this book begin with principles, then turn to specific devices and finally to working systems.

The variety and complexity of digital electronic systems range from simple digital toys and novelties to the most sophisticated computing machines and automated industrial controls. The digital systems described in this book are merely representative of what is being done with TTL and CMOS electronics today. A thorough understanding of a few essential digital systems, however, can be rather easily generalized to include just about any sort of system an engineer or technician will encounter now and in the forseeable future.

Some of the finer details presented here might change with time, but the basic principles and the thought processes engendered by a study of the current state-of-the-art will never become out-of-date.

The language of this book is on what is often loosely termed the "technician level." Most of the discussions, however, are design oriented. The presentations are thus suitable for both technical and engineering courses in practical digital electronics. And if it seems that this approach forces the technician into the role of engineer, so much the better. Knowing how to design a custom digital system quite naturally leads to a high degree of competence in regard to repairing and maintaining existing systems.

The first three chapters deal almost exclusively with the "unchanging" principles of elementary logic circuits and Boolean algebra.

Chapter 4 describes the essential characteristics of TTL and CMOS integrated circuits, and Chapter 5 takes up the vital subject of interfacing those circuits with the outside world.

Chapters 6 and 7 introduce sequential logic circuits that lead to an understanding of basic flip-flops, latches and monostable and astable multivibrators.

Digital counters and counter systems are presented in Chapters 8 and 9. And after a discussion of digital encoding and decoding circuits in Chapter 10, the student will find some compelling and relevant applications of counter systems in Chapter 11.

Chapter 12 ties up some loose ends that are left after describing the basic principles, devices and systems in the first 11 chapters.

Shift registers and basic semiconductor memory systems are discussed in Chapters 13 and 14, respectively.

Chapters 15 and 16 deal with the basic principles and IC devices required for digital numeric and arithmetic operations. Most arithmetic work is now being carried out on a large-scale, making small- and medium-scale TTL and CMOS

devices appear rather awkward by comparison. But in keeping with the general tone of this book, the principles presented in these two chapters lead the student to a greater appreciation of how the larger-scale digital devices work.

DAVID L. HEISERMAN

THE ROLE
OF INTEGRATED CIRCUITS
IN DIGITAL ELECTRONICS

Any digital circuit described in this book can be built from vacuum tubes, relays and incandescent lamps. Many of them were, in fact, built that way during the opening decade of digital electronics—through the 1940's.

Similarly, any circuit in this book can be built from ordinary transistors and neon lamps. Again, this is the way it had to be done through the 1950's and a better share of the 1960's.

Basic digital principles haven't changed much since those first days, but a marked improvement in digital electronic technology is obvious to just about anyone who has lived through most of those years. Only the devices for implementing the basic principles have changed.

Changes in device technology, culminating in modern IC (integrated circuit) technology, have had a powerful influence on system performance, reliability, efficiency and cost factors. Each generation of electronic devices brought with it a marked improvement in all of these vital factors.

One particular vacuum-tube logic module, for instance, sold for about $20 in the mid-1950's. A transistorized version of that same operation, with improved performance, efficiency and reliability, sold for about $5 through the latter part of the 1950's and mid-1960's.

Nowadays an IC does that particular logic operation, but with yet a further quantum jump in performance, efficiency and reliability, for less than 10¢. These cost figures represent a staggering 2000:1 drop in cost per function in less than 30 years!

The real power of modern digital electronics stems from improvements in performance and cost factors brought about by changes in device technology. Engineers in the vacuum-tube days certainly knew how to go about building an electronic digital clock or 12-function calculator for the consumer marketplace. But how many people would be willing to set up the huge, power-gobbling system in their homes, especially since the price tag would carry a figure in excess of $1000?

All of the digital circuits and systems described in this book are built around modern IC technology. This particular technology itself has undergone several generations of change which have brought about corresponding improvements in efficiency and reliability. Some of these IC devices haven't become obsolete, and we are presently going through an era in which several different IC technologies exist at the same time.

It is thus important to understand the essential characteristics of the different types of IC devices and to see how some of them work better than others under a given set of circumstances. Such a study of modern digital ICs begins with a look at how the devices are subdivided and classified into distinct families.

1-1 Organization by Gate Density

Writers of popular scientific literature during the days of vacuum-tube computers estimated that a system having the logic capability of the human brain would occupy a building as large as the Empire State Building and would consume all the electrical energy supplied by the Niagra Falls power complex.

Then through the discrete-transistor days that picture was reduced to the size of a ten-story office building that would require the energy capacity of a medium-sized city.

The coming of IC technology reduced that imaginary computer system to the size of a house; now it is down to the size of a typical room within the house.

Whether or not a computer can actually mimic the full capacity of the human brain isn't the point here. The really valid point is that logic systems have become more densely packed. What once required the space of the Empire State Building now fills just one room in a house. The space occupied by a single logic operation, like the cost, has dropped on the order of 2000:1.

A single logic operation is carried out in a circuit that can be loosely called a *gate*. At one time, there was a 1:1 correspondence between the number of vaccum tubes and gates in a digital system. That 1:1 correspondence is still valid today, but the size of the device for implementing it is on a microscopic level.

Modern IC devices can be classified according to the number of discrete gate functions within a standard-sized package. The general classifications are small-scale, medium-scale and large-scale integration, known more commonly as SSI, MSI and LSI devices.

It is possible to cite actual numbers of gate operations per package, but not many people are anxious to count the thousands of gate operations within a typical

LSI device. Since the whole notion of classifying IC devices by gate density is one of estimating figures, there is a simpler alternative to citing actual gate counts.

An SSI device can be considered one that contains one or more discrete logic gates, each of which is fully accessible from the outside world. A dual SSI package, for example, contains two independent logic gates. The user has the option of using one or both of them.

A quad SSI package, then, is one containing four separate logic gates, each of which can be used independently of the other three.

A few relatively simple logic functions call for connecting together more than one basic gate operation. The connections are made within the IC device itself, giving the user the impression the device is working out a single operation. As long as the device contains relatively few of these second-order logic functions, it can be properly classified as an SSI device.

SSI devices thus rarely contain more than about 6 independent gate functions or more than the equivalent of about 25 transistor-like elements.

Any logic operation, no matter how complex it might be, can be worked out with basic logic gates. There is a long list of relatively complex digital operations that call for interconnecting as many as a hundred basic logic gates. Instead of building up these common circuit functions with dozens of SSI devices, the IC industry has developed a number of MSI devices.

MSI devices contain a relatively large number of discrete logic gates that are internally connected in the more popular patterns. In the context of this analysis of packing density, an MSI device is one that is made up of a number of SSI functions.

MSI devices can contain as many as 8 or 12 separate, higher-order logic functions. The user has access to each one of these functions, but not to the SSI-level connections within the chip.

Thus, an SSI device is one made up of relatively few basic gate operations, while an MSI device is one made up of some internally connected SSI devices. Both are available in the same-sized IC package.

An LSI device is one made up of a number of interconnected MSI devices. LSI devices rarely have the great versatility that characterizes their smaller SSI and MSI counterparts; but the success of modern digital electronics rests heavily on the refinement of LSI technology.

Typical examples of LSI devices are the main IC chips in digital clocks and calculators. An LSI calculator chip must be used in a calculator-type function but in nothing else. MSI and SSI devices, on the other hand, can be used for building up an endless variety of gadgets and systems, including calculators. In other words, LSI devices tend to have dedicated applications, while SSI and MSI devices are more universal in their application.

The value of LSI devices is unquestionable whenever there is such a device available for doing a particular job at hand. But when there are no LSI devices available for a certain application, MSI and SSI devices can be made to do the job.

The circuits described in this book are limited to those built around SSI and MSI devices. The principles involved in analyzing and designing such circuits apply equally well when LSI devices are available.

1-2 Organization by Fabrication Technology

The question of whether a given IC device is SSI, MSI or LSI is often a purely academic one. It is quite possible to work for a lifetime in the digital business without attempting to classify an IC as SSI, MSI or LSI. Digital engineers and technicians, however, must be ever-conscious of whether they are working with RTL, DTL, TTL, CMOS or MOS devices, which are classifications of fabrication techniques.

The procedures for manufacturing an IC determine critical parameters such as operating voltage levels, input and output impedances, drive capabilities and operating speed. Although the members of one particular fabrication family are fully compatible with one another, it is generally difficult to match members of different families.

The RTL (resistor-transistor logic) and DTL (diode-transistor logic) families were the first to realize any great popularity in the digital IC business. These devices used IC equivalents of resistors, diodes and transistors to perform basic gate operations. These two families of fabrication technology are generally considered obsolete today, however, and they are not discussed in particular in this book. (Bear in mind, though, that the basic principles of digital electronics apply through any family of IC technology).

TTL (transistor-transistor logic) is far and away the most popular IC family in the digital business today. TTL circuits, often designated T^2L, are much faster than any other type of IC device, and they have better efficiency figures than their RTL and DTL predecessors.

TTL devices are fabricated around the same general technology that characterizes *NPN* and *PNP* bipolar transistors. It is a strictly silicon technology, as opposed to CMOS and MOS devices which are built around an insulated-gate, field-effect technology.

CMOS devices are intended to replace TTL ICs as far as efficiency is concerned. CMOS packages offer a wider range of operating voltage levels and higher input impedances. CMOS (complementary metal-oxide-silicon) ICs aren't as fast as their TTL counterparts, however; and there is the ever-present problem of making certain that static voltages cannot "blow" the sensitive gate connections.

TTL and CMOS ICs dominate the digital scene as far as SSI and MSI logic devices are concerned. At the present time, TTL circuits are available in a wider selection of digital functions, and the situation will most likely remain that way for a good many years to come. CMOS devices become an invaluable alternative to TTL devices whenever low power consumption and a wider range of supply voltages are prime design considerations.

The IC devices described in this book are thus members of the SSI and MSI and the TTL and CMOS families. Treating these families in a basic textbook makes it is possible for the student to get a good grasp of the entire field of digital electronics.

It turns out that the TTL and CMOS fabrication techniques are wholly unsuitable for LSI systems. It is simply inpossible to cram a sufficient quantity of TTL or CMOS functions into the space required for practical LSI designs. MOS technology has to come to the rescue in this case.

Thus MOS fabrication techniques and LSI devices go hand in hand today. The two expresses are often linked together as MOS/LSI, thus implying that such a device does a big digital job with an insulated-gate, field-effect technique.

The relatively high cost of manufacturing MOS ICs, incidentally, makes it impractical to produce MOS devices in the SSI and MSI categories.

Digital integrated circuits can be classified by relative packaging density (SSI, MSI or LSI) and by the more popular fabrication techniques (TTL, COMS and MOS). Virtually all SSI and MSI circuits on the market and in use today belong to either the TTL or CMOS families, with TTL being the more popular. Any LSI circuit, however, is bound to be a MOS circuit.

The diagram in Fig. 1-1 summarizes the general relationships between the classifications by device packaging density and fabrication technology.

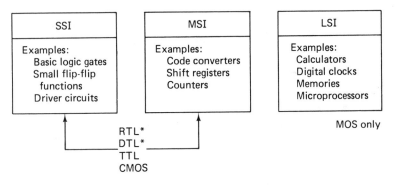

Figure 1-1 General organization of SSI, MSI and LSI digital devices, showing fabrication techniques and some examples in each category.

1-3 The TTL Family

The TTL family (bipolar transistor-transistor logic fabrication) can be subdivided into several distinct groups. Although these groups are all built around a bipolar silicon technology, the different family members have characteristics that make it necessary to exercise some care when attempting to cross "family lines." Figure 1-2 summarizes the names and general characteristics of each family.

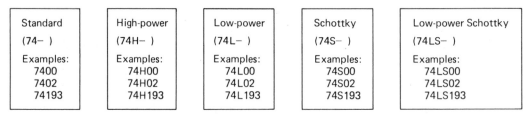

| Standard

(74–)

Examples:
7400
7402
74193 | High-power

(74H–)

Examples:
74H00
74H02
74H193 | Low-power

(74L–)

Examples:
74L00
74L02
74L193 | Schottky

(74S–)

Examples:
74S00
74S02
74S193 | Low-power Schottky

(74LS–)

Examples:
74LS00
74LS02
74LS193 |

Figure 1-2 Summary of the five TTL IC families. Note: IC devices having a 54- prefix indicate TTL devices with better temperature specifications, i.e., −55°C to +125°C. 74- series TTLs are rated at 0°C to 75°C.

The *standard TTL* family is far and away the most popular and widely available subdivision of the TTL logic families. These devices normally carry a 74- prefix, followed by two or three additional digits.

Standard TTL IC devices are mainly characterized by switching or propagation delay times on the order of 10 ns to 12 ns and an average power consumption of about 10 mW per gate. Output drive capability for all but the standard TTL drivers is about 16 mA sink and 400 μA source current.

Because the standard TTL family is the leader in the TTL logic business today, all TTL circuits described in this book belong to that particular family.

The *high-power TTL* is noted for somewhat higher current output drive capacity, but switching times about one-half that of their standard TTL counterparts. Typical switching times in this family are between 5 ns and 10 ns. The trade-off is in power efficiency. Average power consumption in this case is on the order of 22 mW per gate.

High-power TTLs are designated with a 74H- prefix, followed by two or three more digits.

Low-power TTLs trade off switching speed for greater power supply efficiency. Designated with a 74L- prefix, the lowpower TTL devices have typical propagation delay times on the order of 31 ns to 60 ns, but an average power consumption of only 1 mW per gate.

As noted in the fan-out table, Table 1-1, a low-power TTL IC can drive only two standard TTL and only one high-power TTL load. This is a reflection of the fact that the maximum sink and source currents for the 74L- devices are 2 mA and 200 μA, respectively.

Fast-acting Schottky diodes placed across many of the bipolar elements of TTL devices prevent total saturation of these logic elements, thereby giving *Schottky-clamped TTLs* their characteristic ultrahigh switching speeds.

Switching speeds in this instance are between 3 ns and 5 ns, but only at the cost of power supply efficiency. Most of these 74S- devices consume an average of 22 mW per gate. The fan-out table (Table 1-1) shows that Schottky TTL ICs have greater output drive currents than any other member of the TTL generation—sinking up to 20 mA and sourcing as much as 1 mA.

The switching speed of *low-power Schottky TTL* devices is almost identical to that of standard TTL devices, but the average power consumption is much lower. At 2 mW per gate, 74LS- devices represent an "improved" standard TTL line—equal switching or propagation delay times with less power supply drain. Low-power Schottky TTLs cannot drive as many other TTL loads as the standard TTL elements can, however.

All members of the TTL family operate from a supply voltage between $+4.5$V and an absolute maximum of $+7$ V. Typically, the supply voltage $(+V_{cc})$ is specified as 5.5 V. The TTL devices are also wholly compatible with respect to their definitions of high and low logic levels. So the question must arise: Why is it necessary to exercise caution when attempting to mingle different TTL families in the same circuit? The fan-out table, Table 1-1, holds the answer.

Table 1-1   SUMMARY OF TTL FAN-OUT CAPABILITIES

Source (Driving) TTL Device	Load (Driven) TTL Device				
	Standard TTL (74-)	High-Power TTL (74H-)	Low-Power TTL (74L-)	Schottky TTL (74S-)	Low-Power TTL (74LS-)
Standard TTL (74-)	10	8	40	8	20
High-power TTL (74H-)	12	10	50	10	25
Low-power TTL (74L-)	2	1	20	1	10
Schottky TTL (74S-)	12	10	100	10	50
Low-power Schottky TTL (74LS-)	5	4	40	4	20

As far as electrical compatibility is concerned, the main difference between one TTL family and another is in the number of gates each can drive properly. The five major families are listed down the left-hand side and along the top of Table 1-1. The devices doing the driving operation are listed in the vertical column; the devices being driven by each are listed along the top.

A standard 7400 TTL device, for example, is capable of driving ten separate gates of its own family but only eight from the high-power family. The worst case is in the 74L00 family that can drive only two standard TTLs and only one gate from the 74H00 family.

Without taking proper care in selecting and using different members of the TTL family, the designer runs the risk of badly overloading some of the lower-power, high-speed devices.

Then, too, why would one want to drive a higher-power TTL device from a lower-power one anyway? The speed gained by using a low-power TTL gate would be lost in the higher-power stage, and the overall result would be a circuit that switches no faster than the slowest gate in it.

1-4   The CMOS Family

The CMOS devices described in this book have the distinct advantage of having a much wider supply voltage range than any TTL device does—between $+3$ V and $+15$ V. This makes it possible to design digital circuits that are directly compatible with the popular 12-V supply found in the automotive industry and 9-V batteries used in many kinds of consumer electronic equipment.

The engineer has to pay for the wider operating supply voltage range (and lower power consumption as well) with slower switching speeds and poorer current drive capacities. While the power consumption might indeed stand at a very attractive average of 10 ns per gate, typical switching speeds are between 30 ns and 90 ns.

CMOS ICs are thus an attractive option when designing relatively low-speed circuits that are to operate from standard battery power sources. of 6 V, 9 V and 12 V. Where higher speed is essential, however, the choice must come from the TTL families.

It is difficult to subdivide the CMOS family into different families simply because there is little electrical difference between any of the CMOS ICs. The only difference of any note concerns whether or not the inputs and outputs are internally buffered. Buffered CMOS devices have superior input and output driving characteristics to their nonbuffered counterparts.

1-5   Mingling the TTL and CMOS Families

Combining CMOS and TTL IC devices in a single circuit brings up some special interfacing problems that are discussed in greater detail in Sec. 5-3. It is thus sufficient at this point to say that the two families are generally incompatible.

For one thing, the low-power feature of CMOS ICs makes it impossible to drive more than two low-power TTL devices at one time. A CMOS device cannot begin to supply the input current demanded by standard TTLs.

Then there is a matter of logic-level mismatching. What is considered standard output logic-high and logic-low levels for CMOS devices is not compatible with the standards for input high and low levels for TTL. Each interface between a CMOS output and a TTL input calls for some special circuitry, even if the CMOS device is driving a low-power TTL.

Digital designers thus avoid using both CMOS and TTL ICs in the same system whenever possible.

Exercises

1. Obtain a manufacturer's catalog for digital IC devices and compare the variety of devices available in all five of the TTL families and the CMOS family. Draw your own conclusions on which is most useful in the long run.

2. Use this text or other sources to define the following terms: (a) propagation delay; (b) fan out; (c) gate; (d) SSI; (e) MSI; (f) LSI.

3. State the meaning of each of the following IC prefixes, and compare their relative switching speeds, power consumption, fan out and current drive capabilities: (a) 74-; (b) 74H-; (c) 74L-; (d) 74S-; (e) 74LS-; (f) 40-.

THE COMBINATORIAL LOGIC FAMILY

The combinatorial logic family consists of five basic combinatorial logic functions: AND, OR, INVERT, NAND and NOR gates. These functions are the very backbone of modern digital electronics. They are not only combined in different ways to perform an infinite variety of logic operations, but they are also combined to produce flip-flop effects responsible for all sequential logic and memory functions. Any kind of digital circuit—anything from a simple coin-toss game to the most sophisticated computer system—owes its operation to these basic combinatorial logic gates.

This chapter deals with each of the five basic logic gates separately, introducing their Boolean logic equations, schematic symbols and truth tables and describing their essential operating characteristics. The discussions in this chapter also show how each of the basic logic operations can be implemented with discrete-component circuits as well as TTL and CMOS integrated circuits.

2-1 AND Functions

Figure 2-1 summarizes the essential qualities of a basic AND gate function. The logic equation, schematic symbol and truth table in Fig. 2-1(a) can represent the operation of any of the switch, transistor, TTL and CMOS circuits.

2-1.1 An AND Switch Circuit

The switch-equivalent circuit in Fig. 2-1(b) clearly shows that both switches must be depressed or energized at the same time in order produce an energized output—the LED cannot possibly light up if either or both of the input switches A and B are open. If we define an open switch and a turned-off LED as logic-0 states, and if we define a closed switch and lighted LED as logic-1 states, the switch circuit in Fig. 2-1(b) generates the basic AND truth table in Fig. 2-1(a). *An AND function shows a logic-1 output only when all inputs are at logic 1 at the same time.*

2-1.2 AND Logic Equation, Symbol and Truth Table

Figure 2-1(a) shows the basic logic equation, symbol and truth table for 2-input logic gates. The logic operator in an AND logic equation is a dot. This dot is used in a fashion similar to a sign of multiplication in ordinary algebra. And just as the sign of multiplication in ordinary algebra can be omitted without causing any serious confusion, so can the AND operator in Boolean algebra. The Boolean AND logic equation $C = A \cdot B$ (read "C equals A AND B") can thus be written $C = AB$.

In principle at least, an AND gate can have any number of inputs. A 3-input AND gate, for example, can be represented as a switch circuit having three normally open switches in series. The logic symbol in this instance would have three input terminals and one output connection, and the corresponding logic equation could be written $D = ABC$. With three inputs, however, the AND truth table would have to be expanded to show eight different combinations of input 1's and 0's, but it would still show a logic-1 output only where all three inputs are at logic 1 at the same time.

2-1.3 A Switch-and-Transistor AND Circuit

Figure 2-1(c) is a switch-and-transistor version of a 2-input AND circuit. The transistor is normally biased *on* by means of emitter-base current flowing between COMM and $+V_{cc}$ through $R1$; and as long as this transistor is switched *on*, it prevents the LED at output C from lighting. (The emitter-collector saturation voltage of most bipolar transistors is lower than the junction potential of a forward-biased LED).

The only way to turn *off* the transistor in Fig. 2-1(c) is by closing both switches A and B at the same time. Closing both of these switches effectively ties the transistor's base to its emitter, robbing it of its forward-biasing current and switching it *off*. Turning *off* the transistor then allows the LED to conduct (and light up) through resistor R_2.

The LED in Fig. 2-1(c) thus lights up only when both switches are closed at the same time. This is indeed a 2-input AND circuit.

2-1.4 TTL AND Circuits

Figure 2-1(d) shows a pair of TTL circuits for a 2-input AND gate. Both are bipolar equivalents of 2-input AND IC packages on the market today, but the totem-pole version is by far the more popular of the two. The circuits are identical except for the totem-pole output arrangement of D_4, Q_6 and the 130-Ω resistor; and the totem-pole version is more popular because it has a much faster output switching time and does not require an external pull-up resistor.

One of the hallmarks of TTL ICs is the multiple-emitter transistor, Q_1, at the circuit's input. Connecting either input A or B (or both) to COMM forward biases transistor Q_1, causing it to conduct; and the conduction of Q_1 effectively pulls the base connection of Q_2 to COMM. Turning off Q_2 in this fashion interrupts the emitter-base current for Q_3, turning that transistor off as well.

With Q_2 and Q_3 switched off, transistor Q_4 is then forward biased by means of emitter-base current between COMM and $+V_{cc}$ (through the 1-k resistor, D_3 and the 2-k resistor). Turning on Q_4 pulls the base of totem-pole transistor Q_6 near COMM potential to switch it off, but at the same time, turning on Q_4 switches on Q_5, pulling its collector voltage near COMM.

Putting this all together, setting input A, B or both near the 0-V COMM potential ultimately saturates output transistor Q_5, making the circuit show a 0 logic output. Any logic-0 input thus causes a logic-0 output; and that

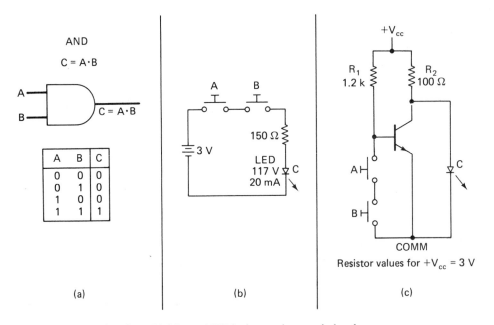

(a)   (b)   (c)

Figure 2-1 AND functions. (a) 2-input AND logic equation, symbol and truth table. (b) A switch-equivalent circuit. (c) A switch-and-transistor AND circuits.

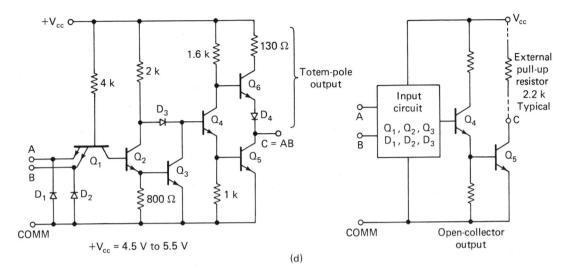

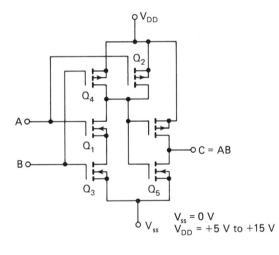

Figure 2.1 (*cont.*) (d) Totem-pole and open-collector TTL and circuits.
(e) A CMOS AND gate circuit.

pattern follows three of the four critical input/output characteristics of a 2-input AND gate.

All that remains is an analysis of the circuit when both inputs are tied near $+V_{cc}$ at the same time. Whenever inputs A and B are both connected near $+V_{cc}$ potential (or not connected to anything at all), transistor Q_1 is switched off because it has no emitter-base current. This transistor is connected as a common-base amplifier, so when it is switched off, forward-biasing current for Q_2 flows through the base-collector junction of Q_1. Q_2 is thus switched on, and that transistor, in turn, switches on Q_3. With Q_2 and Q_3 both saturated, they pull the base connection of Q_4 near COMM potential to turn it off; and turning off Q_4 allows Q_6 to conduct (forward-biasing current through the 1.6-k resistor), but switches off Q_5. Now Q_5 is turned off, and Q_6 is saturated. The overall effect is that the collector voltage of Q_5 rises to a positive potential only slightly less than $+V_{cc}$.

The totem-pole circuit in Fig. 2-1(d) thus shows a logic-1 output only when both inputs are tied near $+V_{cc}$ (or not connected anywhere) at the same time.

Summarizing the action of the totem-pole circuit in Fig. 2-1(d): The output of the circuit is near 0 V whenever either or both of the inputs are connected near 0 V, and the output rises toward $+V_{cc}$ only when both inputs are connected near $+V_{cc}$ (or uncommitted) at the same time. *If we define 0 V as logic 0 and $+V_{cc}$ as logic 1,* this circuit completely satisfies the basic truth table for a 2-input AND gate.

The open-collector version of the 2-input TTL AND gate in Fig. 2-1(d) works in a similar fashion, but without the benefit of an internal totem-pole circuit to pull up the collector voltage of Q_5 when the input logic states call for a logic-1 output. The circuit designer must complete the collector circuit of Q_5 by providing an external pull-up resistor to $+V_{cc}$.

The IC manufacturer can expand the basic TTL AND gates for any number of inputs by simply including more emitters at Q_1. A 3-input AND gate, for example, would look and work exactly like the circuits in Fig. 2-1(d), but it would have three emitter connections at Q_1.

2-1.5 CMOS AND Circuits

Figure 2-1(e) shows the internal structure of a 2-input AND CMOS IC. To see how this circuit works, suppose input B is connected near V_{ss} potential. For all practical purposes, the gate and source connections of Q_3 are thus connected together; and since Q_3 is an enhancement-mode MOSFET, it follows that it will be turned off, acting as an open circuit in the series combination of Q_1 and Q_3.

While the low-input potential at B is turning off Q_3, however, it is switching on the *P*-channel MOSFET, Q_4. Q_4 thus acts as a closed switch in the parallel-circuit combination of Q_4 and Q_2, thus pulling the gates of output transistors Q_6 and Q_5 near the V_{DD} potential. Now the V_{DD} potential on these two gates switches off the *P*-channel FET, Q_6, and turns on the *N*-channel FET, Q_5. The output of the circuit at C is thus very close to the V_{cc} potential.

Putting these steps together, it follows that applying a V_{ss} potential, or logic-low level, to input B guarantees a logic-low level at the output. A similar analysis shows that the C output will be near the V_{ss} potential when A is low or A and B are both pulled down to V_{ss}. In other words, the output of the circuit in Fig. 2-1(e) will be at logic 0 when either or both of the inputs are at logic 0.

The analysis of this circuit will be complete when it is shown that the output is pulled up to logic 1 (near the V_{DD} potential) only when both inputs are pulled up to V_{DD} at the same time.

Connecting inputs A and B to V_{DD} turns on transistors Q_1 and Q_3, but it switches off both Q_4 and Q_2. The gates of Q_6 and Q_5 are thus connected to V_{ss}, turning off Q_5 and turning on Q_6. With Q_6 switched off and Q_5 turned on, output C is essentially connected to V_{DD}—a logic-1 level. And the analysis of this 2-input AND circuit is complete. It follows the basic AND truth table perfectly.

2-2   OR Functions

Figure 2-2 illustrates the essential characteristics and some methods for implementing the basic OR logic operation. The logic equation, schematic symbol and truth table in Fig. 2-2(a) apply to any of the switch, transistor, TTL and CMOS circuits; and the basic notions can be applied to OR circuits having any number of inputs.

2-2.1   An OR Switch Circuit

The simple switch circuit in Fig. 2-2(b) performs the basic 2-input OR function. As long as both switches are open, the LED cannot possibly light up; but closing either one or both of the input switches forward biases the LED and turns it on.

If we define a closed switch contact and a lighted LED as a logic-1 state, and an open switch or turned-off LED as logic 0, the switch circuit generates the basic 2-input OR truth table in Fig. 2-2(a). An OR gate shows a logic-0 output only when all inputs are at logic 0 at the same time; or to state it another way: *An OR function shows a logic-1 output whenever one or more inputs are at logic 1.*

2-2.2   OR Logic Equation, Symbol and Truth Table

Figure 2-2(a) shows the basic logic equation, schematic symbol and truth table for a 2-input OR gate. The logic operator in this instance is a plus sign $(+)$, a symbol used in a manner similar to the sign of addition in ordinary algebra. The Boolean OR equation, however, is read, "C equals A OR B."

It is possible to expand the number of inputs to produce an OR gate having any number of inputs. A 4-input OR gate, for instance, would have a Boolean logic expression following the general pattern, $E = A + B + C + D$, and the corresponding switch-equivalent circuit would show four normally open switches connected in parallel. In any event, the circuit's output goes to a logic-1 state whenever one or more (including all) inputs are set to logic 1.

2-2.3   A Switch-and-Transistor OR Circuit

Figure 2-2(c) is a switch-and-transistor version of the basic 2-input OR gate. As long as both input switches are open, the transistor's emitter-base circuit is forward biased by the connection to $+V_{cc}$ through R_1; and as long as this transistor is switched on in this fashion, the LED cannot light up.

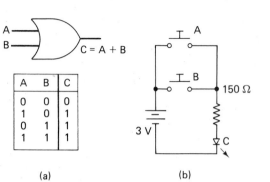

OR
C = A + B

A ─
B ─ C = A + B

A	B	C
0	0	0
1	0	1
0	1	1
1	1	1

(a)

A

B

150 Ω

3 V

C

(b)

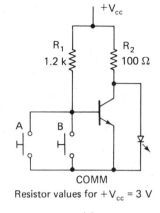

+V$_{cc}$

R$_1$
1.2 k

R$_2$
100 Ω

A

B

COMM

Resistor values for +V$_{cc}$ = 3 V

(c)

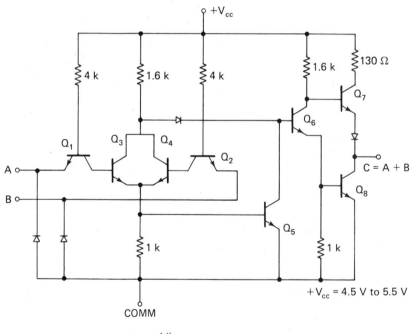

+V$_{cc}$

4 k   1.6 k   4 k   1.6 k   130 Ω

Q$_7$

Q$_6$

Q$_1$ Q$_3$ Q$_4$

Q$_2$

A

B

C = A + B

Q$_8$

Q$_5$

1 k   1 k

+V$_{cc}$ = 4.5 V to 5.5 V

COMM

(d)

Figure 2-2 OR functions. (a) 2-input OR logic equation, symbol and truth table. (b) A switch-equivalent circuit. (c) A switch-and-transistor OR circuit. (d) A TTL OR circuit with totem-pole output.

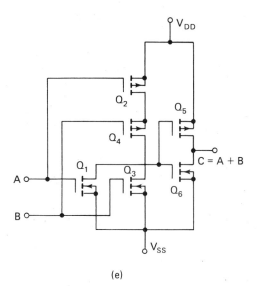

(e)

Figure 2-2 *(cont.)* (e) A CMOS OR gate circuit.

Closing either or both of the input switches grounds the base of the transistor to turn it off. The LED is then free to light, with its forward-biasing current coming from its connection to $+V_{cc}$ through R_2.

The LED in Fig. 2-2(c) thus lights whenever either one or both of the input switches are energized; and this particular set of circumstances follows the basic OR truth table perfectly.

2-2.4 TTL OR Circuits

The circuit in Fig. 2-2(d) represents the internal structure of a TTL 2-input OR gate. The circuit has a totem-pole output configuration made up of D_4, Q_7 and the 130-Ω resistor. Output C is thus set at logic 1 whenever Q_8 is turned off and the totem-pole structure is gated on. On the other hand, the output is set to logic 0 whenever Q_8 is switched on and Q_7 is turned off. This output circuit, in fact, is identical to the output circuit for the totem-pole AND gate in Fig. 2-1(d). The essential differences between a TTL AND and TTL OR circuit must then appear in the preceding circuitry.

To see how the circuit in Fig. 2-2(d) works, suppose input A is tied near the $+V_{cc}$ potential. This condition effectively turns off transistor Q_1 by fixing the emitter and base connections at about the same potential. Turning off Q_1 in this manner allows Q_3 to conduct by virtue of the collector-base current through Q_1 and the 4-k resistor to $+V_{cc}$.

Now the conduction of Q_3 switches on Q_5 which, in turn, pulls the base of Q_6 to COMM potential to turn off that transistor. And with Q_6 switched off, Q_8 cannot possibly conduct because its forward-biasing path to $+V_{cc}$ is interrupted;

but, on the other hand, Q_7 is allowed to conduct because its base connection is raised above ground potential toward $+V_{cc}$ through the 1.6-k resistor.

To summarize this analysis, connecting input A to $+V_{cc}$ (setting the input to logic 1) ultimately turns on Q_7 and switches off Q_8 to yield a logic-1 output. A similar kind of analysis shows the same sort of effect whenever input B is connected to $+V_{cc}$. Output C equals logic 1 whenever either or both of the inputs are set to logic 1. All that remains, then, is to show that the output goes to 0 whenever both inputs are connected to COMM (logic 0) at the same time.

Connecting inputs A and B to COMM at the same time turns on both Q_1 and Q_2. Turning on these two transistors pulls the bases of Q_3 and Q_4 to COMM, thus turning them off; and with these two transistors turned off, Q_5 no longer has a complete path for forward-biasing current; therefore, it also switches off.

Now Q_5 is turned off, allowing the base of Q_6 to be pulled up toward $+V_{cc}$ through D_3 and the 1.6-k resistor at the collectors of Q_3 and Q_4. Q_6 must conduct under these circumstances, and the conduction of that transistor switches off Q_7 and biases on Q_8. The output is thus very close to COMM potential, or logic 0.

This step completes the analysis of this TTL 2-input OR circuit, showing that the output is at logic 0 whenever both inputs are at logic 0 at the same time.

While this OR circuit includes a totem-pole output structure to speed up the transition from logic 0 to logic 1, the very nature of the OR operation precludes the use of a multiple-emitter input transistor. This OR circuit is thus somewhat slower and has a more awkward physical structure than its AND-gate counterpart. Wherever operating speed is critical, then, circuit designers tend to avoid the use of TTL OR gates.

2-2.5 CMOS OR Circuits

Figure 2-2(e) represents the internal structure of a 2-input CMOS OR gate. To see how this circuit works, suppose input A is tied to V_{DD}. A V_{DD} potential on the gate of Q_1 switches it on; but that same potential applied to the gate of the P-channel FET, Q_2, turns that device off. Now Q_1 is part of a parallel circuit made up of Q_1 and Q_3, and Q_2 is part of a series circuit that includes Q_4. Turning on Q_1 thus pulls the gate connections of Q_5 and Q_6 down to V_{ss} potential, and turning off Q_2 disconnects the gates of those output transistors from V_{DD}. And with a V_{SS} potential appearing on the gates of Q_6 and Q_5, Q_6 is switched off and Q_5 is switched on. The output potential at C, then, is very close to V_{DD}.

Connecting input A to V_{DD} thus causes output C to show the V_{DD} potential; and the same sort of analysis applies to input B. In other words, setting either input A or B, or both, to V_{DD} (logic 1) causes the output to show a logic-1 state.

Connecting both inputs to V_{SS} should make the output of an OR gate go to logic 0, or in this case to V_{SS}. And, indeed, setting A and B both to V_{SS} turns off Q_1 and Q_3 and switches on Q_2 and Q_4. The V_{DD} potential then appears on the gates of Q_5 and Q_6, and they respond with Q_5 turning off and Q_6 turning on. The overall result is that output C is pulled down to V_{SS}—a logic-0 output.

Summarizing the operation of this CMOS circuit: Setting either or both of the inputs to logic 1 produces a logic-1 output, and setting both inputs to logic 0 produces a logic-0 output. This format follows the basic OR truth table—it is an OR gate.

2-3   INVERT Functions

As the name implies, the basic function of an INVERT gate is to invert, or switch, a given logic level. Its job is to change a logic-1 state to a logic-0 state, and vice versa. Figure 2-3 summarizes the logic equation, schematic symbol, truth table and various circuits for implementing the INVERT operation.

The INVERT operation might seem too simple or trivial to be of any real importance, but that certainly is not the case. The INVERT function, in fact, can be considered one of the most valuable of all the logic functions.

2-3.1 An INVERT Switch Circuit

The switch-equivalent circuit of an INVERT gate is shown in Fig. 2-3(b). In this particular instance, the LED is normally lit through the forward-biasing path provided by the 150-Ω resistor. Closing switch A, however, short-circuits the LED, turning it off.

If we define a closed switch and turned-on LED as a logic-1 state and an open switch contact and a turned-off LED as a logic-0 state, the little circuit in Fig. 2-3(b) generates the truth table in Fig. 2-3(a). The circuit inverts the input action.

2-3.2 INVERT Logic Equation, Symbol and
Truth Table

The logic equation for the INVERT operation is read, "B equals NOT A." The NOT operator is a bar drawn across the NOT-ed or inverted logic function. Older texts and manuals frequently use an apostrophe or prime (′) notation instead of the bar. The basic NOT operation could be expressed as $B = A'$; but this notation has been abondoned for the most part in favor of the bar notation used throughout this book, e.g., $B = \bar{A}$.

The schematic symbol for an INVERT gate is a triangle with a small circle or "bubble" appearing at the output connection. The circle sometimes appears at the input instead, but common usage tends to favor the diagram in Fig. 2-3(a).

The logic truth table is very simple. Since there is only one input to a logic inverter, there are only two possible input states. And note a simple, but vital fact: The B output is always the complement of the A input.

2-3.3 A Switch-and-Transistor INVERT Circuit

Figure 2-3(c) is a switch-and-transistor equivalent of a logic inverter circuit. The transistor is normally biased off by means of the base connection to COMM through R_3; and as long as this transistor is switched off, the LED conducts (and lights up) through its connection to V_{cc} through R_2.

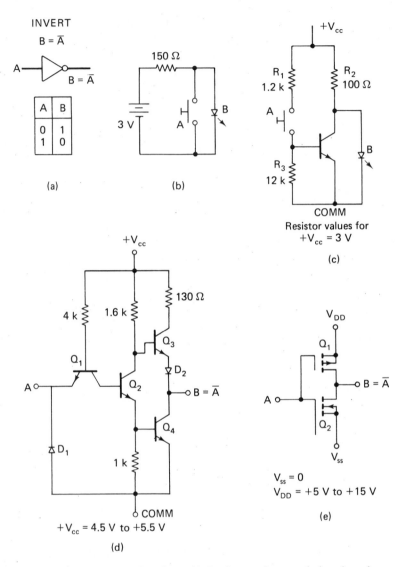

Figure 2-3 INVERT functions. (a) Logic equation, symbol and truth table. (b) A switch-equivalent circuit. (c) A switch-and-transistor INVERT circuit. (d) A TTL INVERT circuit with totem-pole output. (e) A CMOS INVERT gate.

Closing the switch contact, however, turns on the transistor by providing a forward-biasing current path through the emitter-base circuit to $+V_{cc}$. With the transistor thus switched on, the LED goes out because the emitter-collector saturation voltage of the transistor is lower than the conduction threshold voltage of the LED.

Depressing the switch turns out the LED; letting the switch open turns on the

LED. If we define a closed switch and a turned-on LED as logic-1 conditions, the circuit is indeed a logic inverter.

2-3.4 TTL INVERT Circuits

Figure 2-3(d) shows the internal structure of a TTL logic inverter having a totem-pole output configuration. Whenever input A is connected to $+V_{cc}$ or uncommitted (not connected to anything), input transistor Q_1 is switched off; and as long as Q_1 is held nonconducting, Q_2 conducts by virtue of its base connection to $+V_{cc}$ through the collector-base junction of Q_1 and the 4-k resistor.

Turning on Q_2 provides a forward-biasing path for base current to Q_4. Q_4 thus switches on, but Q_3 is held nonconducting by its low-impedance base connection to COMM through Q_2 and the 1-k resistor.

A logic-1 input at A ultimately pulls output B close to COMM—a logic-0 output.

Considering the other half of the INVERT truth table, connecting input A to COMM (or logic 0) forward biases Q_1 to turn it on. Turning on Q_1 in this fashion completes a low-impedance path to COMM for Q_2 to switch off that transistor. And with Q_2 switched off, Q_4 cannot possibly conduct, but Q_3 switches on, getting its forward-biasing base current through the 1.6-k resistor to $+V_{cc}$.

Grounding the input to this circuit thus pulls output B up close to $+V_{cc}$ potential. A logic-0 input yields a logic-1 output.

Open-collector versions of this basic TTL inverter are widely available today. The only difference between the totem-pole and open-collector versions is that the latter does not include the totem-pole components, D_2, Q_3 and the 130-Ω resistor. The circuit designer must provide an external pull-up resistor, usually about 2.2 k, for any open-collector circuit. See the discussion of open-collector TTL outputs in Sec. 2-1.4.

2-3.5 CMOS INVERT Circuits

Figure 2-3(e) shows a CMOS version of the basic logic inverter circuit. Whenever input A is pulled up close to V_{DD}, the N-channel FET Q_2 is switched on, but the P-channel FET Q_1 is switched off. Output B is thus pulled down close to V_{ss}.

Setting input A close to V_{ss}, on the other hand, switches off Q_2 and turns on Q_1. The voltage at output B is thus close to the V_{DD} potential.

The action of this circuit is very simple: A logic-0 input causes a logic-1 output, while a logic-1 input yields a logic-0 output. A close inspection of the AND and OR CMOS circuit in Figs. 2-1(e) and 2-2(e) will show this same inverter circuit appearing at the outputs of the more complex AND and OR circuits.

2-4   NAND Functions

NAND gates have been called the "building-blocks" of digital electronics, and one could certainly make a good case for this point of view, at least as far as TTL and CMOS circuits go. In principle, *any* logic operation—no matter how simple or complex—can be performed by an appropriate combination of NAND gates.

This section introduces the essential characteristics of NAND gates, but their real value won't be apparent until the discussions of implementing gate logic in Chs. 3 and 4.

Figure 2-4 summarizes the essential qualities of a basic 2-input NAND gate function. The logic equation, schematic symbol and truth table in Fig. 2-4(a) apply to any of the switch, transistor, TTL and CMOS circuits.

2-4.1 A NAND Switch Circuit

The switch-equivalent circuit in Fig. 2-4(b) shows that the output LED remains lighted until both input switches are closed at the same time. As long as one or both of the switches are open, there is a clear path for LED current through the 150-Ω resistor. Closing both switches at the same time, however, short-circuits the LED to turn it off.

If we define closed switches and turned-on lights as a logic-1 condition, the simple switch circuit in Fig. 2-4(b) generates the basic NAND gate truth table in Fig. 2-4(a). The essential feature of any NAND gate is that *its output is at logic 1 unless all inputs are at logic 1 at the same time.*

2-4.2 NAND Logic Equation, Symbol and Truth Table

Figure 2-4(a) shows the basic logic equation, symbol and truth table for a 2-input NAND gate. The logic equation expresses the notion that a NAND gate performs a basic AND operation, but that the final result is inverted. The logic equation, $C = \overline{AB}$ is read, "C equals NOT AB," or alternately, "C equals A NAND B." In either case, it is important to note that the NOT bar is drawn across *both* of the AND-ed terms. The expression $C = \overline{A} \cdot \overline{B}$ is entirely different from the correct form, $C = \overline{AB}$. The difference between these often confused expressions is explained in Ch. 3.

The schematic symbol for a NAND gate also reflects the fact that it is a basic AND operation followed by a logic inversion. The symbol is that of an AND gate having an INVERT "bubble" at the output. In effect, a NAND gate first AND-s the inputs, and then it inverts the result.

Carrying this AND-INVERT notion a step further, note that the NAND truth table is very similar to a 2-input AND truth table, with the NAND version showing complemented outputs. The output of a NAND gate is normally high and drops low only when all inputs are high at the same time.

It is little wonder, then, that this circuit is called a NOT-AND or NAND gate.

2-4.3 A Switch-and-Transistor NAND Circuit

Figure 2-4(c) is a switch-and-transistor version of the 2-input NAND circuit. The transistor is normally biased off by the base connection to COMM through R_3; and as long as the transistor is switched off, the LED lights through the path to $+V_{cc}$ provided by R_2. Closing either one of the input switches does not alter the

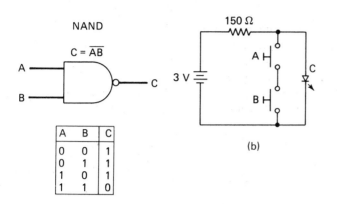

NAND

$C = \overline{AB}$

A	B	C
0	0	1
0	1	1
1	0	1
1	1	0

(a)

150 Ω

3 V

A

B

C

(b)

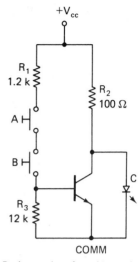

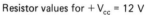

$+V_{cc}$

R_1 1.2 k

R_2 100 Ω

A

B

C

R_3 12 k

COMM

Resistor values for $+V_{cc}$ = 12 V

(c)

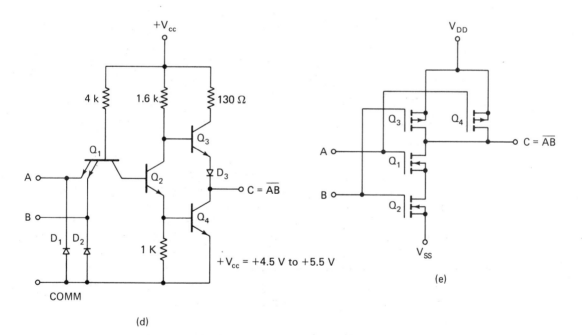

$+V_{cc}$

4 k   1.6 k   130 Ω

Q_1

Q_3

Q_2

D_3

A

B

D_1 D_2

1 K

$C = \overline{AB}$

Q_4

$+V_{cc}$ = +4.5 V to +5.5 V

COMM

(d)

V_{DD}

Q_3 Q_4

A

Q_1

$C = \overline{AB}$

B

Q_2

V_{SS}

(e)

Figure 2-4 NAND functions. (a) Logic equation, symbol and truth table. (b) A switch-equivalent circuit. (c) A switch-and-transistor NAND circuit. (d) A TTL NAND circuit with totem-pole output. (e) A CMOS NAND gate.

23

off state of the transistor, but closing both of the switches at the same time provides a forward-biasing base current that turns on the transistor. Turning on the transistor robs the LED of its forward-biasing junction potential, and it turns off.

The LED in Fig. 2-4(c) is thus normally lighted, going out only when both input switches are closed at the same time. If closed switch contacts and a lighted LED represent the logic-1 state, and if open switches and a turned-off LED represent the logic-0 condition, this circuit generates the basic NAND truth table.

2-4.4 TTL NAND Gates

Figure 2-4(d) shows the internal bipolar structure of a TTL 2-input NAND gate. The circuit is very simple as far as TTL circuits go, but it is a little more complicated than a basic TTL logic inverter. In fact, the only real difference between a TTL NAND gate structure and a TTL INVERT gate is the multi-emitter input on the NAND gates.

As long as either or both of the input connections are connected near the COMM potential, transistor Q_1 is switched on. Turning on Q_1 pulls the base connection of Q_2 near COMM potential, however, to turn off that transistor. (The path to COMM for the base of Q_2 is through the conducting emitter-connections of Q_1 and the logic-0 input.)

As long as Q_2 is held nonconducting in this fashion, Q_4 is held nonconducting by the 1-k resistor to COMM, but Q_3 is switched on by virtue of the base-current path the 1.6-k resistor to $+V_{cc}$.

In other words, setting either or both of the A and B inputs to COMM causes the C output to be pulled up near the $+V_{cc}$ potential. Speaking in terms of logic levels, this means a logic-0 input at either A or B or both yields a logic-1 output, a fact that is consistent with the basic NAND gate truth table in Fig. 2-4(a).

Now run through the analysis of the circuit in Fig. 2-4(d) with both inputs connected near $+V_{cc}$ at the same time. This represents an input logic condition in which both inputs are at logic 1 at the same time; and according to the NAND truth table, output C should show a logic-0 potential.

With both inputs connected near $+V_{cc}$ (or to nothing at all), transistor Q_1 must be switched off. Turning off Q_1 allows Q_2 to conduct by means of forward-biasing base current through Q_1's collector-base junction. The emitter-follower connection between Q_2 and Q_4 forces Q_4 into saturation whenever Q_2 conducts. At the same time, however, the conduction of Q_2 pulls the base of Q_3 near COMM potential to turn off that totem-pole transistor.

Applying a logic-1 level, or making no connection at all, to inputs A and B at the same time produces a logic-0 output, a condition that is consistent with the last line of the NAND gate truth table.

It is possible to expand the number of inputs to the TTL NAND gate in Fig. 2-4(d) by simply providing more emitters at Q_1. No matter how many inputs the NAND gate might have, however, the only way to produce a logic-0 output is by setting all inputs to logic 1 at the same time.

Omitting the internal totem-pole components, D_3, Q_3 and the 130-Ω resistor produces an open-collector NAND gate. When using the open-collector version of the circuit in Fig. 2-4(d), the circuit designer must provide an external pull-up resistor between the collector of Q_4 and $+V_{cc}$. That is the only way to get a true logic-1 output from the circuit. A pull-up resistor for this kind of circuit normally has a value on the order of 2.2 k.

2-4.5 CMOS NAND Circuits

Figure 2-4(e) shows the internal structure of a 2-input NAND CMOS IC. To see how this circuit works, suppose input B is connected near V_{ss} potential. For all practical purposes, the gate and source connections of Q_2 are tied together, making it impossible for this N-channel MOSFET to conduct. Q_2 thus acts as an open switch in the series combination of transistors Q_1 and Q_2.

While the V_{ss} potential of input B is switching off Q_2, it is turning on the P-channel MOSFET, Q_3. The conduction of Q_3 pulls output C up very close to the V_{DD} potential, yielding a logic-1 output.

A logic-0 input at B thus produces a logic-1 output from this circuit; and in a similar fashion, setting input A to V_{ss} produces a logic-1 output close to the positive V_{DD} potential.

The only way output C can be pulled down to a logic-0 V_{ss} potential is by getting Q_1 and Q_2 switched on at the same time while turning off Q_3 and Q_4. This condition is possible only by pulling up inputs A and B near the V_{DD} voltage at the same time. A positive V_{DD} potential appearing on the gates of the N-channel FETs, Q_1 and Q_2, switches them on; but the same potential on the gates of both P-channel FETs, Q_3 and Q_4, turns them off.

The exact conditions for defining a 2-input NAND gate are thus satisfied by the CMOS circuit in Fig. 2-4(e).

2-5   NOR Functions

Figure 2-5 illustrates the essential qualities and some methods for implementing the basic NOR logic operation. The logic equation, schematic symbol and truth table in Fig. 2-5(a) apply to any of the NOR schemes illustrated here.

2-5.1 A NOR Switch Circuit

The switch circuit in Fig. 2-5(b) performs the basic 2-input NOR function. The LED is normally biased on through the 150-Ω resistor connection to the power supply. Closing either or both of the switches, however, extinguishes the LED by shorting it out. Energizing either or both of the inputs to a NOR gate de-energizes the output; or to state it another way: *A NOR gate shows a logic-0 output whenever one or more inputs are at logic 1.*

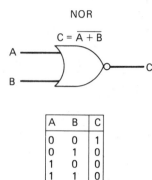

NOR

$C = \overline{A + B}$

A	B	C
0	0	1
0	1	0
1	0	0
1	1	0

(a)

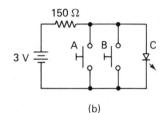

150 Ω

3 V

A B C

(b)

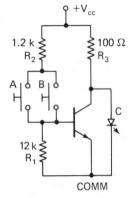

+V_cc

1.2 k R_2   100 Ω R_3

A B

12 k R_1

C

COMM

Resistor values for $V_{cc} = +3$ V

(c)

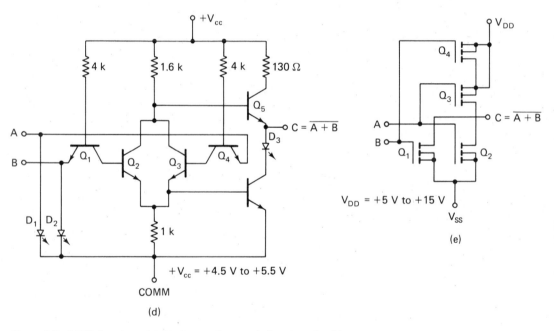

+V_cc

4 k   1.6 k   4 k   130 Ω

Q_5

$C = \overline{A + B}$

D_3

A

B

Q_1   Q_2   Q_3   Q_4

D_1 D_2

1 k

$+V_{cc} = +4.5$ V to $+5.5$ V

COMM

(d)

V_DD

Q_4

Q_3

$C = \overline{A + B}$

A

B

Q_1   Q_2

$V_{DD} = +5$ V to $+15$ V

V_SS

(e)

Figure 2-5 NOR functions. (a) Logic equation, symbol and truth table.
(b) A switch-equivalent circuit. (c) A switch-and-transistor NOR circuit.
(d) A TTL NOR circuit with totem-pole output. (e) A CMOS NOR gate.

2-5.2 NOR Logic Equation, Symbol and Truth Table

Figure 2-5(a) shows the basic logic equation, schematic symbol and truth table
for a 2-input NOR gate. The logic equation clearly reflects the fact that it is a basic
OR function that is inverted. The plus operator indicates an OR-ing operation,
and the bar across the entire expression indicates an inversion of the result.

The logic symbol also implies a NOT-ed OR operation—an OR gate followed by an inverting "bubble." And finally, the truth table is simply an inverted version of the basic OR truth table.

2-5.3 A Switch-and-Transistor NOR Circuit

Figure 2-5(c) is a switch-and-transistor equivalent of the basic 2-input NOR function. The transistor is held normally nonconducting by the base connection to COMM through R_1; and as long as the transistor is switched off this way, the LED lights up through the R_3 connection to $+V_{cc}$.

Closing either or both of the input switches forward biases the transistor to turn it on; and when that transistor is thus saturated, it turns off the LED.

If we define an open switch contact and a turned-off LED at logic 0, and if we define a closed switch and turned-on LED as a logic-1 state, the circuit in Fig. 2-5(c) generates the basic NOR truth table.

2-5.4 TTL NOR Circuits

The TTL NOR circuit shown in Fig. 2-5(d) closely resembles its OR counterpart in Fig. 2-2(d). As might be expected, the NOR version lacks only an internal inverting stage that "uprights" the more natural TTL NOR operation.

Input transistor Q_1 can be turned on only by setting the B input to COMM potential. Any other condition—connecting input B to $+V_{cc}$ or nothing at all—switches Q_1 off. The same general conditions apply to input A and Q_4.

Now, if either of the two input transistors is switched off, the transistor at its collector (either Q_2 or Q_3) must be switched on to pull the base connection of Q_5 down toward COMM potential. That particular state of affairs switches off Q_5.

While the conduction of Q_2 or Q_3 is turning off Q_5, however, it is turning on Q_6 to yield a near 0-V output at C. The conduction of either Q_2 or Q_3 thus causes the circuit to produce a logic-0 output; and since Q_2 and Q_3 are turned on whenever their corresponding input transistors are switched off, it follows that the output of the circuit is at logic 0 whenever either or both of the inputs, A and B, are connected to $+V_{cc}$. This is entirely consistent with the NOR truth table: The output is at logic 0 as long as either or both of the inputs are at logic 1.

To complete the analysis of this circuit, observe the action of the circuit while both inputs are connected to COMM. Under this circumstance, Q_1 and Q_4 are both turned on, and they switch off their respective transistors, Q_2 and Q_3. Turning off both of these parallel-connected transistors switches off Q_6 by letting the 1-k resistor pull its base to COMM potential. At the same time, however, switching off both Q_2 and Q_3 allows Q_5 to conduct by means of forward-biasing base current to $+V_{cc}$ through the 1.6-k resistor.

Grounding both inputs to logic 0 at the same time ultimately turns on Q_5 and switches off Q_6. The overall result is a logic-1 output, a situation that suits the requirements of the NOR truth table in Fig. 2-5(a).

2-5.5 CMOS NOR Circuits

The CMOS circuit in Fig. 2-5(e) represents the internal structure of a typical 2-input NOR gate. Transistors Q_1 and Q_2 are connected in parallel, while Q_3 and Q_4 are in series.

The only way to see a logic-1 output (C pulled up toward V_{DD}) is by switching off both Q_1 and Q_2 while turning on both Q_3 and Q_4. This condition occurs whenever inputs A and B are at logic 0 at the same time. Setting those inputs to V_{ss} switches off the paralleled P-channel FETs and turns on the series-connected FETs, effectively connecting output C to V_{DD}.

The circuit then shows a logic-0 output (C pulled down toward V_{ss} potential) whenever Q_1 or Q_2 is switched on and while either Q_3 or Q_4 is switched off. This particular set of circumstances occurs whenever either or both of the inputs are pulled up toward V_{DD}—logic-1 input states.

The action of the circuit in Fig. 2-5(e) thus follows the basic NOR truth table in Fig. 2-5(a): The output is normally at logic 0, rising up to logic 1 only when both inputs are at logic 0 at the same time.

2-6 Basic Logic IC Packages

Table 2-1 summarizes the basic logic functions available in standard 14-pin DIP (dual in-line plastic) packages. Most of these packages contain more than one gate function, sharing a single set of $+V_{cc}$ and COMM or V_{DD} and V_{ss} power supply connections.

Note that IC packages in the 7400 series belong to the TTL family and those in the 4000 series belong to the CMOS category.

Table 2-1 AVAILABLE BASIC IC PACKAGES

AND Functions		NAND Functions		OR Functions	
7408	Quad 2-input AND; TTL	7400	Quad 2-input NAND; TTL	7432	Quad 2-input OR; TTL
7409	Quad 2-input AND; TTL with open collector	7401	Quad 2-input NAND; TTL with open collector	4071	Quad 2-input OR; CMOS
4081	Quad 2-input AND; CMOS	7403	Quad 2-input NAND; TTL with open collector		*NOR Functions*
		7410	Triple 3-input NAND; TTL	7402	Quad 2-input NOR; TTL
	INVERT Functions	7420	Dual 4-input NAND;	7427	Triple 3-input NOR; TTL
		7430	8-input NAND; TTL	4001	Quad 2-input NOR; CMOS
7404	Hex inverter; TTL	4011	Quad 2-input NAND; CMOS	4002	Dual 4-input NOR; CMOS
				4025	Triple 3-input NOR; CMOS
7405	Hex inverter; TTL with open collector	4012	Dual 4-input NAND; CMOS		
4049	Hex inverter; CMOS	4023	Triple 3-input NAND; CMOS		

Exercises

Attempt to work the following exercises without referring to the material presented in Ch. 2. Check your answers by reviewing this chapter.

1. Show the logic diagram, truth table and logic equation for the following basic combinatorial gates:
   (a) 2-input AND;
   (b) 2-input NAND;
   (c) 2-input OR;
   (d) 2-input NOR;
   (e) INVERT gate.

2. Which kind of basic logic gate shows: (a) a logic 1 only when all inputs are at logic 1? (b) a logic 1 when any or all inputs are at logic 1? (c) a logic 0 when any or all inputs are at logic 0? (d) a logic 0 only when all inputs are at logic 0? (e) a logic 1 only when all inputs are at logic 0? (f) a logic 1 when any or all inputs are at logic 0? (g) a logic 0 when any or all inputs are at logic 1? (h) a logic 0 only when all inputs are at logic 1?

USING THE BUILDING BLOCKS
OF DIGITAL LOGIC

Digital electronics is a building-block technology in the sense that every digital system can be segmented, divided and subdivided into smaller basic elements, ultimately arriving at the basic logic-gate level. The tool for carrying out this kind of building-block analysis is Boolean algebra—a shorthand language and system of logical thinking that can transform the essential operating features of a vast array of semiconductor components into a few concise equations of more manageable proportions.

A digital technician routinely faces this sort of building-block analysis. Given a digital system, he or she must first use the tools of Boolean algebra and his or her familiarity with basic digital circuits to determine exactly how the system is supposed to respond; and then the digital technician looks for any discrepancies between what the circuit is supposed to do and what it actually does. Repairing the system is then a matter of resolving those discrepancies, again, using the tool of Boolean algebra and a knowledge of basic digital circuits.

A digital engineer works with the same building-block idea, but from the opposite point of view. The situation here is one of assembling, or synthesizing, the basic building blocks into a more complex system; but the tools for synthesizing any desired logical operation are still those of Boolean algebra and a knowledge of basic digital logic circuits.

Whether the task is one of repairing or designing a digital logic system, the tools for doing the job are basically the same. A circuit designer might have more choices to make with regard to selecting components and techniques, but a technician and engineer stand on the same level as far as their ability to use Boolean algebra is concerned.

This chapter introduces the basic elements of Boolean algebra, and it shows how Boolean algebra can be applied to the analysis and synthesis of some of the most common and valuable combinations of basic logic gates. The Boolean principles and circuits shown here must be regarded as all-important for both technicians and engineers—they must be recognized on the spot or called up from memory whenever they are needed. This might seem something of a burden at first, but with experience these principles and circuits become a very "natural" way of thinking.

3-1   The Essentials of Boolean Algebra

Table 3-1 summarizes all the essential features of Boolean algebra. Anything more to be said about the subject must have a foundation somewhere in that table. It is tempting, and indeed helpful in most instances, to compare the essentials of Boolean algebra with those of ordinary arithmetic and algebra: The similarities are often as striking as the differences. But in any event, the expressions in Table 3-1 represent the logical foundations of modern digital electronics.

3-1.1 Boolean Postulates

The eight Boolean postulates in Table 3-1 reflect the nature of basic AND, OR and INVERT operations and demonstrate the remarkable self-consistency of a system of logic that permits only two possible states—0 and 1.

Boolean postulates are actually definitions of 0 and 1, AND, OR and INVERT operations. These definitions are subsequently represented by the truth tables in Fig. 3-1.

Postulates 1a and 1b serve two primary purposes: They define the logical meaning and difference between a 1 and a 0 logic state, and they define the basic INVERT operation. According to postulate 1a, any logic condition that is NOT 1 has to be logic 0; and by postulate 1b, any logic condition that is NOT 0 has to be logic 1. There are two, and only two, logic conditions. The truth table in Fig. 3-1(a), an INVERT logic truth table, demonstrates the meaning of postulates 1a and 1b.

The truth table in Fig. 3-1(b) demonstrates the meaning of the three AND postulates, postulates 2a, 3a and 4a. Whenever both inputs to a 2-input AND gate are both 0, for instance, the output must be 0 as well.

In a similar fashion, the OR truth table in Fig. 3-1(c) demonstrates the meaning of postulates 2b, 3b and 4b.

The truth tables in Fig. 3-1 do not prove the truth of the eight basic postulates of Boolean algebra. The postulates themselves cannot be proven. They are definitions that hold together the entire system of Boolean algebra. The truth tables merely demonstrate the basic definitions in tabular form.

Table 3-1 Summary of Boolean Postulates, Properties and Theorems

Boolean Postulates	

1a. $\bar{1} = 0$	1b. $\bar{0} = 1$
2a. $0 \cdot 0 = 0$	2b. $0 + 0 = 0$
3a. $1 \cdot 0 = 0$	3b. $1 + 0 = 1$
4a. $1 \cdot 1 = 1$	4b. $1 + 1 = 1$

Boolean Algebraic Properties

5a. $XY = YX$	5b. $X + Y = Y + X$ Commutative property
6a. $X(YZ) = (XY)Z$	6b. $X + (Y + Z) = (X + Y) + Z$
	Associative property
7 $X(Y + Z) = XY + XZ$ Distributive property	

Boolean Theorems

8a. $X \cdot 0 = 0$	8b. $X + 0 = X$
9a. $X \cdot 1 = X$	9b. $X + 1 = 1$
10a. $X \cdot X = X$	10b. $X + X = X$
11a. $X \cdot \bar{X} = 0$	11b. $X + \bar{X} = 1$
12 $\bar{\bar{X}} = X$	
13a. $\overline{X \cdot Y} = \bar{X} + \bar{Y}$	13b. $\overline{X + Y} = \bar{X}\bar{Y}$ DeMorgan's theorems
14 $(X + Y)(\bar{X} + Z) = XZ + \bar{X}Y$ Law of consensus	

Figure 3-1 Truth table demonstrations of the eight Boolean postulates. (a) The INVERT postulates. (b) The AND postulates. (c) The OR postulates.

3-1.2 Algebraic Properties of Boolean Algebra

The algebraic properties of Boolean algebra, listed as items 5a through 7 in Table 3-1, are identical to those of ordinary algebra and arithmetic. Although the commutative, associative and distributive properties cannot be proven in a formal sense, they can be clearly demonstrated by means of truth tables and the application of some "common sense" logic.

The commutative properties of Boolean AND and OR operations (properties 5a and 5b) merely indicate that logic inputs are not order sensitive. It makes no difference whether we say input X is being AND-ed with Y, or Y is being AND-ed with X; the result is the same in either case. And in the same way, it makes no difference whether one thinks of input X as being OR-ed with Y, or Y with X.

Properties 6a and 6b in Table 3-1 show that groups of AND-ed or groups of OR-ed operations can be subgrouped in any arbitrary fashion. One can, for example, AND an X term with previously AND-ed terms Y AND Z to get the same result as AND-ing the term X AND Y with a third input, Z. As long as the terms in an expression are all AND-ed (associative property 6a) or all OR-ed (associative property 6b), they can be grouped in any convenient manner.

While the commutative and associative properties of Boolean algebra might be considered obvious or almost too trivial for formal consideration, they are nevertheless applied routinely, usually without conscious effort.

The distributive property of Boolean algebra, listed as item 7 in Table 3-1, often seems to have more power as far as day-to-day digital work is concerned; at least it is used more consciously than the others. The distributive property combines loggic AND and OR operations; and when read from left to right, it defines a form of Boolean "multiplication" that is identical to the more familiar process of multiplying a monomial times a polynomial. Read from right to left, the distributive property defines Boolean "factoring", a process that is quite familiar to any student of ordinary algebra.

Figure 3-2 shows a truth table that demonstrates the validity of the distributive property of Boolean algebra. Columns 1, 2 and 3 in this truth table show all eight possible combinations of 1 and 0 logic states for the three inputs, X, Y and Z. The fourth column shows the results of OR-ing the Y and Z inputs (OR-ing together columns 2 and 3). Column 5 then shows what happens when the results in column 4 are AND-ed with input X. Column 5, in other words, is the result of performing the operation $X(Y + Z)$ under the input conditions specified in Columns 1, 2, and 3.

Column 6 then begins another logic operation, AND-ing together inputs X and Y; and column 7 shows the results of AND-ing inputs X and Z. OR-ing together these two columns produces the results shown in column 8, a column representing the operation $XY + XZ$.

Given the same sets of inputs X, Y and Z, a comparison of columns 5 and 8 show identical results. The conclusion is that $X(Y + Z)$ is indeed the same as $XY + XZ$.

This is an example of a truth-table demonstration of the truth or validity of a Boolean logic equation; or as some logicians would prefer to put it, it is a *proof by perfect induction*.

While the truth-table analysis in Fig. 3-2(a) demonstrates the validity of the distributive property of Boolean algebra, the circuits in Fig. 3-2(b) show how the property can be implemented with simple 2-input AND and OR gates. Both of these circuits behave in exactly the same fashion, responding to their X, Y and Z inputs according to the truth table. The circuit implementing the expression $X(Y + Z)$, however, is somewhat simpler than the one for working the expression

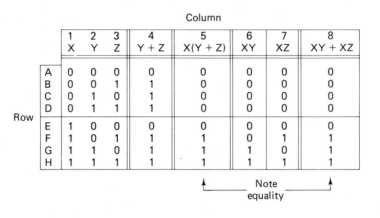

Column

		1	2	3	4	5	6	7	8
		X	Y	Z	Y + Z	X(Y + Z)	XY	XZ	XY + XZ
Row	A	0	0	0	0	0	0	0	0
	B	0	0	1	1	0	0	0	0
	C	0	1	0	1	0	0	0	0
	D	0	1	1	1	0	0	0	0
	E	1	0	0	0	0	0	0	0
	F	1	0	1	1	1	0	1	1
	G	1	1	0	1	1	1	0	1
	H	1	1	1	1	1	1	1	1

Note
equality

X(Y + Z) = XY + XZ

(a)

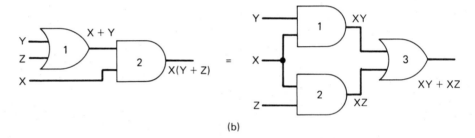

(b)

Figure 3-2 The distributive property of Boolean algebra. (a) A truth table demonstration. (b) Equivalent logic circuits.

$XY + XZ$; and that particular observation foreshadows some of the powerful design applications of Boolean algebra in discussions to come.

3-1.3 Boolean Theorems

The Boolean theorems listed in Table 3-1 grow quite naturally from the basic postulates and properties. They are generalizations of the eight basic postulates and follow the commutative, associative and distributive properties of Boolean algebra.

The truth table and circuits in Fig. 3-3 illustrate the Boolean AND theorems 8a, 9a, 10a and 11a.

The circuit in Fig. 3-3(b) shows how fixing one of the two inputs of a 2-input AND gate to logic 0 always yields a logic-0 output. Note from the truth table that $W = 0$ whenever input Y is fixed at 0, no matter what the logic state of the X input might be. Setting one of the inputs to logic 0, in effect, turns off the gate as far as the X input is concerned. And that notion can be extended to AND gates

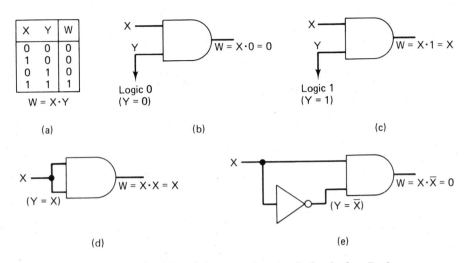

X	Y	W
0	0	0
1	0	0
0	1	0
1	1	1

$W = X \cdot Y$

(a)

$W = X \cdot 0 = 0$

Logic 0
(Y = 0)

(b)

$W = X \cdot 1 = X$

Logic 1
(Y = 1)

(c)

$W = X \cdot X = X$

(Y = X)

(d)

$W = X \cdot \overline{X} = 0$

(Y = $\overline{X}$)

(e)

Figure 3-3   Truth table and demonstration circuits for the four Boolean AND theorems.

having any number of inputs: As long as at least one input is at logic 0, none of the others has any influence on the output.

Figure 3-3(c) shows how theorem 9a can be implemented with a 2-input AND gate. One input in this instance is fixed at logic 1, and the output is equal to whatever the other input logic input level might be. As long as $Y = 1$, the W output is equal to the X input. This effect shows up clearly on the last two lines of the truth table in Fig. 3-3(a). Here Y is equal to 1 in both instances, and the W output follows the X input.

Theorem 10a shows what happens whenever the inputs of an AND gate are identical: The output equals the inputs. This can be easily demonstrated by connecting together all the inputs of an AND gate as shown in Fig. 3-3(d). Whatever the data input source does under these circumstances, the W output follows suit. The first and last lines of the truth table in Fig. 3-3(a) also demonstrate how the W output follows the inputs when the inputs are all exactly the same.

While AND-ing like inputs produces an output equal to those inputs, AND-ing complemented terms always produces a logic-0 output. The inverter in Fig. 3-3(e) guarantees a complemented version of X at the Y input; so when $X = 0$, it follows that $Y = 1$; and when $X = 1$, it follows that $Y = 0$. In either case, one of the two inputs is always at logic 0; and theorem 8a has already shown that any input with a logic-0 state always generates a logic-0 output. Hence $X \cdot \overline{X} = 0$.

The truth table and circuits in Fig. 3-4 demonstrate the Boolean OR theorems 8b, 9b, 10b and 11b. OR-ing any term with a logic 0, for instance, always produces an output equal to that variable input. See theorem 8b and Fig. 3-4(b), and compare them with the first two lines of the truth table in Fig. 3-4(a).

Stated in words, theorem 9b says OR-ing any expression with a logic-1 level always produces a logic-1 result. The truth table in Fig. 3-4(a) demonstrates this

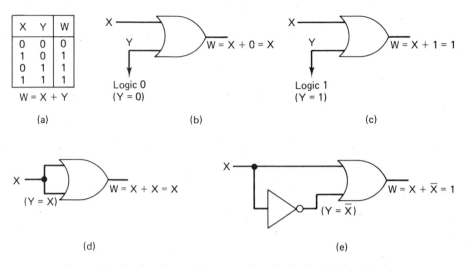

X	Y	W
0	0	0
1	0	1
0	1	1
1	1	1

W = X + Y

(a)

Logic 0
(Y = 0)

W = X + 0 = X

(b)

Logic 1
(Y = 1)

W = X + 1 = 1

(c)

(Y = X)

W = X + X = X

(d)

(Y = X̄)

W = X + X̄ = 1

(e)

Figure 3-4 Truth table and demonstration circuits for the four Boolean OR theorems.

fact in its last two lines. Figure 3-4(c) then shows how it can be implemented with a 2-input OR gate.

OR-ing any logic expression with itself always produces that same expression. See the first and last lines of the truth table and the circuit in Fig. 3-4(d). The idea is expressed as a logic equation in item 9b in Table 3-1: $X + X = X$.

While OR-ing any logic state with itself produces that same logic level, OR-ing any logic state with its complement always generates a logic-1 output. This fact can be demonstrated by the little circuit in Fig. 3-4(e) and the two middle lines of the OR truth table in Fig. 3-4(a). Indeed, $X + \bar{X} = 1$.

And, finally, theorem 12 shows that any double-NOT-ed term is equal to that term itself. Passing a logic level through two consecutive INVERT gates, for example, always yields the original logic state. See the circuit in Fig. 3-5.

The double-NOT theorem works very much like the double-negative sign situation in ordinary algebra. A double-negative term such as $--3$, for instance, is actually the same as $+3$, while $---3$ can be better expressed as -3. In a similar fashion, $\bar{\bar{1}}$ in logic terms is equal to 1, while $\bar{\bar{\bar{1}}} = \bar{1} = 0$.

Boolean theorems 8a through 12 in Table 3-1 represent the most basic and straightforward of all the theorems, and after some experience with them, they become almost intuitively obvious. Theorems 13a, 13b and 14, however, call for some special consideration.

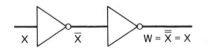

X   X̄   W = X̄̄ = X

Figure 3-5 Demonstration circuit for the Boolean double-NOT theorem.

3-1.4 DeMorgan's Theorems

DeMorgan's theorems, shown as theorems 13a and 13b in Table 3-1, represent powerful logical bridges between AND and OR expressions. It is important to understand exactly what these theorems are saying from the outset. Some vital points that are often confused at the beginning make the difference between understanding a lot of basic digital electronics and being totally lost.

DeMorgan's theorem 13a (which we will call *DeMorgan's NAND theorem* throughout this book) states that X NAND-ed with Y is equal to NOT X OR NOT Y. Expressing this in the form of an equation, $\overline{XY} = \bar{X} + \bar{Y}$. In a sense, splitting the bar across the XY term changes the sign of operation. The bar still exists, but it is divided between the OR-ed X and Y terms.

Then DeMorgan's theorem 13b, called *DeMorgan's NOR theorem* for the sake of convenience, states that X NOR-ed with Y is equal to NOT X AND NOT Y. In Boolean form: $\overline{X + Y} = \bar{X} \cdot \bar{Y}$. Again, splitting the bar changes the sign of operation.

These statements of DeMorgan's theorems are not intuitively obvious. There seems to be no rational explanation for why they work. The fact is, however, that they do work, and both of them can be demonstrated by truth-table analyses.

Figure 3-6 shows the truth-table "proof" and equivalent logic circuits for

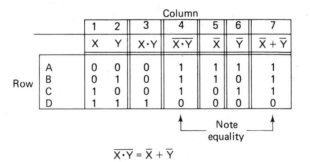

	Column							
		1	2	3	4	5	6	7
		X	Y	X·Y	$\overline{X \cdot Y}$	$\bar{X}$	$\bar{Y}$	$\bar{X} + \bar{Y}$
Row	A	0	0	0	1	1	1	1
	B	0	1	0	1	1	0	1
	C	1	0	0	1	0	1	1
	D	1	1	1	0	0	0	0

Note equality

$$\overline{X \cdot Y} = \bar{X} + \bar{Y}$$

(a)

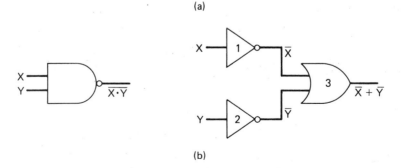

(b)

Figure 3-6 DeMorgan's NAND theorem. (a) Truth-table demonstration. (b) Equivalent logic circuits.

DeMorgan's NAND theorem. Columns 1 and 2 in that truth table merely show all four possible combinations of 1's and 0's for inputs X and Y. The third column then shows the results of AND-ing the X and Y inputs, and column 4 is an inverted version of column 3. Column 4 is actually a NAND result achieved with the inputs X and Y.

Columns 5 and 6 in the truth table in Fig. 3-6(a) show inverted versions of the X and Y inputs; column 7 is the result of OR-ing those two inverted inputs.

Since the data in column 4 represent $\overline{XY}$ and the data in column 7 represent $\bar{X} + \bar{Y}$, it follows that these two expressions are equal since they generate exactly the same truth-table results.

Although the origin of DeMorgan's NAND theorem might not be intuitively or even logically clear, we are forced to accept its validity from the truth-table analysis.

The two circuits in Fig. 3-6(b) thus perform exactly the same basic logic operation. The NAND gate version yields $\overline{XY}$, while the OR-gate circuit having pre-inverted inputs yields $\bar{X} + \bar{Y}$. But in spite of their differences in appearance, the circuits are identical from a functional point of view.

Figure 3-6 shows a truth-table analysis and equivalent logic circuits for DeMorgan's NOR theorem: $\overline{X + Y} = \bar{X} \cdot \bar{Y}$. Column 4 in Fig. 3-6(a) shows the

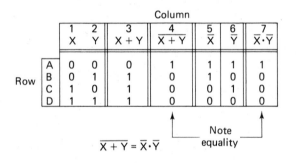

(a)

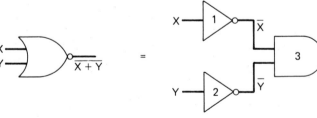

(b)

Figure 3-7 DeMorgan's NOR theorem. (a) Truth-table demonstration. (b) Equivalent logic circuits.

results of NOR-ing inputs X and Y, while column 7 represents $\bar{X} \cdot \bar{Y}$. Since the data in columns 4 and 7 are identical, we are forced to accept the validity of DeMorgan's NOR theorem.

The circuits in Fig. 3-7(b) demonstrate how two vastly different logic elements can work according to DeMorgan's theorem to perform the same basic logic function.

The key to remembering DeMorgan's theorems is to *break the bar and change the sign of operation*. Although this is not a genuine logic statement, it is a workable trick for recalling the essence of the two theorems. Of course, they are used in reverse order as often as they are used from left to right.

Exercises at the end of this chapter offer many opportunities for working with DeMorgan's theorems from different points of view.

3-1.5 The Law of Consensus

The law of consensus, like DeMorgan's theorems, has some features that tend to defy any logical origin. The validity of this law, however, is easily proven by a truth-table analysis. See Fig. 3-8.

The truth table in Fig. 3-8(a) breaks down the equation for the law of consensus

| | | | | | Column | | | | | |
| | 1 | 2 | 3 | 4 | 5 | 6 | 7 | 8 | 9 | 10 |
	X	Y	Z	$\bar{X}$	X + Y	$\bar{X}$ + Z	(X + Y)($\bar{X}$ + Z)	XZ	$\bar{X}$Y	XZ + $\bar{X}$Y
A	0	0	0	1	0	1	0	0	0	0
B	0	0	1	1	0	1	0	0	0	0
C	0	1	0	1	1	1	1	0	1	1
D	0	1	1	1	1	1	1	0	1	1
E	1	0	0	0	1	0	0	0	0	0
F	1	0	1	0	1	1	1	1	0	0
G	1	1	0	0	1	0	0	0	0	1
H	1	1	1	0	1	1	1	1	0	1

Row (label to the left of the table)

————Note equality————

$$(X + Y)(\bar{X} + Z) = XZ + \bar{X}Y$$

(a)

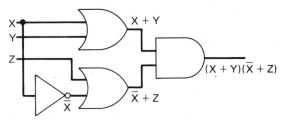

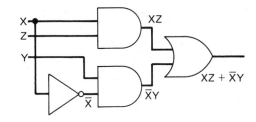

(b)

Figure 3-8 The law of consensus. (a) Truth-table demonstration. (b) Equivalent logic circuits.

into its elementary parts, ultimately showing that $(X + Y)(\bar{X} + Z) = XZ + \bar{X}Y$ (note that column 7 is identical to column 10).

The circuits in Fig. 3-8(b) are the equivalent logic circuits for the two components of this logic equation.

3-2 Building Inverters from NAND and NOR Gates

This section introduces the powerful notion that any sort of logic can be built from other kinds of gates. In this particular instance, it is possible to construct an INVERT operation from single NAND or NOR gates.

Figure 3-9 summarizes three different approaches to an INVERT logic operation. The diagrams in Fig. 3-9(a) show the most straightforward approach to achieving an INVERT operation. The inverter symbol having the circle at the apex of the triangle is the more common symbol for a logic inverter, although the second one is an acceptable alternative. In either case, the circuit performs the basic INVERT function, $B = \bar{A}$.

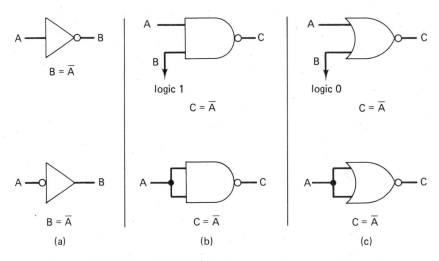

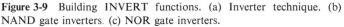

Figure 3-9 Building INVERT functions. (a) Inverter technique. (b) NAND gate inverters. (c) NOR gate inverters.

Figure 3-9(b) shows two different ways to get an INVERT logic function from a 2-input NAND gate. In the first instance, the B input is permanently tied to a logic-1 source; and the overall result is that $C = \bar{A}$—a basic INVERT function. The principle behind this technique rests upon the theorem that says $X \cdot 1 = X$. Generalizing this theorem to a NAND function, we get: $\overline{X \cdot 1} = \bar{X}$.

The second diagram in Fig. 3-9(b) shows the inputs of the NAND gate tied together. Here the B input is actually equal to A; and from the theorem that states $X \cdot X = X$, it follows that $\overline{X \cdot X} = \bar{X}$.

Both of the circuits in Fig. 3-9(b) thus show how to achieve an INVERT function by using 2-input NAND gates. Either technique is acceptable under most practical circumstances, and the same basic ideas can be extended to include 3- and 4-input NAND gates. When a 4-input NAND gate is used, for example, three of the inputs can be tied to a logic-1 source or all four of them can be connected together to the data to be inverted.

To see the practical significance of using a NAND gate to perform the task of an inverter, consider the following situation: A circuit designer finds he or she must use one more INVERT operation to complete the circuit. There are no unused inverters left in the basic design, but there is at least one unused NAND gate in one of the IC packages already in the circuit. Rather than specifying an additional hex inverter IC package, the designer can use the spare NAND gate to do the job, thus avoiding the additional complexity, inconvenience and cost of another inverter IC.

The circuits in Fig. 3-9(c) show how to build an inverter from a NOR gate. The general approach is the same as for the NAND gate version, but the first approach calls for tying all the unused inputs to logic 0 instead of to logic 1. The logic-equation analyses accompanying the figures justify their validity. Again, the only reason one would use a NOR gate to do the job of an inverter is when a spare NOR gate is available and an inverter is not.

3-3   Expanding and Building AND Functions

It is easy to draw an AND gate having any desired number of inputs, but it is difficult to find an AND-gate IC package having more than two inputs. Table 2-1 shows only three AND ICs, and all three of them have only two inputs. This is not meant to imply that AND functions aren't used very often or that most AND functions call for only two inputs. The fact of the matter is that basic AND functions are so easily implemented with other, more popular function gates. Figure 3-10 shows how it is possible to use more than one AND gate to build up AND functions having three or four inputs; but more importantly, Figs. 3-10(d) through 3-10(i) show how to build AND functions from the more popular and versatile NAND and NOR gates.

3-3.1 Expanding AND Gate Functions

Figure 3-10(a) shows a basic 2-input AND gate such as the one available in 7408 TTL and 4081 CMOS IC packages. Figure 3-10(b) then shows how to use two of these 2-input AND gates to build up a 3-input AND function. The basic idea is to AND together two of the inputs and then AND the result with the third input. The output of AND gate 1, for instance, is AB; and after AND-ing this term with input C at AND gate 2, the result is $(AB)C$—or ABC since the parentheses can be omitted without changing the expression or causing any confusion.

Figure 3-10(c) shows how to implement three 2-input AND gates to achieve the effect of a 4-input AND function. The analysis of this circuit follows the same

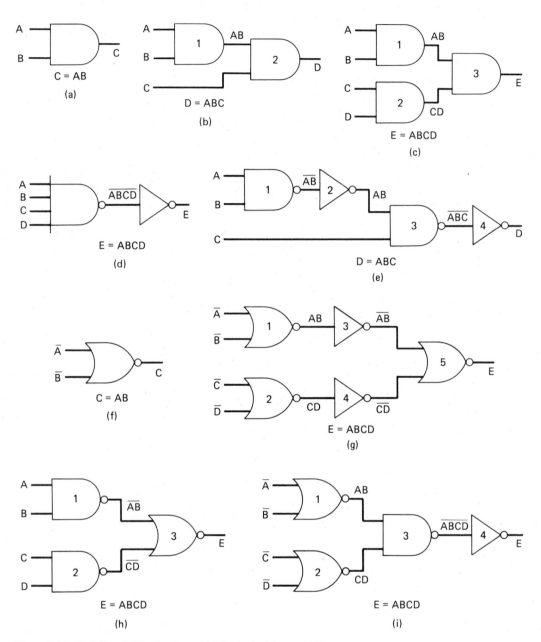

Figure 3-10 Building AND functions. (a) The basic 2-input AND gate. (b) A 3-input AND function from AND gates. (c) A 3-input AND function from AND gates. (d) A 4-input AND function from a NAND gate and inverter. (e) A 3-input AND function from 2-input NAND gates and inverters. (f) A 2-input AND function from a 2-input NOR gate. (g) A 4-input AND function from 2-input NOR gates and inverters. (h) A 4-input AND function from NAND and NOR gates requiring non-inverted inputs. (i) A 4-input AND function from NAND and NOR gates requiring inverted inputs.

42

general procedure described for the 3-input version in Fig. 3-10(b), with the final result being *ABCD* in this instance.

The basic AND function can be expanded to any number of inputs by using one less 2-input AND gate than there are inputs. The 3-input AND function in Fig. 3-10(b), for example, uses two 2-input AND gates; the 4-input version in Fig. 3-10(c) calls for three 2-input AND gates. An 8-input AND function, using this procedure, would then call for seven 2-input AND gates.

3-3.2 Building AND Functions from NAND Gates

Figure 3-10(d) and 3-10(e) illustrate two techniques for building AND functions from NAND gates. The circuit in Fig. 3-10(d) takes advantage of the fact that a NAND function is simply an inverted AND function—*changing a NAND function to an AND operation is a simple matter of inverting the output of the NAND gate.*

NAND gates are readily available with 2, 3, 4 and even 8 inputs, and if any of these are followed by a logic inverter, the overall result is the creation of 2-, 3-, 4- or 8-input AND functions. In principle, this is the simplest approach to building AND functions having more than two inputs; and in a practical sense, it is usually the most economical approach because NAND gates are more widely used and readily available as off-the-shelf items.

The circuit in Fig. 3-10(e) performs the function of a 3-input AND circuit. The job could be done with a 3-input NAND gate followed by an inverter; but this particular circuit is built up from 2-input NAND gates and inverters. Even the inverters in this circuit could be replaced with NAND gate versions as shown in Fig. 3-9(b). Now why would a circuit designer go to all this trouble when a 3-input NAND gate followed by an inverter could do it? Perhaps it is because the designer finds he or she has some spare 2-input NAND gates and inverters left over in IC packages already specified for other operations in the system; and in such instances it is more economical and efficient to use up these spare gates than it is to add another IC package such as a 7410 triple 3-input NAND IC.

3-3.3 Building AND Functions from NOR Gates

Figures 3-10(f) and 3-10(g) show how it is possible to build up AND functions from NOR gates. Carefully note that the inputs to these NOR-gate equivalents are all inverted. While the 2-input NOR gate in Fig. 3-10(f) does perform the logic operation $C = AB$, the A and B inputs must be inverted before they are applied to the circuit. If they aren't inverted first, the NOR gate generates its natural function, $C = \overline{A + B}$.

The idea of using NOR gates to perform AND operations rests with DeMorgan's NOR theorem and the double-NOT theorem. See Table 3-1. A direct analysis of the circuit in Fig. 3-10(f) shows that $C = \overline{A} + \overline{B}$. Applying DeMorgan's NOR theorem, we see that the expression becomes $C = \overline{\overline{A}} \cdot \overline{\overline{B}}$; and applying the double-NOT theorem produces the result, $C = AB$.

NOR gates thus provide a convenient medium for performing AND functions—*building an AND function from NOR gates is a matter of inverting the inputs before applying them to the NOR gates.*

The process of pre-inverting the inputs might seem rather awkward, but it often happens that previous operations on the inputs produce inverted logic levels anyway. If the inputs are already inverted, NOR gates provide the most efficient means for building an AND function; but if these inputs aren't already inverted, it is usually more efficient to use the NAND gate technique.

NOR gates are available in the TTL family with 2 or 3 inputs, making it rather simple to perform 2- or 3-input AND functions with preinverted inputs. The CMOS family provides NOR gates having 2, 3 and 4 inputs. But if the circuit designer is stuck with only 2-input NOR gates, he or she can use the technique shown in Fig. 3-10(g) to build up NOR-gate AND functions having any number of inputs. Of course, it is possible to replace the inverters in Fig. 3-10(g) with the NAND or NOR gate versions shown in Figs. 3-9(a) and 3-9(b).

3-3.4 Building AND Functions from NAND and NOR Gates

Figures 3-10(h) and 3-10(i) show how to build up 4-input AND functions from combinations of 2-input NAND and NOR gates. The circuit in Fig. 3-10(h) is more desirable when the four terms to be AND-ed are not already inverted by some previous operations. The circuit in Fig. 3-10(i), however, is better when the inputs are already inverted. Study both of these circuits carefully, especially noting how the NOR gates take advantage of DeMorgan's NOR theorem to perform AND operations.

3-4 Expanding and Building OR Functions

OR functions appear in digital electronics about as often as any of the other basic functions. A look at the list of available OR IC packages in Table 2-1, however, shows that there are relatively few (only two) OR gate ICs on the market. The primary reason for the scarcity of OR gate packages is the fact that the basic OR function can be easily built up from the more popular NAND and NOR gates.

3-4.1 Expanding OR Gate Functions

Figure 3-11(a) shows a basic 2-input OR gate such as the one available in the 7432 TTL package or the 4071 CMOS IC package. In either case, the circuit has only two inputs; so a circuit designer wanting to use OR gates to perform basic OR functions calling for more than two inputs must expand the simpler gates as shown in Figs. 3-11(b) and 3-11(c).

The circuit in Fig. 3-11(b) shows how it is possible to build a 3-input OR function from a pair of 2-input OR gates. The idea is to first OR two of the inputs in one gate and then OR the result with the third input. In this particular example,

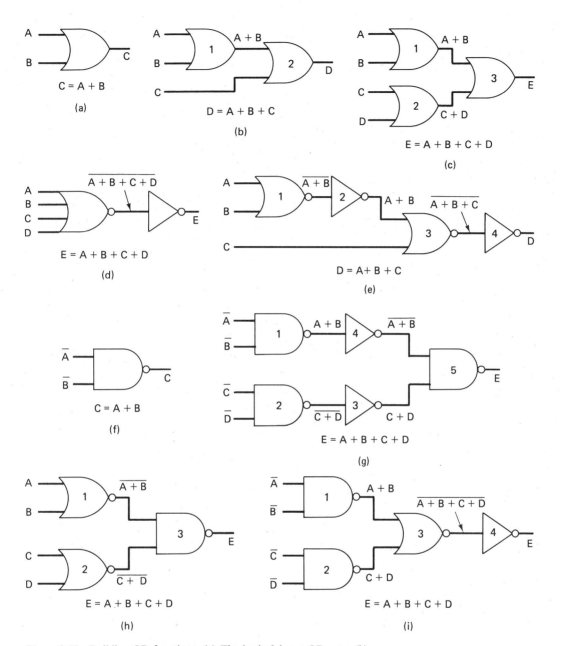

Figure 3-11 Building OR functions. (a) The basic 2-input OR gate. (b) A 3-input OR function from 2-input OR gates. (c) A 4-input OR function from 2-input OR gates. (d) A 3-input OR function from a NOR gate and inverter. (e) A 3-input OR function from 2-input NOR gates and inverters. (f) A 2-input OR function from a 2-input NAND gate. (g) A 4-input OR function from 2-input NAND gates and inverters. (h) A 4-input OR function from NAND and NOR gates requiring non-inverted inputs. (i) A 4-input OR function from NAND and NOR gates requiring inverted inputs.

OR gate 1 yields $A + B$, and when this result is OR-ed with input C in gate 2, the final outcome is $D = A + B + C$.

The circuit in Fig. 3-11(c) then extends this expansion technique to work out the effect of a 4-input OR function. The outputs of OR gates 1 and 2 are $A + B$ and $C + D$, respectively; and after OR-ing these terms in gate 3, the final result is $D = A + B + C + D$.

It is possible to carry this procedure as far as necessary, adding one more 2-input OR gate for each additional term to be OR-ed. It turns out that the number of gates required to do the job is always one less than the number of inputs.

3-4.2 Building OR Functions from NOR Gates

Figures 3-11(d) and 3-11(e) illustrate two basic techniques for building up OR functions from NOR gates and inverters. Of course, the inverters can be replaced by their NAND or NOR gate versions whenever necessary or feasible [see Figs. 3-9(b) and 3-9(c)].

The simple circuit in Fig. 3-11(d) takes advantage of the fact that *inverting the output of a NOR function yields an OR operation*. This principle takes advantage of the double-NOT theorem in Table 3-1.

NOR gates having 2 and 3 inputs are available in both the TTL and CMOS families, but the CMOS family offers the advantage of a 4-input NOR gate. Following any one of these NOR gates with an inverter produces the effects of an OR gate having a corresponding number of inputs.

Circuit designers, however, often face a situation in which they have some spare 2-input NOR gates available from other operations in the system, and they feel obligated to use them instead of adding another IC just to work out a simple 3- or 4-input OR operation. Figure 3-11(e) shows how to combine 2-input NOR gates and inverters (or NAND and NOR versions of them) to produce a 3-input OR function. This circuit is actually a modified version of the OR-gate circuit in Fig. 3-11(b); it replaces the basic 2-input OR gates with NOR/INVERT combinations that do the same job.

3-4.3 Building OR Functions from NAND Gates

Figures 3-11(f) and 3-11(g) show how to build basic OR functions from NAND gates. Carefully note that the inputs to these NAND circuits are inverted; the inputs to the circuit in Fig. 3-11(f), for example, are $\bar{A}$ and $\bar{B}$. If these inputs were not pre-inverted, the NAND gate would perform its natural function, producing the output $C = \overline{AB}$. Inverting the inputs before applying them to the NAND gate, however, directly yields $C = \overline{\bar{A} \cdot \bar{B}}$, which, by DeMorgan's NAND theorem, is equal to $\bar{\bar{A}} + \bar{\bar{B}}$. And by the double-NOT theorem, the equation reduces to $C = A + B$.

Building an OR function from a NAND gate is thus a matter of inverting the inputs before applying them to the NAND gate. Since NAND gates with 2, 3 and 4 inputs are readily available in both the TTL and CMOS families, they provide a popular means of performing basic OR functions, provided the inputs are inverted. Four inverted inputs applied to one of the 4-input NAND gates in a 4012 CMOS package, for example, yields the effect of a 4-input OR circuit.

Whenever it is undesirable or impractical to add a 3- or 4-input NAND gate to achieve the effect of a 3- or 4-input OR function, it is possible to build up the OR function from 2-input NAND gates and inverters. See the 4-input OR function in Fig. 3-11(g). Gates 1 and 2 perform an OR function on their respective inputs according to DeMorgan's NAND theorem. The outputs from gates 1 and 2 are then inverted and applied to NAND gate 5 where they are OR-ed, again by DeMorgan's NAND theorem.

3-4.4 Building OR Functions from NAND and NOR Gates

Figures 3-11(h) and 3-11(i) show two techniques for combining NAND and NOR gates to perform and expand basic OR functions. These particular examples apply to 4-input OR functions, but their general patterns can be reduced or expanded to yield OR functions with any number of inputs.

In both instances, the NAND gates see inverted logic levels, indicating the application of DeMorgan's NAND theorem for performing OR operations with NAND gates. The circuit in Fig. 3-11(h) uses a pair of NOR gates to generate terms $A + B$ and $C + D$ from sets of non-inverted inputs. The outputs of these NOR gates are already inverted, making the use of a NAND gate OR-ing operation a very convenient one.

The circuit in Fig. 3-11(i) is more useful when the four terms to be OR-ed are already inverted by some previous logic operations (the circuit would otherwise call for four inverters at the input). The two NAND gates thus perform a DeMorgan-based OR function, yielding $A + B$ and $C + D$ at their outputs. These non-inverted terms are then NOR-ed by gate 3 and are finally inverted by inverter 4 to generate the basic OR function, $E = A + B + C + D$.

3-5 Expanding and Building NAND and NOR Functions

NAND and NOR functions are the most "natural" logic functions in the sense that they are the simplest to implement in TTL and CMOS IC structures. Their popularity, however, actually arises from the fact that they can be used for synthesizing the basic AND, OR and INVERT functions, as well as expanded versions of themselves. And if there is any question about the great popularity of NAND and NOR gates, a brief analysis of the list of basic gate packages in Table 2-1 ought to clear it up: There are 14 different varieties of NAND and NOR gates, and there are only 8 AND, OR and INVERT IC packages.

3-5.1 Expanding and Building NAND Functions

Figure 3-12 shows some samples of how NAND gate functions can be expanded or built up from combinations of NAND, NOR and INVERT gates. The circuit in Fig. 3-12(a) shows an elementary 2-input NAND gate that directly generates the NAND function, $C = \overline{AB}$. The circuit in Fig. 3-12(b) then shows how it is possible to build up a 4-input NAND function from 2-input NAND gates and inverters (or inverters made from NAND gates). Of course, it would be simpler to perform a 4-input NAND function with one of the common 4-input NAND gates, such as the 7420 TTL dual 4-input NAND package or 4012 CMOS version of the same thing. But as expressed many times in this chapter, circuit designers often find themselves in the position of wanting to use up some spare 2-input gates instead of adding another IC to do the same task.

The NAND-expansion circuit in Fig. 3-12(b) can be modified to reduce or extend the number of input terms. Using a pair of 4-input NAND gates in place

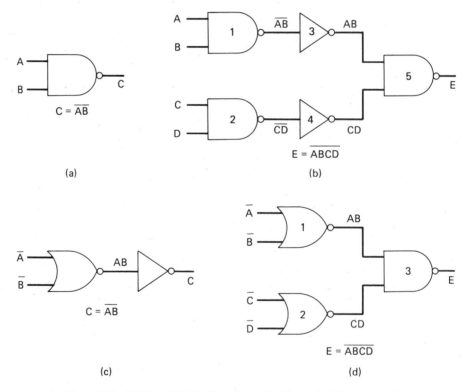

(a)

(b)

(c)

(d)

Figure 3-12 Building NAND functions. (a) The basic 2-input NAND gate. (b) A 4-input NAND function from 2-input NAND gates and inverters. (c) A 2-input NAND function from a NOR gate and inverter. (d) A 4-input NAND function from NAND and NOR gates requiring inverted inputs.

of gates 1 and 2, for example, would produce an 8-input NAND gate; and to carry this notion to an extreme, replacing gates 1 and 2 with 7430 8-input NAND gates would extend the number of inputs to 16.

Figure 3-12(c) shows how to build a simple 2-input NAND function from a 2-input NOR gate and an inverter. The NOR gate has inverted inputs; thus by DeMorgan's NOR theorem it acts as an AND gate. The output of the NOR gate is then AB; and after this result is inverted, the circuit yields $C = \overline{AB}$. The number of possible inputs is dictated by the number of inputs available at the NOR gate; and although it is possible to expand this sort of circuit to accommodate any number of inputs, the resulting complexity of the scheme makes it rather unpopular. Expanding this function is a matter of building up an expanded AND function [see Fig. 3-10(g)] and following it with an inverter.

The circuit in Fig. 3-12(d) is one example of building up a 4-input NAND function with 2-input NOR and NAND gates. As long as the inputs to the NOR gates are inverted as shown in the figure, they yield AND-ed outputs AB and CD. The NAND gate then simply NANDs these terms to generate the function $E = \overline{ABCD}$. This technique is especially useful when working with input terms that are already inverted by some previous logic operations; otherwise, the circuit in Fig. 3-12(b) is somewhat more desirable from a practical point of view.

3-5.2 Expanding and Building NOR Functions

The circuits in Fig. 3-13 demonstrate some common techniques for expanding and building up NOR functions from NAND and NOR gates and inverters.

The circuit in Fig. 3-13(a) simply shows the basic NOR function as performed by a single 2-input NOR gate. NOR gate ICs having 2 and 3 inputs are available in both the TTL and CMOS families, and the CMOS family offers a 4-input NOR gate. This straightforward approach is often the most desirable (and certainly the simplest) way to carry out 2-, 3- and 4-input NOR functions.

Whenever it is impractical to add a separate NOR package, however, it is possible to build up a 4-input NOR function from combinations of 2-input NOR gates and inverters. See Fig. 3-13(b). The inverters can be NAND or NOR gate versions as shown in Figs. 3-9(b) and 3-9(c).

This technique for expanding NOR functions can be extended to accommodate any number of inputs. Replacing gates 1 and 2 with 3-input NOR gates, for instance, produces the effect of a 6-input NOR function.

The little circuit in Fig. 3-13(c) shows how one can use a NAND gate followed by an inverter to produce a NOR function. NAND configurations such as this one actually make it possible to design entire digital systems with NAND gates only, provided the inverter is replaced with a NAND-gate equivalent.

Take note of the important fact that the inputs to the circuit in Fig. 3-13(c) must be inverted. This feature allows the NAND gate to perform an OR function (based on DeMorgan's NAND theorem). The number of possible inputs is dictated by the number of inputs available at the NAND gate, running as high as eight inputs for the 7430 TTL 8-input NAND package.

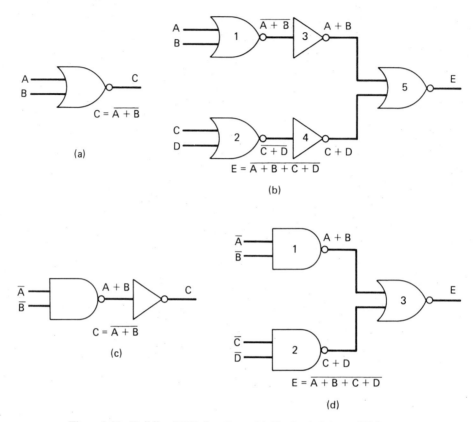

Figure 3-13 Building NOR functions. (a) The basic 2-input NOR gate. (b) A 4-input NOR function from 2-input NOR gates and inverters. (c) A 2-input NOR function from a NAND gate and inverter. (d) A 4-input NOR function from NAND and NOR gates requiring inverted inputs.

Using combinations of NAND and NOR gates permits a convenient procedure for expanding NOR functions to any number. As long as the inputs are first inverted, they are OR-ed by the input NAND gates and then they are NOR-ed by gate 3. Using a 7420 dual 4-input NAND IC in place of gates 1 and 2 in Fig. 3-13(d), for instance, would build an 8-input NOR function.

Comparing the NOR circuits in Figs. 3-13(b) and 3-13(d), we would see that the one in Fig. 3-13(d) would be more desirable when the inputs are not already inverted and that the circuit in Fig. 3-13(d) would be better in cases in which the inputs are already inverted by some previous operations.

Any logic function can, indeed, be expanded and built from combinations of NAND, NOR and INVERT gates. The one most desirable under a given set of conditions depends on several important factors:

1. Whether or not the inputs are already inverted
2. Whether or not the outputs should be inverted
3. The number and types of gates available in IC packages already in use
4. The availability of gates required for doing the job in the most straight-forward manner

3-6 DeMorgan Symbols

The technique of generating basic AND and OR functions from NOR and NAND gates having inverted inputs is a very popular one. The technique is so popular, in fact, that many circuit designers use a special set of logic symbols to indicate its application.

Figure 3-14 shows two kinds of inverted-input circuits and their equivalent DeMorgan symbols. In Fig. 3-14(a) the situation is one in which the designer is using inverted inputs to a standard NAND-gate package to generate an OR operation. The equivalent DeMorgan symbol shows an OR gate having inverting "bub-

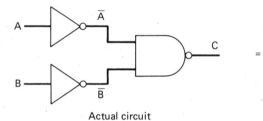
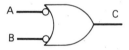

Actual circuit
$C = A + B$

DeMorgan symbol
$C = A + B$

(a)

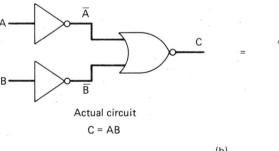

Actual circuit
$C = AB$

DeMorgan symbol
$C = AB$

(b)

Figure 3-14 Equivalent DeMorgan symbols. (a) OR function from a NAND gate having inverted inputs. (b) AND function from a NOR gate having inverted inputs.

bles" at the input. The OR symbol implies an OR-ing operation, while the bubbles denote an inversion of inputs *A* and *B*.

Figure 3-14(b) shows a 2-input AND operation derived from a basic NOR gate having inverted inputs. The same circuit, using DeMorgan symbol notation, is an AND symbol preceded by a set of inverting bubbles.

Since any NAND device can yield either a NAND or OR function, depending on whether the inputs are non-inverted or inverted, they are often described as positive NAND/negative OR devices. By the same line of thinking, a NOR device can be described as a positive NOR/negative AND circuit.

Exercises

1. State the eight basic postulates of Boolean algebra. Check your answers against items la through 4b in Table 3-1.

2. According to the distributive property of Boolean algebra, $X(Y + Z) = XY + XZ$. Extend this basic equation by "multiplying" the following terms: $(W + X)(Y + Z)$ = _____ .

3. State the eight basic AND and OR theorems of Boolean algebra. Check your answers against items 8a through 11b in Table 3-1.

4. Write the two forms of DeMorgan's theorem: Check your results with items 13a and 13b in Table 3-1.

5. Use a truth table to show that (a) $\bar{X}\cdot\bar{Y}$ does not equal $\overline{XY}$ and (b) $\bar{X} + \bar{Y}$ does not equal $\overline{X + Y}$.

6. Name the basic functions performed by the circuits in Fig. E-3.6.

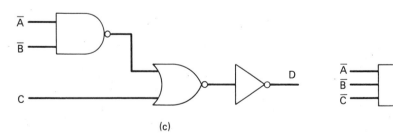

(a)

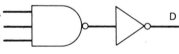

(b)

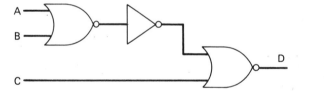

(c)

(d)

Figure E-3.6

7. Draw circuits for the following functions by using only 2-input NAND gates:
  (a)   3-input AND;
  (b)   3-input OR;
  (c)   3-input NAND.
  Hint: Build inverters by connecting together the inputs of the NAND gates; assume
  non-inverted inputs.

4 ANALYZING AND DESIGNING DIGITAL LOGIC FUNCTIONS

Whether one is attempting to analyze an existing digital logic circuit or design one, the procedures involved are much the same. Analyzing an existing circuit is generally a matter of generating a truth table or writing logic equations that express the circuit's main operating characteristics in some meaningful way. By the same token, designing digital logic circuits is a matter of generating truth tables and logic equations and then selecting the logic gates that can do the specified job.

In either case—whether anlyzing or designing logic circuits—a good working knowledge of logic equations and truth-table techniques is essential. This chapter extends and refines some of the techniques already described in previous chapters. The overall objective is to round out your understanding of basic combinatorial logic circuits.

4-1   Boolean Logic Identities

A logic identity is any logic equation that expresses an equality between two terms that often look quite different. Either one of the two terms of an identity, however, can be either expanded or reduced by algebraic techniques to yield a statement of obvious equality. Table 4-1 shows four very common and especially useful Boolean identities.

The application of Boolean identities is important to circuit analysis because it leads to simpler and more obvious expressions of a circuit's overall operating

Table 4-1 Four Common Boolean Identities

1. $A(A + B) = A$
2. $A + AB = A$
3. $A(\bar{A} + B) = AB$
4. $A + \bar{A}B = A + B$

characteristics. And as far as circuit design procedures are concerned, identities provide a convenient means for simplifying the circuits and exploring alternate ways to do a particular job.

In principle, there is an infinite variety of valid Boolean identities; therefore, it is virtually impossible to list more than a few of the most useful ones. It is up to the individual working with digital logic circuits to perfect his or her insight into deriving, recognizing and proving the validity of identity equations of all kinds.

Proving Boolean identities is a necessary part of working with digital logic equations. Anyone with some experience in digital electronics learns to identify certain identity relationships and take their validity for granted. Situations continually arise, however, where logic equations do not appear to fall into a familiar pattern; in such instances, it is wise to prove the equality of the two terms.

There are two basic approaches to proving the validity of an identity equation: by algebraic manipulation or by means of a truth-table proof. Sometimes both approaches are necessary for revealing the real essence of a logic expression.

Identity 1 in Table 4-1 is actually a specific application of the distributive property of Boolean algebra, showing how a relatively complex AND/OR relationship can be reduced to a single term. Figure 4-1(a) shows the algebraic and truth-table proofs as well as a set of equivalent circuits.

The first step in the algebraic proof is to state the identity to be proven: $A(A + B) = A$. The objective is to expand or reduce the more complicated of the two terms until it is equal to the simpler one. Applying the distributive property to the original statement yields the expression $AA + AB = A$; and taking advantage of the theorem that says any term AND-ed with itself yields that term, we have $A + AB = A$. Now applying the distributive property in the factoring sense yields $A(1 + B) = A$; but since any term OR-ed with 1 is equal to 1, the expression reduces to $A \cdot 1 = A$. And, finally, applying the theorem that says anything AND-ed with 1 is equal to that term itself, we get $A = A$, an obvious statement of equality that proves the validity of the original statement.

The truth table in Fig. 4-1(a) shows a different kind of approach to proving the same identity, $A(A + B) = A$. The first two columns, labeled A and B, merely show all four possible combinations of 1's and 0's for the circuit's two inputs. The next column shows the results of OR-ing columns A and B; and, finally, the last column shows the results of AND-ing the $A + B$ column with input A—a truth-table expression of the term $A(A + B)$. From that point, it doesn't take much study to see that the entire expression $A(A + B)$ can be replaced with the simpler term, A.

The B term in identity 1 is thus totally irrelevant. No matter what the B input to the circuit might be, the output always follows the A input. The AND/OR circuit in Fig.4-1(a) can thus be replaced with a straight wire; and using a straight piece of wire is certainly more economical than using the equivalent AND/OR circuit that generates the expression $A(A + B)$.

Identity 2 in Table 4-1 is similar to identity 1 in that they are both specific applications of the Boolean distributive property. In identity 2, however, the proof calls for "factoring," rather than expanding, the more complex term. Figure 4-1(b)

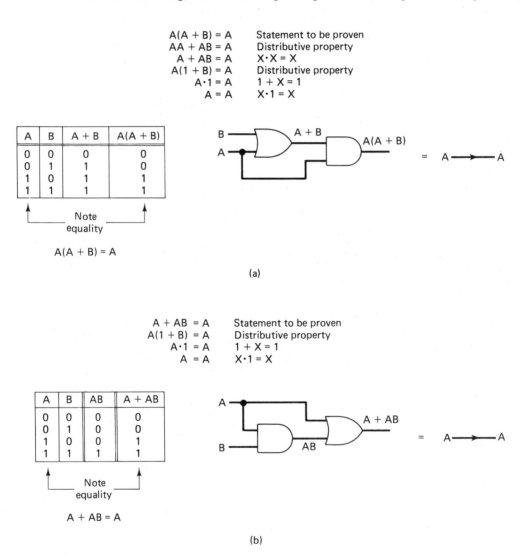

(a)

(b)

Figure 4-1 Boolean identities: algebraic and truth-table proofs and equivalent circuits. (a) $A(A + B) = A$. (b) $A + AB = A$.

$$A(\overline{A} + B) = AB \qquad \text{Statement to be proven}$$
$$A\overline{A} + AB = AB \qquad \text{Distributive property}$$
$$0 + AB = AB \qquad X \cdot \overline{X} = 0$$
$$AB = AB \qquad 0 + X = X$$

A	B	$\overline{A}$	$\overline{A}$ +B	$A(\overline{A} + B)$	AB
0	0	1	1	0	0
0	1	1	1	0	0
1	0	0	0	0	0
1	1	0	1	1	1

$$A(\overline{A} + B) = AB$$

Note equality

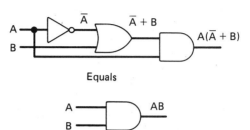

Equals

(c)

$$A + \overline{A}B = A + B \qquad \text{Statement to be proven}$$
$$\overline{\overline{A + \overline{A}B}} = A + B \qquad X = \overline{\overline{X}}$$
$$\overline{\overline{A}(\overline{\overline{A}B})} = A + B \qquad \overline{X + Y} = \overline{X} \cdot \overline{Y}$$
$$\overline{\overline{A}(\overline{\overline{A}} + \overline{B})} = A + B \qquad \overline{XY} = \overline{X} + \overline{Y}$$
$$\overline{\overline{A}(A + \overline{B})} = A + B \qquad \overline{\overline{X}} = X$$
$$\overline{\overline{A}A + \overline{A}\overline{B}} = A + B \qquad \text{Distributive property}$$
$$\overline{0 + \overline{A}\overline{B}} = A + B \qquad \overline{X}X = 0$$
$$\overline{\overline{A}\,\overline{B}} = A + B \qquad 0 + X = X$$
$$\overline{\overline{A}} + \overline{\overline{B}} = A + B \qquad \overline{XY} = \overline{X} + \overline{Y}$$
$$A + B = A + B \qquad \overline{\overline{X}} = X$$

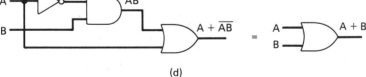

(d)

Figure 4-1   (*cont.*) (c) $A(\overline{A} + B) = AB$. (d) $A + \overline{A}B = A + B$.

shows the algebraic and truth-table proofs of identity 2 as well as equivalent circuits for carrying out the prescribed logic operations.

The first step in the proof is to factor out the A's in the left-hand side of the equation. This yields $A(1 + B) = A$. Applying the theorem that says $1 + X = 1$, we see that $A \cdot 1 = A$, which is certainly true from the theorem, $X \cdot 1 = X$. The final result is that $A = A$, an obviously true statement that proves the identity.

The truth-table proof in Fig. 4-1(b) first shows all possible combinations of 1's and 0's for the A and B inputs; then it shows the results of AND-ing inputs A

and B. The final step, OR-ing A with AB, appears in the last column. And since the pattern of 1's and 0's in this last column is identical to the pattern in the A column, it follows that $A + AB = A$.

These two proofs of identity 2 imply that the B input is irrelevant; no matter what the B input does, the output always follows the A input. And once again, a two-gate logic circuit can be replaced with a straight piece of wire as shown in the logic circuits in Fig. 4-1(b).

Identity relationships sometimes contain surprises. The left-hand side of identity 3 in Table 4-1, for example, looks very much like the left-hand side of the equation listed as identity 1. The inverted A term in identity 3 might not seem very important at first. It wouldn't seem that using this one inverted A term would make the B term relevant, but it does. Figure 4-1(c) summarizes the algebraic and truth-table proof for identity 3.

The proof begins with an expansion of the left-hand side of the equation by using the principle of distributive "multiplication": $A(\bar{A} + B) = A\bar{A} + AB$. Since $A\bar{A} = 0$ (from the theorem $X\bar{X} = 0$), the equation further reduces to $0 + AB = AB$; and the statement is finally proven by recognizing the fact that $0 + AB = AB$.

The truth-table proof in Fig. 4-1(c) merely confirms the fact that $A(\bar{A} + B)$ is actually equal to the simple AND function, AB; and it is certainly easier to implement the function AB than it is to set up the logic gates for working out $A(\bar{A} + B)$. Compare the logic circuits in Fig. 4-1(c).

As far as circuit analysis is concerned, applying Boolean identities greatly simplifies logic expressions, reducing an otherwise complicated and perhaps confusing expression to its simplest and clearest form. And for anyone designing logic circuits, the proper application of Boolean identities can reduce the amount of logic circuitry required for carrying out a specified logic operation. Compare the equivalent circuits in Figs. 4-1(a), 4-1(b) and 4-1(c).

Now there is one important question that often arises at this point: How do you know what to do first when trying to prove a Boolean identity? The answer is generally quite simple: Do whatever you *can* do first, that is, apply whatever postulate, property or theorem of Boolean algebra you can. In the case of proving identities 1 and 3, the first step is to expand the left-hand side of the equation by distributive expansion ("multiplying" A by the terms in parentheses). Actually, there are no other postulates, properties or Boolean theorems that can be applied with any real meaning at first. Do whatever you can do first; and after taking that first step, observe how opportunities for applying other theorems and properties crop up one at a time, *leading* you to the final step, one logical step at a time.

Any valid identity can be proven by an appropriate sequence of steps. Of course, there are times when it is extremely difficult to see what can be done next; but experience is the key to reducing the number of tough-dog proofs to a minimum. And if all else fails, a truth-table proof always works.

Identity 4 in Table 4-1 is an example of an identity that is rather tricky to prove because it is not obvious what should be done first. It appears that none of the basic Boolean properties and theorems apply. The distributive property doesn't apply to the left-hand side of the equation because A and $\bar{A}$ are entirely different terms. In such instances, try double NOT-ing the entire expression.

Double NOT-ing one side of a logic equation doesn't change its "value" at all, but it does provide an opportunity to apply one of DeMorgan's theorems. Sometimes this double NOT-ing procedure is the only way to get an algebraic proof underway. Carefully study the rather lengthy, but workable, proof for identity 4 in Fig. 4-1(d). A truth-table proof of this identity is left as an exercise at the end of this chapter.

4-2 Simplifying Boolean Logic Expressions

Whether designing a circuit or analyzing an existing one, there is nearly always a need for simplifying Boolean logic expressions for the circuits. There are several important reasons for simplifying logic expression. For one, a logic expression reduced to its simplest terms is easier to "read"; it is easier to see the essential operating features of the circuit it represents. Then, too, a reduced or simplified logic expression lends itself to simpler and more direct truth-table proof and investigation. And as far as circuit design is concerned, there is a general rule of thumb that says *the simplest circuits come from the simplest equations.*

In principle, it is possible to reduce any logic expression to its simplest form by applying the postulates, properties, theorems and identities of Boolean algebra. Reducing logic expressions by algebraic manipulation calls for the same kind of insight and skill required for proving Boolean identities. In fact, reducing logic equations is much like "proving" a Boolean identity that has a term on just one side of the equal sign. It is up to the user to provide the other term in its most basic and straightforward form.

Just about any logic expression containing the same factor in more than one place can be simplified. The expression $A(A + B)$, for example, can be simplified because the A appears more than once. Applying identity 1 to this particular expression reduces it to A, an expression that is certainly simpler than the original one.

By the same token, the expression $AB + AC$ can also be simplified. Factoring out the A terms reduces it to the simpler form $A(B + C)$. The latter expression cannot be simplified any further, since each term appears only once.

Can an expression such as $ABC + ABD + AB$ be simplified? It certainly can because the AB term appears more than once in that expression. Factoring an AB from the three terms yields $AB(C + D + 1)$; and since any expression OR-ed with 1 is equal to 1, the expression further reduces to $AB \cdot 1$, which is even further reducible to AB.

Figure 4-2 compares the logic circuits for working equivalent circuits $ABC + ABD + AB$ and AB. The value of reducing logic expressions to their simplest terms *before* designing a circuit is clearly demonstrated in this particular example.

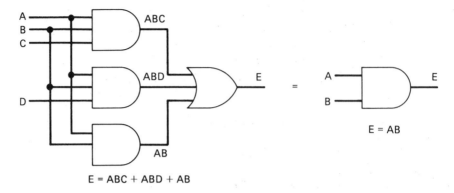

E = ABC + ABD + AB

E = AB

Figure 4-2 Results of simplifying $ABC + ABD + AB$.

Logic expressions containing more than one like term are generally reducible to a simpler form; and anyone working with such equations, whether for the purposes of circuit analysis or design, ought to make the effort to go through the simplification procedures.

Expressions containing a term and an inverted version of itself are often reducible. Take, for instance, the expression $AB(\bar{A} + C)$. An A and an $\bar{A}$ appear in the same expression, hinting at the fact that the expression might be reduced further. It is always worth a try, anyway. In this case, Boolean "multiplication" leads to $AB\bar{A} + ABC$; and since $A\bar{A} = 0$, the expression becomes $B \cdot 0 + ABC$. Now, anything AND-ed with 0 is equal to zero, and the final and simplest form of $AB(\bar{A} + C)$ is thus ABC.

Not all expressions containing a term plus an inverted version of itself are reducible. The expression $A\bar{B} + \bar{A}B$ is one example. This expression can be altered in form as shown in Fig. 4-3, but it cannot be really simplified to remove some of the terms.

A third indication that a Boolean logic expression can be simplified is an overabundance of NOT signs. Suppose the analysis of an existing logic circuit yields the expression $\overline{A + \overline{BC}}$. Here there are two NOT bars, one within the other. It is always possible to reduce the number of NOT bars, or at least to eliminate the confusing NOT-within-NOT situation.

One of DeMorgan's theorems is usually appropriate for getting rid of excessive NOT signs in a logic expression. Where the expression has the form $\overline{A + \overline{BC}}$, for example, using DeMorgan's NOR theorem yields $\bar{A} \cdot \overline{\overline{BC}}$; and applying the double-NOT theorem produces the simplest possible form, $\bar{A}BC$. Thus $\overline{A + \overline{BC}} = \bar{A}BC$, and a truth-table analysis can prove the validity of this simplification procedure.

Any algebraic simplification of a logic expression actually generates a Boolean identity: One side of the identity equation is the original expression, and the other side is the simplified version. And like any Boolean identity, the relationship can be

$$A\overline{B} + \overline{A}B$$
$$= (A + B)(\overline{A} + \overline{B}) \quad \text{Law of consensus}$$
$$= (A + B)\overline{AB} \quad \text{DeMorgan's NAND theorem}$$

$$A\overline{B} + \overline{A}B = (A + B)\overline{AB}$$

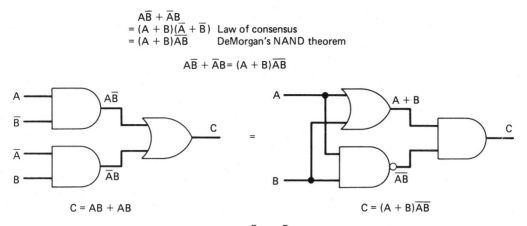

Figure 4-3 Equivalent circuits for the expression $\overline{A}B + \overline{A}B$.

doublechecked and proven to the user's satisfaction by means of a truth-table analysis.

The signs of a reducible logic equation are thus:

1. A term or set of terms appearing more than once
2. A term and an inverted version of itself appearing in the same expression (except for the special case $\overline{A}B + A\overline{B}$)
3. An overabundance of NOT bars, especially NOT-within-NOT situations

4-3 Analyzing Combinatorial Logic Circuits

The whole purpose of analyzing an existing logic circuit is to uncover its fundamental operating characteristics, thereby getting some insight into what it is used for and how it works. Experience with digital circuitry makes it possible to recognize many popular operations at a glance; and in such instances, formal procedures for analyzing the circuit aren't necessary. There is a rather simple procedure, however, that always leads to a complete analysis of any combinatorial logic circuit, and that procedure is the subject of this section. (Analyzing logic circuits containing sequential logic components such as timers and flip-flops call for a slightly different approach described in a later chapter of this book).

The first step in analyzing a logic circuit is to assign labels to the outputs of *all* the gates. The selection of labels is arbitrary, but it is sometimes helpful to select labels that reflect the nature of the outputs; but in any event, it is important to identify the output of each gate in some convenient fashion.

Figure 4-4 shows a combinatorial logic circuit made up of two NAND gates and a pair of inverters. Labels A, B and C are assigned to the circuit's inputs and D is assigned to the output. Labels E, F and G are then assigned to the outputs of all the intermediate stages.

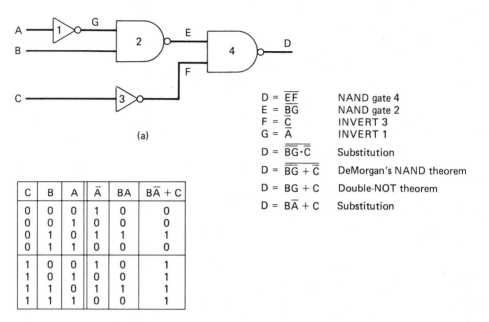

(a)

D = $\overline{EF}$	NAND gate 4
E = $\overline{BG}$	NAND gate 2
F = $\overline{C}$	INVERT 3
G = $\overline{A}$	INVERT 1
D = $\overline{\overline{BG} \cdot \overline{C}}$	Substitution
D = $\overline{\overline{BG}} + \overline{\overline{C}}$	DeMorgan's NAND theorem
D = BG + C	Double-NOT theorem
D = $B\overline{A} + C$	Substitution

C	B	A	$\overline{A}$	BA	$B\overline{A}$ + C
0	0	0	1	0	0
0	0	1	0	0	0
0	1	0	1	1	1
0	1	1	0	0	0
1	0	0	1	0	1
1	0	1	0	0	1
1	1	0	1	1	1
1	1	1	0	0	1

(b)

Figure 4-4   Analysis of a NAND circuit.

After all the inputs, outputs and intermediate stages are assigned labels, the next step in the analysis is to write the fundamental logic equation for each gate by using the assigned labels. The output of NAND gate 4 in Fig. 4-4, for instance, is $\overline{EF}$; the output of NAND gate 2 is $\overline{BG}$. The same idea holds for the inverters: $G = \overline{A}$ and $F = \overline{C}$.

Since there are four gates in this particular circuit, there have to be four preliminary logic equations expressing the outputs of each one of them.

The whole idea of this analysis procedure is to derive a single logic equation that expresses the output in terms of the inputs—the intermediate expressions should not appear in the final equation. So the next step is to eliminate the intermediate expressions for E, F and G.

Beginning with the output equation $D = \overline{EF}$, eliminate the E and F terms by substituting the preliminary expressions for E and F. By substitution, then, $D = \overline{\overline{BG} \cdot \overline{C}}$. This expression can be simplified by first applying DeMorgan's NAND theorem, followed by the double-NOT theorem. The result is $D = BG + C$; but it still contains one of the intermediate terms, G. The final step, then, is to eliminate the G term by substituting its equivalent value, $\overline{A}$. The overall result is the final equation, $D = \overline{A}B + C$.

The analysis of the circuit in Fig. 4-4 might be complete at this point as far as an experienced digital technician is concerned. The final equation states the fact that the D output will be at logic 1 whenever $C = 1$ OR when $A = 0$ AND $B = 1$ at the same time. (Substituting those patterns of 1's and 0's into the equation and applying the basic postulates of Boolean algebra yield $D = 1$. Otherwise, $D = 0$).

It is possible, and often desirable, to extend this analysis to include a truth table. Truth tables can reflect the essential character of a circuit more clearly.

The truth table in Fig. 4-4 first shows all eight possible combinations of inputs A, B and C, followed by columns representing $\bar{A}$, $B\bar{A}$ and the final result $B\bar{A} + C$. A study of this truth table shows that the circuit's output is at logic 1 when $C = 0$, $B = 1$ and $A = 0$, and also whenever $C = 1$. The final conclusion here is the same as that "read" from the logic equation, but the truth table shows the results in a somewhat clearer tabular form.

A digital technician can then use this truth table as a basis for testing and troubleshooting the circuit.

The procedure for analyzing any combinatorial logic circuit can be summarized as follows:

1. Assign labels to all inputs and outputs, including the intermediate stages.
2. Write the basic logic equation for each gate in the circuit.
3. Use the process of substitution to eliminate expressions for the intermediate stages, simplifying the results whenever possible.
4. Use the final form of the logic equation for generating a truth table.

This procedure applies to any combinatorial logic circuit, including the simple three-gate circuit in Fig. 4-5. This particular example, however, points out the value of trying several truth tables to find a scheme that most clearly shows the essence of the circuit.

After arbitrary lables are assigned and the basic gate equations are written, the process of substitution and some simplification yields the expression $D = A(B + C)$. Anyone learning to read logic equations can see that $D = 1$, provided $A = 1$ while B OR $C = 1$ at the same time. Or to put it another way: This is an OR circuit that OR's inputs B and C, but only if $A = 1$. It is a *gated OR circuit*. A beginner might not reconize this fact, however; so the next reasonable step in the analysis is to generate a truth table, a truth table that hopefully sheds some light on the real function of the circuit.

Figure 4-5(b) shows one version of a truth table for this circuit. The table shows all eight possible combinations of inputs A, B and C, and the final column shows the result, $D = A(B + C)$. The truth table in Fig. 4-5(b) is indeed a valid representation of the circuit in Fig. 4-5(a), but it doesn't really show the gated-OR feature very clearly.

Whenever a truth table doesn't seem to say anything really meaningful, it is a good idea to draw another one using a different sequence of inputs. The truth table in Fig. 4-5(c) is also a valid one for the circuit in Fig. 4-5(a), but the sequence of inputs has been reversed compared to the first truth table.

Note that the 1's and 0's in the final column in Fig. 4-5(c) are grouped together. This is a good sign that the table more clearly reflects the main operating charac-

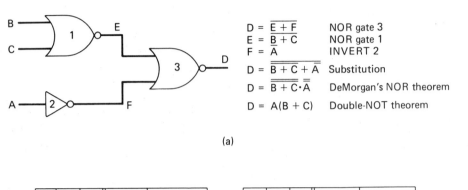

(a)

C	B	A	B + C	A(B + C)
0	0	0	0	0
0	0	1	0	0
0	1	0	1	0
0	1	1	1	1
1	0	0	1	0
1	0	1	1	1
1	1	0	1	0
1	1	1	1	1

(b)

A	B	C	B + C	A(B + C)
0	0	0	0	0
0	0	1	1	0
0	1	0	1	0
0	1	1	1	0
1	0	0	0	0
1	0	1	1	1
1	1	0	1	1
1	1	1	1	1

(c)

Figure 4-5 Analysis of a NOR circuit. (a) Circuit diagram and algebraic analysis. (b) One possible truth table. (c) A better truth table.

teristics of the circuit at hand. In this instance, it is quite clear that the *A* input must be equal to 1 before the output can be equal to 1; and by looking at just the lower half of the truth table (where $A = 1$), it becomes apparent that it is following a simple $B + C$ pattern. In other words, the table in Fig. 4-5(c) shows that $D = B + C$, but only if $A = 1$. Otherwise, the output is 0.

Being able to interpret the true function of a logic circuit from its logic equation is a valuable asset to any digital technician; but when this fails, it is possible to glean the same understanding from a good truth-table representation. Sometimes the input pattern for the truth table has to be shuffled around a few times to see the real essence of the circuit, but there is always a *best* truth table, one that most clearly reflects the essential operating features of the circuit at hand.

4-4 Designing Combinatorial Logic Circuits

Logic equations and truth tables play vital roles in the analysis of any given combinatorial logic circuit; and by the same token, these tools are valuable for designing logic circuits. In a sense, the procedure for designing a logic circuit is just the reverse of analyzing an existing circuit. The process usually begins with a truth table, proceeds to a logic equation built from the truth table, and then goes to the actual circuit design based on the logic equation. The only real difference between the operations involved in analyzing and designing combinatorial logic circuits is that the circuit designer has more latitude as far as selecting the actual gates for implementing the logic operation.

Any logic design begins with defining the basic concept of what the circuit is supposed to do. The designer might be able to verbalize the idea, but in many instances a truth table can show what must be done more clearly. In either case, the truth table in Fig. 4-6(a) expresses the following notion: The output of the circuit should be at logic 1 whenever $C = 0$ AND A and B are equal ($A = B = 0$ or $A = B = 1$). An experienced circuit designer could go directly to a logic equation from this notion, but a truth table is in order whenever there is any doubt about what the logic equation should look like.

Once the truth table has been set up, the next step is to write a logic equation that expresses all conditions where $D = 1$. In this particular instance, $D = 1$ at two places in the truth table: where all three inputs are at logic 0 at the same time and where $C = 0$ and $B = A = 1$. Expressing these two conditions as a logic equation, $D = \bar{A}\bar{B}\bar{C} + AB\bar{C}$. Note that logic-0 terms at the inputs of the truth table appear as NOT-ed terms in the preliminary logic equation, while logic-1 terms appear as non-inverted expressions.

The preliminary equation drawn directly from the truth table can be simplified by first "factoring" out the $\bar{C}$ terms and then applying DeMorgan's NAND theorem to the $\bar{A}\bar{B}$ term. The final result is the equation, $D = \bar{C}(\overline{A + B} + AB)$.

With a reduced logic equation at hand, the next step is to devise a logic circuit that does the job. Inspection of the logic equation shows that the circuit will require at least four logic gates: one for AND-ing A and B, one for NOR-ing A and B, one for the OR function between $\overline{A + B}$ and AB, and finally one for AND-ing $\bar{C}$ with the rest of the terms.

The circuits in Fig. 4-6(b) show how to develop the logic equation into a working circuit by the most straightforward approach. It is a matter of realizing that the equation is basically an AND operation, AND-ing together $\bar{C}$ and $(\overline{A + B} + AB)$. So the last step in the logic circuit should do this particular job. See step 1 in Fig. 4-6(b).

Completing the circuit is then a matter of building circuits to generate the inputs required for AND gate 1. An OR gate connected to one input of AND gate 1, for instance, can take care of the $\overline{A + B} + AB$ operation (step 2). A NOR gate then provides the operation $\overline{A + B}$, while another AND gate does the A AND B job. The circuit is thus complete as shown in step 4, using four logic gates as predicted at the outset.

AND and OR gates are not widely used in digital technology these days, however; therefore, it is necessary to build up NAND and NOR equivalents of these basic gate functions.

The circuits in Fig. 4-6(c) show how to do the job by using only NOR gates.

C	B	A	D
0	0	0	1
0	0	1	0
0	1	0	0
0	1	1	1
1	0	0	0
1	0	1	0
1	1	0	0
1	1	1	0

$D = \overline{A}\,\overline{B}\,\overline{C} + AB\overline{C}$ From truth table
$D = \overline{C}(\overline{A}\,\overline{B} + AB)$ Distributive property
$D = \overline{C}(\overline{A + B} + AB)$ DeMorgan's NAND theorem

(a)

Step 1

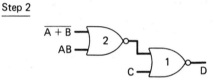

Step 2

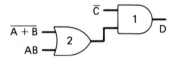

Step 3

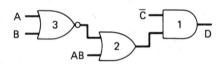

Step 4

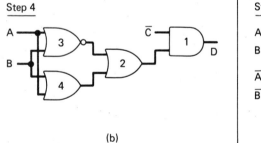

(b)

Step 1

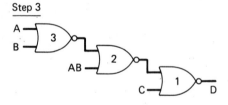

Step 2

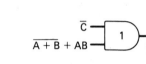

Step 3

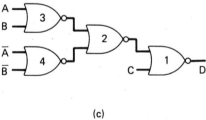

Step 4

(c)

Figure 4-6 Design of a logic circuit. (a) Required truth table and resulting Boolian expression. (b) A straight-forward, but impractical, design. (c) A NOR gate design.

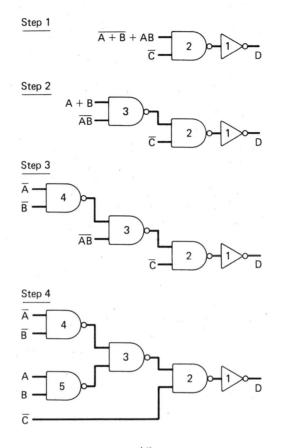

Figure 4-6   (*cont.*) (d) A NAND gate design.

Step 1 in Fig. 4-6(c) takes advantage of the fact that inverted inputs to a NOR gate yield an AND function (see Sec. 3-3.3). Another NOR gate naturally provides the NOR function required at one input of NOR gate 1; NOR gate 3 then provides the $\overline{A + B}$ function at one input of NOR gate 2. See step 3 in Fig. 4-6(c). NOR gate 4 puts the final touch on the circuit, doing the job of an AND gate by having inverted inputs applied to it.

This circuit, like the more straightforward version, requires only four logic gates, but it has the disadvantage of requiring a set of inverted inputs. If these inverted inputs aren't readily available from some previous logic operations, the designer must include a pair of inverters to get those inverted inputs for NOR gate 4.

The sequence of design steps in Fig. 4-6(d) shows how to build the circuit from NAND gates and one inverter. NAND gate 2 and the inverter in step 1 perform the function of an AND gate. Step 2 in Fig. 4-6(d) then shows how a NAND gate with inverted inputs can be used for carrying out a basic OR function. Also see Sec. 3-4.3.

NAND gate 4 in step 3 OR's inputs A and B, while NAND gate 5 in step 4 naturally provides the prescribed $\overline{AB}$ input to NAND gate 3.

The NAND gate version of this circuit, like its NOR gate counterpart in Fig. 4-6(c), requires a set of inverted inputs. The NAND gate version also requires an additional INVERT gate at its output. Although it would appear that the NAND gate version is less desirable in terms of the amount of required gate functions, it is not necessarily the worst choice in the long run. The choice of circuits depends a great deal on certain factors not introduced here. The inverter required for the NAND gate design, for instance, might be readily available in a partly used IC package elsewhere in the system. And there is also the simple fact that circuits cost less when they use as many identical gate functions as possible, and the designer might be using a large quantity of NAND gates throughout the design.

The circuit in Fig. 4-6(b) is certainly the least desirable from a practical point of view because it calls for using three different kinds of IC logic packages. It makes a good first design, however. After that, the choice of the NOR or NAND version depends on many other considerations that aren't directly related to this particular design example.

A good procedure for designing combinatorial logic circuits is as follows:

1. Build a truth table that expresses the desired relationships between the output and all possible combinations of 1's and 0's at the inputs.
2. Derive a logic equation from the truth table, first OR-ing together all of the input logic levels that yield a logic-1 output and then simplifying the results.
3. Devise a straightforward logic circuit using AND, OR, NAND, NOR and INVERT gates as directly indicated in the simplified logic equation.
4. Build an equivalent circuit from NAND, NOR and INVERT gates as necessary and desirable.

There are certainly numerous shortcuts for basic logic circuit design, and many of them will be described at appropriate places throughout this book. The basic procedure just described here, however, always works for combinatorial logic design situations (designing circuits around sequential logic elements such as timers and flip-flops calls for slightly different techniques).

The circuit design procedure illustrated in Fig. 4-7 clearly shows the importance of being able to detect possible identity relationships in logic equations. The basic design problem in this instance is to devise a logic circuit that performs the function of a three-way switch, a switching scheme that lets the user turn a single light on and off from any one of three switch stations.

C	B	A	D
0	0	0	0
0	0	1	1
0	1	0	1
0	1	1	0
1	0	0	1
1	0	1	0
1	1	0	0
1	1	1	1

(a)

$D = A\overline{B}\,\overline{C} + \overline{A}B\overline{C} + \overline{A}\,\overline{B}C + ABC$ From truth table

$D = \overline{C}(A\overline{B} + \overline{A}B) + C(\overline{A}\,\overline{B} + AB)$ Distributive property

Identify proof

$$A\overline{B} + \overline{A}B \overset{?}{=} \overline{\overline{A}\,\overline{B} + AB}$$ Possible Identity

$$= \overline{\overline{A} + B} + \overline{AB}$$

$$= \overline{\overline{A} + B}\,(\overline{AB})$$

$$= (A + B)(\overline{A} + \overline{B})$$

$$= A\overline{A} + A\overline{B} + \overline{A}B + B\overline{B}$$

$$A\overline{B} + \overline{A}B = A\overline{B} + \overline{A}B$$ Identity proven

$$\text{Let } U = A\overline{B} + \overline{A}B$$

$$\therefore D = \overline{C}U + C\overline{U}$$

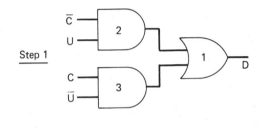

Step 1

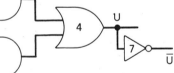

Step 2

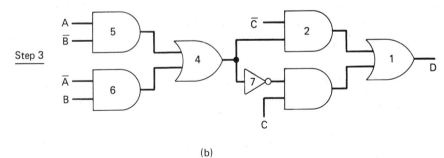

Step 3

(b)

Figure 4-7 Design of a logic circuit. (a) Required truth table and the resulting Boolean expression. (b) A straightforward, but impractical, design.

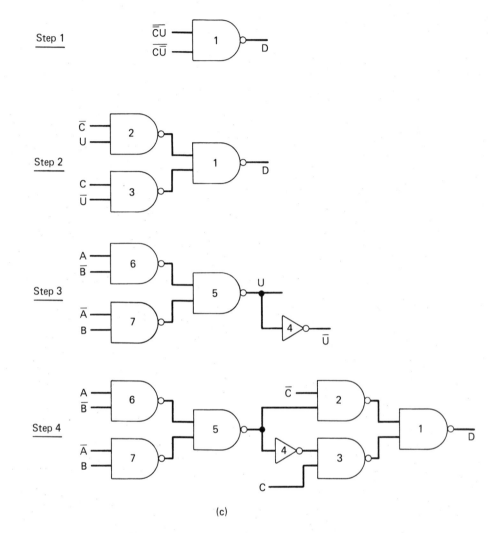

Figure 4-7 (cont.) (c) A practical NAND gate design.

Deciding to define an *on* switch or lamp as a logic-1 state and an *off* switch and lamp as a logic-0 state, the user can work out the truth table in Fig. 4-7(a). Note that the light is off ($D = 0$) whenever all three switches are off ($A = B = C = 0$), but the light goes on whenever any *one* or all three of the switches are on. The light also goes off whenever any two switches are on. Study the truth table carefully, convincing yourself that it actually does the intended job—allowing a user to control a single lamp from any one of three different switching stations.

Once the truth table looks good, work out a preliminary logic equation that shows all combinations of input conditions that produce a logic-1 output. There are four conditions that satisfy this requirement; therefore, there are four OR-ed elements in the first logic equation in Fig. 4-7(a).

Simplifying the preliminary logic equation is a matter of seeing that a $\bar{C}$ can be "factored out" of the first two terms and that a C can be "factored out" of the second two. This application of the Boolean distributive property generates the equation $\bar{C}(A\bar{B} + \bar{A}B) + C(\bar{A}\bar{B} + AB)$. Now this is a rather complicated equation that contains several like terms. It should be possible to simplify it a great deal, but trying to simplify this particular equation as it stands turns out to be a very cumbersome task.

It is at this point a circuit designer calls on his or her past experience and knowledge of Boolean identities. Is it possible that there is some direct identity relationship between the terms $(A\bar{B} + \bar{A}B)$ and $(\bar{A}\bar{B} + AB)$? They look very much alike and they contain the same series of expressions. Is it possible that one of these terms is merely an inverted version of the other? To check this idea, express $\overline{A\bar{B} + \bar{A}B} = \bar{A}\bar{B} + AB$ as an identity; then attempt to prove it. The proof of this particular identity is shown in Fig. 4-7(a); and, indeed, the "guess" pays off.

Since these two A and B expressions are merely inverted versions of one another, it is possible to simplify the design procedure by letting $U = A\bar{B} + \bar{A}B$. The basic equation then becomes $D = \bar{C}U + C\bar{U}$, a rather simple relationship that requires only three logic gates. See step 1 in Fig. 4-7(b).

Getting the U input is then a matter of working out the expression $U = A\bar{B} + \bar{A}B$; and this can be done with three more gates as shown in step 2 of Fig. 4-7(b). Step 3 then combines the two basic circuits to complete the preliminary cirucit design.

The circuits in Fig. 4-7(c) then show a NAND gate version of the circuit. Gate 1 performs the basic OR-ing function by having inverted inputs, while NAND gates 2 and 3 provide the necessary NAND functions for the inputs of gate 1. The remainder of the steps in Fig. 4-7(c) follow the same general pattern, ultimately arriving at a complete NAND gate version of the desired circuit. It is left to you to work out a comparable NOR gate version.

Exercises

1. Prove the following Boolean identities by using both algebraic and truth table methods:

   (a) $AB(B + C) = AB$
   (b) $A + AB(A + B) = A$
   (c) $A + \bar{A}B = A + B$
   (d) $AB(\overline{AB} + BC) = ABC$
   (e) $AB(\bar{A} + \bar{B}) = 0$
   (f) $ABC + AB\bar{C} = AB$
   (g) $A + \overline{A + B} = A + \bar{B}$

2. Simplify the following Boolean logic expressions:

   (a) $ST(S + TU) = $ _____

(b) $WX(\bar{X}Y + \bar{W}Z) =$ _____

(c) $\bar{A}(B + C) =$ _____

(d) $(S + T\bar{U})(\bar{S} + T + \bar{U}) =$ _____

(e) $\bar{X}YZ + W =$ _____

(f) $ABC(\bar{D}EF + DEF) + A(\bar{D}A + DE + D\bar{E}) =$ _____

3. Write a simplified logic expression for each of the circuits shown in Fig. E-4.3.

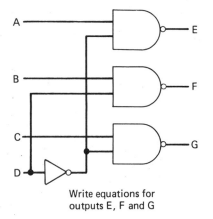

(a)

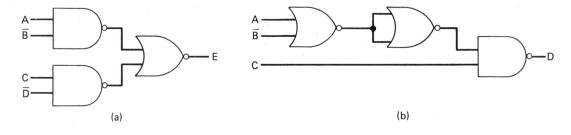

(b)

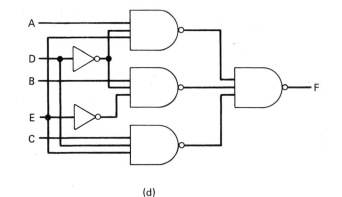

(d)

Write equations for
outputs E, F and G

(c)

Figure E-4.3

4. Design a logic circuit for performing the logic equations listed below or the truth tables shown in Fig. E-4.4. Use only NAND gates and inverters, and assume all inputs must be true (non-inverted).

(e) Use the expression in Ex. 2a.

(f) Use the expression in Ex. 2c.

(g) Use the expression in Ex. 2d.

(h) Use the expression in Ex. 2f.

A	B	C	D
0	0	0	0
0	0	1	0
0	1	0	1
0	1	1	1
1	0	0	0
1	0	1	1
1	1	0	0
1	1	1	1

(a)

A	B	C	D
0	0	0	0
0	0	1	0
0	1	0	0
0	1	1	1
1	0	0	0
1	0	1	0
1	1	0	1
1	1	1	0

(b)

A	B	C	D
0	0	0	0
0	0	1	0
0	1	0	0
0	1	1	1
1	0	0	1
1	0	1	1
1	1	0	0
1	1	1	0

(c)

A	B	C	D	E
0	0	0	0	0
0	0	0	1	0
0	0	1	0	0
0	0	1	1	0
0	1	0	0	0
0	1	0	1	0
0	1	1	0	0
0	1	1	1	1
1	0	0	0	0
1	0	0	1	0
1	0	1	0	0
1	0	1	1	0
1	1	0	0	0
1	1	0	1	1
1	1	1	0	1
1	1	1	1	1

(d)

Figure E-4.4

5
INTERFACING DIGITAL
LOGIC CIRCUITS

In a practical sense, digital logic circuits are useless without some provisions for properly interfacing them with the outside world. There must be provisions for getting logic data into them and out of them in some meaningful fashion that is compatible with the voltage and power specifications of the logic ICs. Then, too, there is a growing need nowadays for interfacing one logic family with another— interfacing TTL, CMOS and PMOS large-scale circuits.

The entire subject of interfacing digital logic circuits rests wholly on the electrical characteristics of the digital circuits at hand. Although the members of any given logic family are directly compatible and require no special interfacing consideration, a circuit designer must take special care when interfacing them with the outside world or attempting to work members of different logic families into the same digital system.

The electrical characteristics that are especially relevant for interfacing purposes are as follows:

1. Input/output voltage levels
    (a) Maximum voltage level defining the circuit's logic-0 input, V_{IL}
    (b) Maximum voltage level defining the circuit's logic-0 output, V_{OL}
    (c) Minimum voltage level defining the circuit's logic-1 input, V_{IH}
    (d) Minimum voltage level defining the circuit's logic-1 output, V_{OH}
2. Input/output current levels
    (a) Maximum input source current at logic 0, I_{IL}

(b) Maximum output sink current at logic 0, I_{OL}
(c) Maximum input sink current at logic 1, I_{IH}
(d) Maximum output source current at logic 1, I_{OH}

3. Absolute maximum and minimum input voltage levels (as determined by the power supply voltages)
4. Input noise margins, generally the difference between V_{IH} and V_{IL}.

Table 5-1 lists the input and output voltage and current specifications for the TTL and CMOS logic IC families as determined for a supply voltage between $+4.5$ V and $+5.5$ V. This happens to be the usual supply voltage for TTL circuits, but they can be operated to an absolute maximum of $+7$ V.

The CMOS family has a much wider supply voltage range, running between $+3$ V and $+15$ V. The input and output specifications for CMOS circuits vary widely with the supply voltage, and they are shown here at TTL levels to illustrate the contrast between TTL and CMOS operating specifications.

Table 5-1 COMPARISON OF TTL AND CMOS INPUT AND OUTPUT SPECIFICATIONS WHEN THE POSITIVE SUPPLY VOLTAGE IS BETWEEN $+4.5$ V AND $+5.5$ V

| | Input/Output Voltage Levels | | | |
| | Inputs | | Outputs | |
	V_{IL}	V_{IH}	V_{OL}	V_{OH}
7400-series TTL	0.8 V max.	2 V min.	0.4 V max.	2.4 V min.
4000-series CMOS	1.5 V max.	3.5 V min.	0.5 V max.	4.5 V min.

Supply voltage = $+4.5$ V to 5.5 V
COMM = 0 V

| | Input/Output Current Levels | | | |
| | Inputs | | Outputs | |
	I_{IL}	I_{IH}	I_{OL}	I_{OH}
7400-series TTL	-1.6 mA max.	40 μA max.	16 mA max.	-400 μA max.
4000-series CMOS	-1 μA max.	1 μA max.	1.7 mA max.	-1.7 mA max.

Supply voltage = $+4.5$ V to 5.5 V
COMM = 0 V

5-1   Input Interfacing

Virtually all digital logic circuits have at least one input point that is capable of accepting logic levels from the outside world. A mechanical switch is the most common means of providing such input information, although a great many digital systems accept data from a wide variety of transducers such as phototransistors, strain gauges, piezoelectric materials and so on.

The whole business of interfacing digital inputs with the outside world can be summarized in a single expression: input signal conditioning. A circuit designer must make certain that any data entering a digital system from the outside is suitably conditioned for the kinds of ICs used in that system.

5-1.1 Mechanical Switch Inputs

Mechanical switches provide the simplest and most straightforward means for getting outside information into a digital circuit. The switch can be a user-operated manual switch or a set of relay contacts. In either case, the interfacing techniques are identical.

The basic design problem in this instance is to make certain the switch scheme pulls the IC's input voltage below the V_{IL} (maximum guaranteed logic-0 level) for a logic-0 input, and pulls the input voltage above V_{IH} (minimum guaranteed logic-1 level) for a logic-1 input. If the switching scheme meets these input voltage specifications and also provides the required amount of input current, it automatically meets the circuit's noisemargin specifications.

Figure 5-1 shows two techniques for interfacing a mechanical switch contact with any TTL or CMOS input. In Fig. 5-1(a) the resistor pulls down the circuit's input level to a logic 0 as long as the switch contact is open. Closing the switch contact pulls up this point to $+5$ V to guarantee a logic-1 input to the circuit.

The resistor to ground in Fig. 5-1(a) is absolutely necessary. If it is omitted, the circuit's input is left floating whenever the switch contact is open; and since an uncommitted logic input acts as a logic 1, it follows that the circuit would never see a logic-0 input, whether the switch contacts are closed or not. In other words, the resistor to ground provides the only means for introducing a logic-0 level to the circuit's input. Then, too, an uncommitted input is susceptible to outside noise and thus false switching. This noise condition is especially critical in the case of CMOS circuits.

Pulling the circuit's input down with a resistor to ground serves two important functions: It provides the necessary logic-0 input whenever the switch contact is open, and it eliminates the possibility of false switching due to external electrical noise while the switch is open. Obviously, the logic input in Fig. 5-1(a) cannot be connected directly to ground because such a connection would cause a short circuit across the power supply whenever the switch contacts are closed.

The values shown for the pull-down resistor in Fig. 5-1(a) are typical maximum values. The actual maximum values for this resistor are 500 Ω for TTL inputs and 1.5 M for CMOS circuits, as determined by the equation $R = V_{IL}/I_{IL}$. The values can be much less than those shown in the diagram, but only at the cost of reduced power efficiency of the input circuit.

The circuit in Fig. 5-1(b) shows as input switch circuit that provides a logic-1 level to the gate as long as the switch contact is open. Closing the switch then pulls the gate's input down to logic 0.

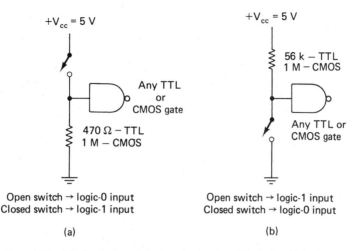

Open switch → logic-0 input
Closed switch → logic-1 input

(a)

Open switch → logic-1 input
Closed switch → logic-0 input

(b)

Figure 5-1 Mechanical switch input circuits. (a) Active-high input. (b) Active-low input.

The pull-up resistor from the gate input to $+V_{cc}$ in this instance provides the necessary voltage and current levels to guarantee a logic-1 input whenever the switch contact is open. The values for this resistor in Fig. 5-1(b) are typical maximum values, and many circuit designers select pull-up resistors as low as 2.2 k for both TTL and CMOS inputs. The actual maximum allowable value can be determined by the equation

$$R = \frac{V_{cc} - V_{IH}}{I_{IH}}$$

or about 1.5 M for CMOS and 75 k for TTL.

Technically speaking, the pull-up resistor in Fig. 5-1(b) could be omitted, taking advantage of the fact that the gate would "see" a logic-1 input whenever the switch contact is open, thereby leaving the gate input uncommitted. The possibility of false switching due to noise pickup, however, makes the pull-up resistor a necessary part of the input circuit.

The choice of the circuit in Fig. 5-1(a) or 5-1(b) depends on whether the designer wants the gate to see true or inverted inputs. If a closed switch and a high logic level are defined as logic 1, and an open switch and low logic levle as logic 0, it follows that the circuit in Fig. 5-1(a) has a true (non-inverted) logic input. The circuit in Fig. 5-1(b), however, provides inverted logic to the gate. If the circuit calls for inverted inputs, it is certainly more economical to arrange the position of the switch as in Fig. 5-1(b) than it is to use the circuit in Fig. 5-1(a) followed by a logic inverter.

The normally open switches can be replaced with normally closed versions, thereby reversing the logic-level format—giving the circuit in Fig. 5-1(a) an inverted-logic format and the one in Fig. 5-1(b) a non-inverted format. Using normally closed switches does not call for any change in the values of the pull-down and pull-up resistors.

Before leaving the subject of mechanical switch inputs, a word about contact bounce is in order. Whenever two pieces of metal strike one another, there is bound to be some bouncing effects. Closing any switch contact, then, is bound to generate some *contact-bounce noise* that can last anywhere between 1 ms and 10 ms, depending on the construction and quality of the switch. The input to the gate circuit in Fig. 5-1(a), for instance, is normally at logic 0; and when the switch contacts close, the input rises up to +5 V. But as long as the contacts are bouncing, the logic level alternates rapidly between logic 1 and logic 0. Such input noise is not critical in low-performance combinatorial logic circuits; but when high operating speed is a critical factor, or when the signal is being delivered to a sequential logic circuit such as a flip-flop, the bouncing effect can play havoc with the system.

There are two popular techniques for eliminating contact-closure bounce, and they will be fully described later in this book. For the time being, it is sufficient to be aware of contact-bounce noise when using mechanical switch inputs to logic circuits.

5-1.2 Discrete-Component Inputs

Input situations occasionally call for interfacing discrete semiconductors to the inputs of digital logic gates. The discrete devices are usually bipolar transistors in such instances, but they can also be JFETs, MOSFETs, low-power SCRs or any other semiconductor device that can be operated in a switching mode. The input signal might even come from a linear IC device such as a voltage comparator or operational amplifier. In any case, the idea is to provide the digital input with the voltage and input current drive levels necessary for fast, reliable and noise-free switching.

The two circuits in Fig. 5-2 show how discrete switching components can be properly interfaced with any digital gate. Note from the outset that the circuit in Fig. 5-2(a) is suitable for both TTL and CMOS gates, while the one in Fig. 5-2(b) is recommended only for CMOS gates.

The circuits shown here use bipolar transistors, but they can be replaced with any of the devices mentioned in the opening paragraph, provided the values of R_1 and R_2 are adjusted to ensure clean switching between saturation and complete cutoff.

Working as a switch, Q_2 in Fig. 5-2(a) pulls the input potential of the gate circuit up to about +4.7 V whenever it is switched on, and then it allows R_3 to pull the input down to less than 0.8 V whenever it is switched off. These input voltage levels are adequate for defining input logic-1 and logic-0 states for either TTL or CMOS logic gates. In a sense, Q_2 simply replaces the mechanical switch in Fig. 5-1(a).

Q_1 in Fig. 5-2(a) makes it possible to reference the circuit's input to ground. Whenever Q_1 is switched on by forward biasing it with a positive potential at its

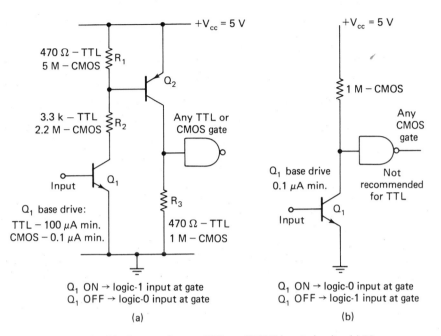

Figure 5-2 Bipolar transistor to TTL or CMOS input circuits. (a) Non-inverting input interface. (b) Inverting input interface.

base connection, its collector current forward biases Q_2 to turn it on as well. Using the supply voltage and resistor values shown on that diagram, the turn-on drive current for Q_1 must be at least 100 μA when using TTL logic circuits, or at least 0.1 μA in the case of CMOS gates. Using any less base drive current runs the risk of failing to saturate either transistor, causing unreliable switching of the logic gate element.

Turning off Q_1 by bringing its base potential close to ground turns off both transistors, letting R_3 pull down the gate's input potential to a logic-0 level.

Forward biasing Q_1 thus provides the gate circuit with a logic-1 input, while switching off Q_1 provides a logic-0 input.

The circuit in Fig. 5-2(b) is simpler and less expensive than the one in Fig. 5-2(a). The basic idea here is to provide the logic gate with a logic-0 input whenever Q_1 is switched on and then let the pull-up resistor raise the input to a logic-1 level whenever Q_1 is turned off. Q_1 in this instance works as the mechanical switch contact in Fig. 5-1(b).

This circuit works quite well for CMOS logic gates, but it is not recommended for use with TTL gates. The problem as far as TTL gates are concerned stems from the fact that bipolar transistors and most other semiconductor switching devices have an emitter-to-collector saturation voltage that is barely within the TTL definition of logic 0. A saturation voltage of 0.7 V is typical for many silicon bipolar transistors, and that is too close to the 0.8 V max. definition of TTL input logic 0.

Any kind of semiconductor devices can be interfaced with the inputs of TTL and CMOS logic gates as long as (1) the input scheme operates in an on/off switching mode and (2) the voltage levels and drive currents are within the margins specified for the logic circuits' inputs.

5-1.3 Schmitt-Trigger Inputs

Schmitt-trigger inputs to digital logic circuits must be considered one of the most reliable and useful of all possible input schemes. Schmitt-trigger gates are widely available in both the TTL and CMOS families, having the common 14-pin DIP package that makes them electrically and mechanically compatible with most logic circuits.

Figure 5-3 shows the logic symbol for a Schmitt-trigger inverter and a diagram that illustrates its basic principle of operation. Note that the symbol is that of a logic inverter that has a hysteresis-loop figure drawn in its center.

The *voltage* axis in the diagram in Fig. 5-3(b) shows the 0 V and $+V_{cc}$ levels, plus two special Schmitt-trigger levels designated V_{T+} and V_{T-}. The V_{T+} voltage level is the input level at which a positive-going signal causes the inverter's output to switch from logic 1 to logic 0. This particular transition takes place at point A in the diagram. The V_{T-} level is the input level at which a negative-going signal causes the inverter's output to switch from logic 0 to logic 1 again.

The Schmitt-trigger inverter thus produces a logic-0 output whenever the input voltage is below V_{T-} and a logic-1 output whenever the input is above the V_{T+} level. What happens between these two levels—in the *hysteresis* interval—depends on the state of the output as the input signal enters that interval. If the output is at logic 1 when the input signal enters the hysteresis interval, the output tends to remain at logic 1 until the input signal exceeds the V_{T+} transition level; and if the output is at logic 0 while the input voltage enters the hysteresis interval, the output remains at logic 0 until the signal drops below the V_{T-} transition level.

The overall effect of a Schmitt-trigger input gate is to clean up noisy or irregular signals before applying them to standard logic ICs. The irregular input signal shown in Fig. 5-3(c), for example, might be a digital pulse that has been sent to the logic circuit from a noisy environment through long cables. The logic levels emerging from the Schmitt-trigger inverter, however, are perfectly clean and ready for application to standard logic gates of other types.

Schmitt-trigger logic gates were designed specifically for cleaning up noisy or irregular digital signals, but they can be used for translating analog-type voltage levels into 1 or 0 logic levels that are wholly compatible with the input requirements of digital logic circuits. The input waveform shown in Fig. 5-3(a), for instance, could represent a triangular input waveform; and whenever this input signal exceeds the V_{T+} level, the circuit's output drops to logic 0, remaining there until the waveform falls below the V_{T-} level.

Now this is not an analog-to-digital converter, but rather a simple scheme for translating analog-like voltage level changes into a 1-or-0 binary format.

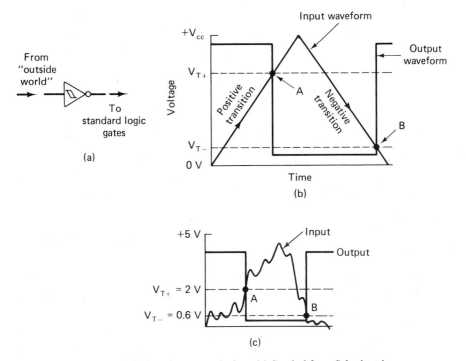

Figure 5-3   Schmitt-trigger gate devices. (a) Symbol for a Schmitt-trigger inverter. (b) Schmitt-trigger output response to a triangular waveform input, showing the effect of hysteresis. (c) Response of a Schmitt-trigger inverter to an irregular input waveform.

The circuit in Fig. 5-4 uses a Schmitt-trigger inverter to obtain 120-Hz digital pulses from the standard 60-Hz utility line. Transformer T_1 steps down the 120-VAC input to about 6.3 VAC and, at the same time, isolates the digital system from utility-system ground. The full-wave bridge rectifier assembly then converts the 60-Hz sinusoidal waveform into 120-Hz pulsating dc. See the waveform for point S is Fig. 5-4(b).

The 220-Ω resistor and 5.1-V zener diode clip the peaks of the pulsating dc, making certain the input voltage to the Schmitt-trigger inverter does not exceed the logic circuit's supply voltage. (Using a CMOS Schmitt trigger and a +12-V supply, for instance, would call for using a 10-V or 12-V zener diode.) The Schmitt-trigger inverter then responds to the clipped pulsating dc by producing the square waveform shown for point U in Fig. 5-4(b).

The threshold voltages of the Schmitt-trigger logic gates are not adjustable; nevertheless, it is possible to simulate the effects of an adjustable Schmitt-trigger threshold device by using a simple voltage divider at the input. See Fig. 5-5.

Resistor R_2 in Fig. 5-5 is fixed at either 470 Ω or 1 M, depending on whether the Schmitt-trigger gate belongs to the TTL or CMOS family. The value of R_1 is

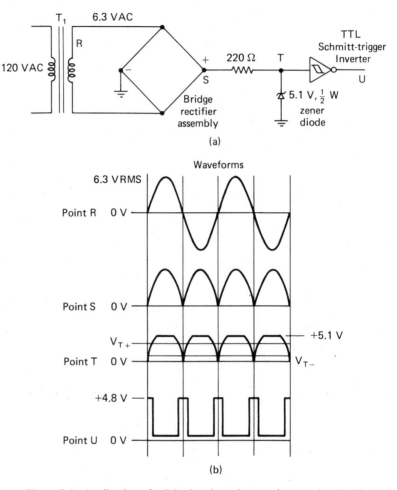

Figure 5-4 Application of a Schmitt-trigger inverter in a precise 120-Hz source of clock pulses. (a) Schematic diagram. (b) Circuit waveforms.

then selected according to the value of the positive input signal that is to set the inverter's output to its logic-0 state. The equation for determining the value of R_1 is $R_1 = (V_s R_2)/V_{T+}$, where V_s is the desired trigger point of the input waveform, R_2 is the fixed value of 470 Ω or 1 M, and V_{T+} is the positive-transition trigger voltage for the Schmitt-trigger gate. If the gate is supposed to respond with a logic-0 output when the input waveform exceeds $+12$ V, for example, the calculated value of R_1 is 2820 Ω, or about 2.7 k.

The zener diode at the input prevents the gate's input voltage from exceeding the logic supply voltage as described in connection with the circuit in Fig. 5-4.

Of course V_s must be greater than the inverter's maximum V_{T+} rating, and the

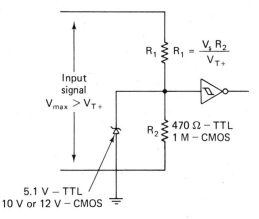

Figure 5-5   Simulating the effect of a Schmitt-trigger circuit with an adjustable threshold.

input signal must drop very close to 0 V to make certain the inverter can be switched to a logic-1 output again.

It is possible to modify the circuit in Fig. 5-5 to accommodate input signals that are smaller than the normal threshold ratings of the Schmitt-trigger device. The trick here is to amplify the signal first with an analog amplifier such as an LM3900 comparator or any operational amplifier that uses a single supply voltage. Once the signal is amplified above the gate's positive threshold level, it can be fed to the gate through the same circuit shown in Fig. 5-5.

None of the Schmitt-trigger devices described here are precision devices—they aren't intended to be. Their main purpose is to clean up noisy or irregular digital pulses, and their usefulness as analog-to-pulse generators is simply an extra feature. Any system calling for precision Schmitt triggering must use a high-performance operational amplifier to do the main switching job. The digital Schmitt-trigger gate can then be used for setting the pulse to the proper voltage and impedance levels for the digital system.

Figure 5-6 lists the avilable TTL and CMOS Schmitt-trigger devices and their maximum and minimum input transition levels.

The NAND Schmitt-trigger gates perform the usual NAND function on any input signals that meet the specifications for transition voltage levels, including the outputs from standard logic gates.

5-2   Output Interfacing

While all input interface circuits must be designed around the general input voltage and current specifications for the logic system at hand, output interface circuits must take into consideration the devices' output voltage and current-loading specifications. The diagrams in Fig. 5-7 completely summarize these output param-

7413   Dual 4-input NAND schmitt-trigger – TTL, totem-pole output
74132 Quad 2-input NAND schmitt-trigger – TTL, totem-pole output
7414   Hex schmitt-trigger inverter – TTL, totem-pole output

40106 Hex schmitt-trigger inverter – CMOS

(a)

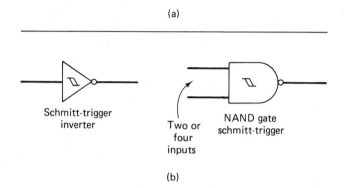

| Schmitt-trigger inverter | Two or four inputs | NAND gate schmitt-trigger |

(b)

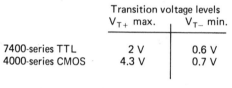

	Transition voltage levels	
	V_{T+} max.	V_{T-} min.
7400-series TTL	2 V	0.6 V
4000-series CMOS	4.3 V	0.7 V

Supply voltage = +4.5 V to 5.5 V
COMM = 0 V

Outputs are compatible with the devices' respective families
(See table 5.1).

(c)

Figure 5-6 Summary of TTL and CMOS Schmitt-trigger devices. (a) Device numbers and general descriptions. (b) Schematic symbols. (c) Transition voltage levels.

eters as they apply to interfacing TTL and CMOS logic gates with other types of output devices.

Unless stated otherwise, all of the discussions in this section deal with standard TTL totem-pole or CMOS ICs. The special advantages of open-collector TTL outputs will become quite apparent from examples scattered throughout the remainder of this chapter, however. See Chapter 2 for complete discussions of totem-pole and open-collector outputs.

In principle, any digital logic circuit can be loaded from its output to common or from its output to the positive supply voltage. See Figs. 5-7(a) and 5-7(b), respectively.

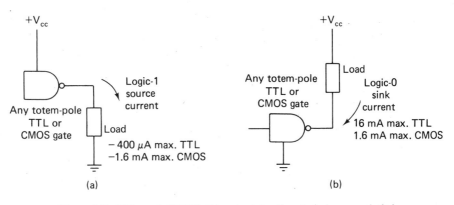

Figure 5-7 TTL and CMOS IC output loading techniques and their essential characteristics. (a) Source loading: load from device output to COMM. (b) Sink loading: load from device output to V_{CC} (or V_{DD} for CMOS).

Whenever a logic circuit is loaded from its output to ground as in Fig. 5-7(a), maximum load current flows whenever the logic gate generates its logic-1 output. Under this particular loading condition, the totem-pole elements of a TTL gate or the *P*-channel FET in a CMOS gate conduct to pull up the output voltage near the $+V_{cc}$ supply potential. The device thus sources current (conventional current flow terminology) through the load to ground. Note that CMOS devices are capable of sourcing up to 1.6 mA, while TTL gates can source only 400 μA. While the gate circuit is generating a logic-0 output, however, the load current is negligible for both TTL and CMOS circuits.

Figure 5-7(b) shows a logic gate loaded from its output to the positive supply voltage. In this case, maximum loading occurs whenever the gate is generating a logic-0 output. The logic device sinks load current from the positive supply and through the lower output transistors in the IC to ground. While a CMOS gate can sink 1.6 mA this way, a TTL can drive the load with up to 16 mA. Loading is negligible with a logic-1 output.

In comparing the two loading methods in Fig. 5-7, it is important to note that the load in Fig. 5-7(a) is energized whenever the logic gate generates a logic-1 output; in Fig. 5-7(b) the load is energized with a logic-0 output. The circuit in Fig. 5-7(b) thus inverts the logic scheme; the one in Fig. 5-7(a) does not.

A comparison of the two loading methods also shows that the current-driving capability for CMOS gates is the same in both instances: 1.6 mA. TTL gates with a totem-pole output, however, have a far greater current-driving capacity when the load is connected to the positive power supply voltage than when it is connected to ground: 16 mA as opposed to a mere 400 μA.

Looking at these loading specifications from a practical point of view, we see that a CMOS gate can be loaded either way, but it is generally better to load a TTL gate from its output to the positive supply voltage. In fact, the practice of loading totem-pole TTL gates from output to ground ought to be avoided whenever possible. The totem-pole elements simply do not have the power-handling capacity of the lower output transistor in TTL IC packages.

5-2.1 Driving LEDs

One of the most popular output devices for digital circuits is a light-emitting diode (LED). An LED emits light energy whenever it is forward biased with a sufficient amount of drive current and voltage, normally requiring $+1.7$ V forward voltage drop with currents up to 20 mA. The LED load is often a single indicator lamp. but it can also be one element in a 7-segment LED numeric display assembly.

Figure 5-8 shows the recommended procedures for driving LEDs with TTL and CMOS logic ICs. Note in Fig. 5-8(a) that a TTL gate is capable of delivering the required power whenever the LED is connected through a limiting resistor to the postive supply voltage. Whenever the logic gate shows a logic-0 output, then, the gate output is near ground potential to complete a path for current flow through the limiting resistor and LED. The LED lights up whenever the gate shows a logic-0 output. A logic-1 output from the gate, however, pulls the gate's output potential near $+5$ V, depriving the LED of its required forward junction potential of about 1.7 V. Thus the LED cannot light whenever the gate shows a logic-1 output.

Connecting the load from the output of the logic circuit to the positive supply voltage inverts the logic relationship between the gate output and the LED: The LED lights up with a logic-0 output, and it goes out as the gate's output switches to logic 1.

The resistor in Fig. 5-8(a) performs two basic functions whenever the LED lights. The resistor limits the logic-0 sink current to an acceptable value, and it takes up that portion of the supply voltage not occupied by the forward drop of the LED and the logic-0 output level of the logic circuit. For TTL totem-pole gates, this limiting resistor has a typical value of 330 Ω, thereby limiting the gate's gate's output sink current to about 7 mA, a value well within the 16-mA limit recommended for logic-0 output currents for TTL totem-pole configurations.

It is possible to drive the LED to brighter levels with TTL sink currents as high as 16 mA. The required limiting-resistor value can be determined by the equation

$$R = \frac{V_{cc} - V_{OL}}{I_{OL}}$$

where I_{OL} is any desired LED-driving sink current up to 16 mA.

The fact that the LED circuit in Fig. 5-8(a) inverts the logic relationship between the gate's output and the lighting of the LED can be something of a nuisance at times. The more natural tendency is to design output circuits that respond with a true or non-inverted logic relationship. It would be nice if the circuit designer had the option of getting a true output response by connecting the limiting resistor and

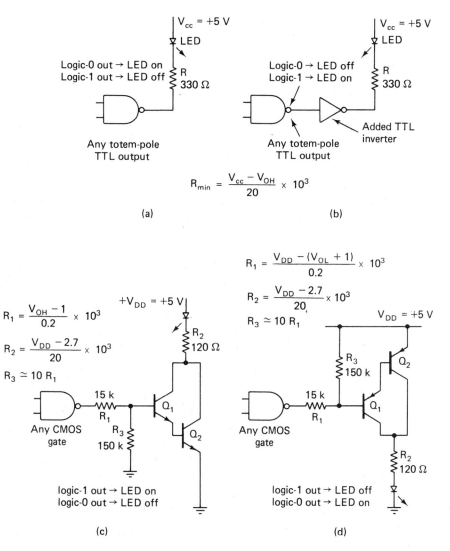

$$R_{min} = \frac{V_{cc} - V_{OH}}{20} \times 10^3$$

(a)

(b)

$$R_1 = \frac{V_{DD} - (V_{OL} + 1)}{0.2} \times 10^3$$

$$R_2 = \frac{V_{DD} - 2.7}{20} \times 10^3$$

$$R_3 \simeq 10\,R_1$$

$$R_1 = \frac{V_{OH} - 1}{0.2} \times 10^3$$

$$R_2 = \frac{V_{DD} - 2.7}{20} \times 10^3$$

$$R_3 \simeq 10\,R_1$$

(c)

(d)

Figure 5-8   LED outputs. (a) TTL-to-LED with logic inversion. (b) TTL-to-LED without logic inversion effects. (c) CMOS-to-LED with logic inversion. (d) CMOS-to-LED without logic inversion.

LED from the gate's output to ground. That way, the LED would light up whenever the gate showed a logic-1 output. Unfortunately, since a totem-pole TTL output is incapable of sourcing enough current to light the LED with a logic-1 output, the designer is stuck with the inverting, output-to-V_{cc} configuration.

Of course, it is possible to compensate for this inverting situation by designing output stages of the logic system so that the final outputs are inverted; but when this is not practical, the only alternative is to add an inverter stage as shown in

Fig. 5-8(b). The NAND gate, actually representing any final-stage TTL device, has a true output. The output is then inverted by means of a logic inverter (or any convenient equivalent as described in Sec. 3-2) before applying the logic levels to the LED circuit.

In short, the circuit in Fig. 5-8(a) yields an inverted relationship between the TTL output and the response of the LED. The circuit in Fig. 5-8(b) can compensate for this inversion effect, effectively producing a true relationship by first inverting the output signal.

The basic CMOS IC gates have the advantage of equal sink and source currents, making it possible to load them equally well to V_{DD} or ground. The implication is that the circuit designer has the option of generating an inverted or non-inverted relationship between the gate's output logic levels and the response of the LED. There is a catch, however: CMOS gates cannot supply enough current to light an LED directly.

The circuits in Figs. 5-8(c) and 5-8(d) show how to take advantage of the high current gain characteristics of Darlington pairs of transistors to boost the current output of CMOS ICs. The circuit in Fig. 5-8(c) has an inverted relationship between the gate's output logic levels and the response of the LED—the LED lights whenever the logic circuit generates a logic-0 output, and the LED switches off whenever the logic circuit shows a logic-1 output. Resistor R_1 limits the CMOS source current for logic-1 outputs, while resistor R_3 ensures the turn-off of Q_1 and Q_2 whenever the gate's output drops to logic 0. R_2 limits the forward-biasing LED current to about 20 mA.

The resistor values shown in Fig. 5-8(c) assume an operating supply voltage of $+5$ V. The equations accompanying the diagram can be used for determining the values of the resistors at other supply voltage levels.

The circuit in Fig. 5-8(d) yields a true logic relationship between the CMOS device's output and the response of the LED. In this instance, a logic-1 output from the gate turns on the LED, while a logic-0 output turns it off. The resistors in Fig. 5-8(d) have the same values and perform the same roles as the corresponding resistors in Fig. 5-8(c). The equations for determining the value of R_2 are different for the two circuits; but even so, it often turns out that they have much the same values in either case.

The Darlington amplifier circuits in Figs. 5-8(c) and 5-8(d) use discrete bipolar transistors. Using discrete components in this fashion can be something of a problem whenever the design is to be used in a high-volume production situation. Many IC manufacturers now supply CMOS compatible LED driver ICs in their lines of proprietary devices. While the discrete-transistor circuits shown here are quite adequate for one-of-a-kind or low-volume production purposes, a circuit designer anticipating large-volume production of the circuit will do well to investigate the possibility of using LED driver ICs.

The foregoing discussion applies to the problems of driving discrete LED lamps. The need for 7-segment LED numeric displays, however, has generated a demand for 7-segment LED driver circuits that are compatible with TTL and

CMOS circuitry. These 7-segment drivers usually appear as part of a decoder or code-converter package, however; and they will be described in greater detail in Chapter 10.

5-2.2 Driving Higher-Power Loads

The principles involved in driving LED loads apply to the more general problem of driving higher-current loads such as relay coils, incandescent lamps and dc motors. The only real difference is that the higher current output interfacing circuits call for larger amounts of current gain, and perhaps some voltage gain as well.

Figure 5-9(a) shows the only acceptable circuit for interfacing the output of a TTL gate to a discrete-transistor amplifier circuit. The circuit takes advantage of the fact that the gate has a relatively high logic-0 sink current level, more than enough current output to switch Q_1 into saturation. If it is assumed that the transistor has a beta of at least 10 (a safe assumption in a vast majority of cases), it follows that this circuit can drive loads up to 160 mA—ten times the maximum sink current output of the TTL gate.

R_1 in Fig. 5-9(a) serves as a limiting resistor for holding the gate's output current to 16 mA or less. Generally, the value of R_1 is selected by the equation

$$R_1 = \frac{V_{cc} - (V_{OL} + V_{eb})}{I_{sink}}$$

where V_{eb} is the forward-biasing emitter-base voltage of Q_1 (about 0.5 V) and I_{sink} is the desired base-drive current (16 mA maximum). Since the amplifier will certainly have a minimum current gain of 10, $I_L = 10 I_{sink}$; and that makes it possible to determine the value of R_1 in terms of the desired amount of load current:

$$R_1 = \frac{V_{cc} - (V_{OL} + V_{eb})}{0.1 I_L}$$

As a design example, suppose the problem is to drive a relay coil having a current rating of 50 mA. Using the circuit in Fig. 5-9(a),

$$R_1 = \frac{5.5 - (0.4 + 0.5)}{(0.1)(50)} \times 10^3 \text{ or } 920 \text{ } \Omega.$$

That particular value isn't available as a standard resistor, so the circuit designer has the option of using a standard 1-k or 680-Ω resistor. The choice in this instance can favor the 1-k resistor. Although this choice will reduce the base-drive current level somewhat, the transistor most likely has a beta greater than 10, and it will saturate properly whenever the gate pulls the base potential close to common.

The reverse diode, incidentally, should always be connected in this reverse-biased fashion across inductive loads such as relay coils and dc motor windings. The purpose is to shunt out inductive "kinkback" potentials that occur whenever the current to the load is switched off. Such potentials do not occur when using resistive or capacitive loads.

Note in Fig. 5-9(a) that there is an inverse relationship between the logic output level from the TTL gate circuit and the operation of the load device. Whenever the gate shows a logic-0 output, Q_1 is biased on and applies full power to the load. Switching the output of the gate circuit to logic 1 turns off Q_1 and opens the circuit to the load.

The load device can be yet another transistor; this brings up the possibility of

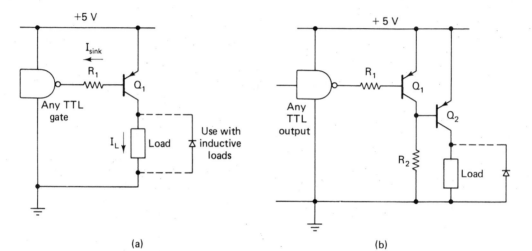

(a)   (b)

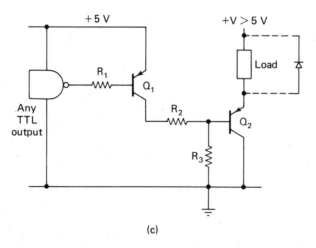

(c)

Figure 5-9 Interfacing digital ICs to discrete bipolar transistor amplifiers. (a) Current gain of 10 with logic inversion between gate output and load power. (b) Current gain of 100 with no logic inversion between gage output and application of load power. (c) Both current and voltage gain with logic inversion.

constructing a non-inverting power amplifier or one that has greater current and voltage gain.

The circuit in Fig. 5-9(b) uses a pair of *PNP* transistors to produce a true relationship between the logic level from the gate circuit and the energizing of the load and, at the same time, permit higher current gain.

Whenever the output of the gate circuit is at logic 0, Q_1 is switched on by means of the gate's sink current through the emitter-base junction of Q_1 and limiting resistor R_1. With Q_1 thus saturated, the emitter-base junction of Q_2 sees a low impedance that is adequate for turning off Q_2. No current is applied to the load, then, as long as the output of the gate circuit is at logic 0.

Switching the output of the gate circuit to logic 1 turns off Q_1; but then Q_2 can pick up forward-biasing emitter-base current through R_2, thereby saturating that transistor and applying full power to the load device.

In short, a logic 0 from the gate turns off the load power, while a logic-1 level from the gate turns on the load power.

The circuit in Fig. 5-9(b) also has an assumed maximum current gain of 100, a gain of 10 for each transistor amplifier stage. If the circuit is run at its maximum current level, the system can operate loads as high as 1.6 A ($100 \times I_{OL\,max}$).

Resistor R_2 is the key to the operation of this circuit. Whenever Q_1 is switched on, R_2 serves as its resistive load; and when Q_2 is switched on, R_2 serves as the forward-biasing limiting resistor for that transistor. Resistor R_2 thus plays a dual role, and both roles must be born in mind when selecting the values of R_1 and R_2.

As a design example, suppose the load current is 1 A maximum. This means that Q_2 should be selected so that it can properly saturate whenever its collector current is 1 A (a transistor power rating between 0.6 W and 1 W is satisfactory). If a gain of 10 is assumed for Q_2, its forward-biasing base current should be 100 mA, thereby dictating a value of R_2 in the neighborhood of 47 Ω.

Now the load resistance for Q_1 is fixed at 47 Ω, and that translates into a maximum load current of about 100 mA. If a gain of 10 is assumed for Q_1, the base-drive current for that transistor ought to be at least 10 mA; and by using the design procedure described in connection with the circuit in Fig. 5-9(a), R_1 becomes about 470 Ω.

The current gain of any output circuit can be extended even further by replacing the load in Fig. 5-9(b) with yet another current amplifier stage.

The circuits in Figs. 5-9(a) and 5-9(b) assume that the load can operate properly from the +5-V supply voltage provided for the TTL circuits. Suppose, however, the load is rated at 12 V instead. This situation calls for both current and voltage gain.

The only problem associated with stepping up the voltage level to the load is isolating the higher supply voltage to the load from the output of the TTL circuit. Figure 5-9(c) illustrates how this voltage-translation problem can be solved.

In Fig. 5-9(c) the output of the TTL gate is interfaced to a *PNP* transistor, Q_1, in the usual fashion. The collector circuit of Q_1, however, is used for driving the base circuit of Q_2. Transistor Q_1 must operate from the same power supply as the

TTL gate, but Q_2 can operate from any desired power supply voltage greater than $+5$ V. The TTL gate is effectively isolated from the higher-voltage supply potential by virtue of the fact that the emitter-base voltage of Q_2 never exceeds $+0.5$ V, an approximate figure for the maximum emitter-base saturation potential for Q_2.

Whenever the output of the TTL gate is at logic 0, transistor Q_1 is biased on; and as long as Q_1 is switched on, it provides forward-biasing base current to Q_2. Q_2 thus applies full power to the load whenever the TTL gate generates its logic-0 output.

Changing the output of the TTL circuit to logic 1 switches off Q_1 and deprives Q_2 of its forward-biasing current. As a result, Q_2 turns off to remove power to the load.

Resistors R_1 and R_2 limit base current to Q_1 and Q_2, respectively. Resistor R_3 merely ensures a low-impedance, noise-inhibiting impedance between the base and emitter of Q_2 whenever it is switched off.

If a current gain of 10 is assumed for each amplifier stage, the amplifier in Fig. 5-9(c) has a maximum recommended current gain of 100. The voltage gain is equal to the ratio of the higher supply voltage to the TTL supply voltage. If the higher supply voltage, $+V$, happens to be 12 V and the TTL power supply is at $+5.5$ V, the overall voltage gain of the circuit is on the order of 2.2.

Considering a design example, suppose the load is rated at 12 V, 500 mA. Transistor Q_2 can then be just about any medium-power *NPN* transistor. If a beta of 10 is assumed, its base-drive current should be about 50 mA; and as a general rule of thumb, the non-critical value of R_3 can be set so that it carries about 0.1 times the base-drive current for Q_2—5 mA in this instance.

Transistor Q_1 must then be able to supply a total of 55 mA whenever Q_2 is to be switched on. This current level is adjusted by means of R_2. Now the voltage across R_2 will be equal to the TTL supply voltage, less the combined forward voltage drops of the emitter-base junction of Q_2 and the emitter-collector circuit of Q_1. These two voltages are usually in the neighborhood of 0.5 V and 0.7 V, respectively; so the voltage across R_2 when Q_1 is conducting is about 3.8 V. Ohm's law then shows that the value of R_2 should be 3.8 V/55 mA $= 69$ k. Or to use a standard-value resistor, R_2 can be selected as 68 k.

Resistor R_3 ought to carry 5 mA as described above whenever its voltage is clamped at 0.5 V by the emitter-base junction of Q_2. That particular resistor should thus have a value of about 100 Ω.

If the maximum load current for Q_1 is known to be 55 mA, it follows from the procedure outlined earlier in this section that the TTL gate must provide about 5.5 mA of drive current for the base circuit of Q_1. This figure fixes the value of R_1 at about 1 k.

Although using discrete-component amplifiers provides the most universal approach to the problems of interfacing TTL outputs with load devices having higher current and voltage ratings, there is a family of open-collector TTL buffer/drivers that can supply up to 40 mA of sink current to loads operating as high as $+30$ V. Figure 5-10 lists some of these devices and shows how they are normally implemented.

Device		Maximum Load Voltage	Maximum Sink Current
7406	Hex inverting buffer/driver	30 V	40 mA
7407	Hex non-inverting buffer/driver	30 V	40 mA
7416	Hex inverting buffer/driver	15 V	40 mA
7417	Hex non-inverting buffer/driver	15 V	40 mA
7426	Quad 2-input NAND buffer	15 V	16 mA
7438	Quad 2-input NAND buffer	5.5 V	48 mA

(All open-collector TTL devices)

(a)

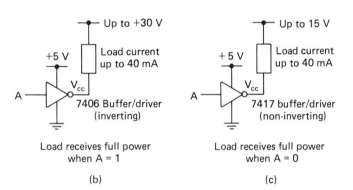

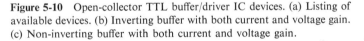

(b)   (c)

Figure 5-10   Open-collector TTL buffer/driver IC devices. (a) Listing of available devices. (b) Inverting buffer with both current and voltage gain. (c) Non-inverting buffer with both current and voltage gain.

Interfacing CMOS gates to current and voltage amplifiers follows much the same procedures already outlined here for totem-pole TTLs. The primary difference between interfacing TTL and CMOS devices to an external load concerns the differences in supply voltages and output drive capabilities. One of the fortunate features of CMOS gates is that they can be loaded equally well to V_{DD} with a *PNP* transistor or to V_{SS} with an *NPN* transistor.

Table 5-2 shows the output drive characteristics of CMSO gates as they apply to interfacing bipolar transistor driver circuits. The figures are approximations, but they are close enough to actual values to make the design scheme outlined here a workable one.

The table lists the drive characteristics at the three most popular supply voltage levels: 5 V, 9 V and 12 V. The 9-V supply is normally associated with small battery-operated devices such as electronic calculators and digital games and novelties of all sorts. The 12-V supply is a logical choice when designing equipment to be used in automobiles and boats.

The V_{OH} column in Table 5-2 shows the definitions of logic-1 output levels that are relevant when driving an *NPN* base circuit with an output source current. The V_{OL} column is important when desiging an interface in which the CMOS gate drives a *PNP* transistor with a sink current.

To appreciate the usefulness of this table, suppose there is a need to build a driver circuit such as the one in Fig. 5-9(a) that uses a CMOS gate and a supply voltage of +12 V. If the load is rated at 50 mA, assume Q_1 has a gain of 10, and specify a base drive current of 5 mA. Since a CMOS gate can sink up 10 mA when operated from a 12-V supply (see the I_O column in Table 5-2), the gate can handle the job easily.

Table 5-2   CMOS Output Drive Specifications

Supply Voltage	V_{OH}	V_{OL}	I_O (Sink or Source)
+5 V	4.5 V	0.5 V	1.7 mA
+9 V	8.2 V	0.8 V	10 mA
+12 V	11 V	1.0 V	10 mA

All that remains to be done is to calculate the value of R_1. Using the equation already cited for calculating R_1,

$$R_1 = \frac{12 - (1 + 0.5)}{5} \times 10^3 \text{ or } 2.1\text{-k}$$

In this instance, $V_{OL} = 1.0$ V as shown in Table 5-2. A 2.2-k resistor would be a reasonable value for R_1.

The circuits and procedures described for Figs. 5-9(b) and 5-9(c) apply equally well to CMOS gates that drive a *PNP* base to $+V_{DD}$. The procedures can be easily generalized to CMOS outputs that drive *NPN* base-emitter junctions to COMM or V_{SS}. In such instances, the value of the limiting resistor is found by the equation

$$R_1 = \frac{V_{OH} - V_{eb}}{I_{\text{source}}}$$

where V_{OH} is found from Table 5-2, V_{eb} is the emitter-base forward voltage drop of the *NPN* transistor (usually about 0.5 V), and I_{source} is the desired output source current from the gate (not to exceed the I_O figure cited in Table 5-2).

5-3 TTL-to-CMOS Interfacing and CMOS-to-TTL Interfacing

The best way to deal with any problems of interfacing TTL and CMOS circuits within one digital system is to avoid using the two families: Select a TTL or CMOS design and use that technology throughout. The need for combining TTL and CMOS circuits frequently arises, however, whenever one wants to couple together two entirely different digital subsystems. If one of the systems has already been engineered around a TTL technology and the other has already been assembled from CMOS ICs, an understanding of proper interfacing techniques becomes essential.

5-3.1 TTL-to-CMOS Interfacing

At first thought, there might seem to be no problems associated with using a relatively high-power TTL device to drive a relatively low-power CMOS device. The table in Fig. 5-11 summarizes the output characteristics of a typical TTL device and compares these outputs with the input characteristics of a CMOS IC.

It turns out that a CMOS gate can, indeed, be driven directly from a TTL IC; but there is one qualifying factor—the TTL gate defines a logic-1 level as 2.4 V or more, while a CMOS input defines logic 1 as 3.5 V or more. There is a good chance that the logic-1 level from the TTL output will not be close enough to +3.5 V to ensure that the CMOS gate will interpret it as logic 1.

If the two kinds of devices are operated from the same 5-V power supply, the cure for this particular mismatch of logic-1 definitions is rather simple. Figure 5-11(a) shows a 22-k *pull-up resistor* connected between the output of the TTL

TTL-to-CMOS ($V_{cc} = V_{DD} = +5$ V)

TTL output		CMOS input		Notes
V_{OL} = 0.4 V max		V_{IL} = 1.5 V max		Compatible
I_{OL} = 16 mA max		I_{IL} = 1 μA max		Compatible
V_{OH} = 2.4 V min		V_{IH} = 3.5 V min		Possible trouble
I_{OH} = -400 μA max		I_{IH} = 1 μA max		Compatible

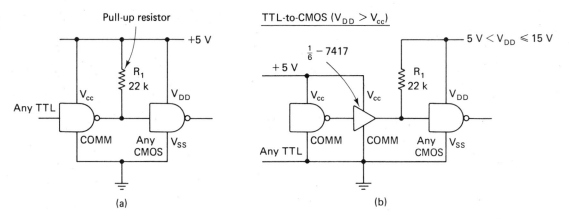

Figure 5-11 Interfacing any TTL device to CMOS input. (a) Interfacing circuit when both power sources are +5 V. (b) Interfacing when the CMOS power supply is greater than the TTL supply level.

device and $+V_{cc}$. This resistor makes certain that the logic-1 level from the TTL gate will exceed the 3.5-V minimum for logic 1 required by the CMOS gate.

Whenever a TTL IC is driving a CMOS IC, and they are both operating from the same supply voltage, properly interfacing the two is a simple matter of including the pull-up resistor, R_1.

If the CMOS system is operating from a higher supply voltage, the situation is somewhat more complicated. Recall that TTL circuits must operate from a supply voltage no less than $+4.5$ V and absolutely no more than $+7$ V. CMOS devices, however, can operate from a supply voltage anywhere between $+3$ V and $+15$ V. It is thus quite possible that the CMOS system will be operating at a higher voltage level than the TTL system driving it.

This is a voltage-level translation problem that is rather easily solved by connecting a non-inverting, open-collector TTL buffer between the TTL and CMOS gates. See Fig. 5-11(b). While the TTL gate and 7414 buffer are both operated from the same $+5$-V source, resistor R_1 serves as the load resistance for the open-collector output of the buffer device. And since this particular TTL buffer can withstand open-collector voltages as high as 15 V, it is fully compatible with the higher-voltage supply for the CMOS gate.

The output drive characteristics of the buffer IC are equal or better than that of a typical totem-pole TTL gate; therefore, there are no problems as far as current-drive capability is concerned.

5-3.2 CMOS-to-TTL Interfacing

The table in Fig. 5-12 shows that the output of CMOS gates is fully compatible with the input characteristics of TTL gates on all points except one: current drive at logic 0. The CMOS device can deliver a maximum of 1.7 mA when V_{DD} is at $+5$ V, but the TTL input requires -1.6 mA. A difference of only 0.1 mA is too close for comfort. A current driver ought to be inserted between a CMOS device that is to operate a TTL input.

The most popular CMOS-to-TTL buffer IC is the 4010 or 4009. The 4010 is a hex non-inverting, CMOS-oriented buffer. The 4009 is just an inverting version of the same circuit.

These CMOS buffers have two different supply-voltage connections: V_{DD} and V_{cc}. As one might expect, the V_{DD} terminal is connected to the positive supply voltage for the CMOS device, while the V_{cc} terminal goes to the positive supply point for the TTL IC.

If the CMOS and TTL circuits are operated from the same 5-V supply as shown in Fig. 5-12(a), the V_{DD} and V_{cc} terminals of the buffer are simply tied together. But if there is a need to translate the higher V_{DD} level to a TTL-oriented V_{cc} level of 5 V, the buffer's V_{DD} and V_{cc} terminals must be connected to their their respective power supplies. See the example in Fig. 5-12(b).

CMOS-to-TTL (V_{cc} = V_{DD} = 5 V)

CMUS output		TTL input		Notes
V_{OH}	= 4.5 V min	V_{IH}	= 2 V min	Compatible
I_{OH}	= −1.7 mA max	I_{IH}	= 40 μA max	Compatible
V_{OL}	= 0.5 V max	V_{IL}	= 0.8 V max	Compatible
I_{OH}	= 1.7 mA max	I_{IL}	= 1.6 mA max	Possible trouble

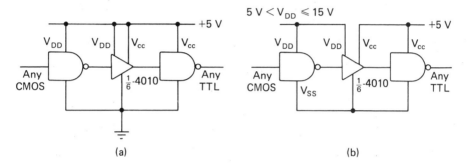

Figure 5-12 Interfacing any CMOS device to TTL input. (a) Interfacing circuit when both power supplies are +5 V. (b) Interfacing when the CMOS power supply is greater than the TTL supply voltage.

Exercises

1. What is the primary purpose of a switch-debouncing circuit?

2. What is the general definition of *noise margin* as it applies to TTL and CMOS digital circuits? Using the data in Table 5-1, what is the noise margin for TTL gates? For CMOS gates operating from a 5-V power supply?

3. Whenever a TTL circuit is interfaced to a discrete component, such as a transistor or LED, why is it necessary to connect the discrete component from the output of the TTL circuit to $+V_{cc}$?

4. What are the general definitions of *sink current* and *source current* as they apply to digital IC outputs? To digital IC inputs?

5. Why is there an effective logic inversion between the logic level from a TTL IC and an LED load it drives?

6. Which output specification for TTL and input specification for CMOS make a pull-up resistor necessary when interfacing TTL to CMOS?

7. Which output specifications for CMOS and input specification for TTL make it necessary to use a buffer/driver circuit when interfacing CMOS to TTL?

6 BASIC FLIP-FLOP CIRCUITS

AND, OR, NAND, NOR and INVERT gates make up the family of basic combinatorial logic circuits. Their job is to combine logic input levels to yield some prescribed output level, and that output logic level is strictly determined by the prevailing input conditions. Input conditions that might have existed previously have absolutely no effect on the operation of the circuit. Combinatorial logic circuits are not sequence sensitive.

By contrast, sequential logic circuits are sequence sensitive: The output logic states are determined by the sequence of input logic conditions. The circuits introduced in this chapter all have at least one operating mode in which the output logic state is strictly determined by the sequence of input logic states.

6-1 The Basic Latch Circuit

Figure 6-1(a) shows a rather simple-looking NAND gate circuit and truth table. Note that inputs A and B each go to one input of a NAND gate, while the output of each gate is fed back to an input of the opposite gate. This cross-coupled feedback configuration is the hallmark of all sequential logic circuits, including digital latches and flip-flops.

If the input logic levels specified in the truth table are used, an analysis of this circuit is fairly straightforward, at least down to the last line where $A = B = 1$. Whenever A and B are both at logic 0, for instance, both NAND gates are forced

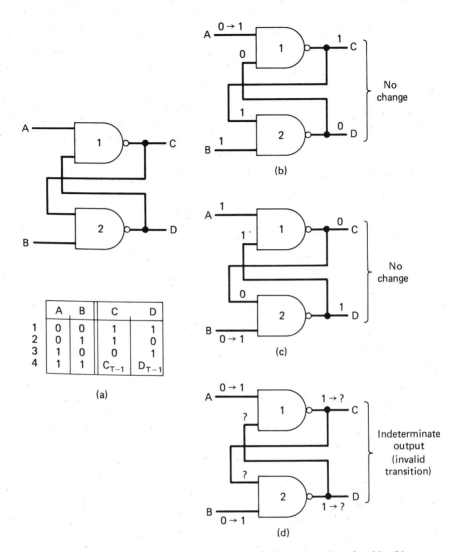

Figure 6-1   A basic latch circuit. (a) Logic diagram and truth table. (b) Entering the memory mode from $A = 0$, $B = 1$. (c) Entering the memory mode from $A = 1$, $B = 0$. (d) The invalid transition from $A = B = 0$ to $A = B = 1$.

to generate logic-1 outputs (applying a logic 0 to any input of a NAND gate guarantees a logic-1 output). The fact that a logic-1 level is then fed back to the second inputs of each gate does not alter the fact that the NAND gates produce logic-1 outputs. By the first line in the truth table, then, $A = B = 0$ yields outputs $C = D = 1$ in a combinatorial-logic fashion.

Changing input B to a logic 1 as shown in the second line of the truth table changes output D to logic 0. To see how this happens, note that input A is still at logic 0; so the output of NAND gate 1 is still forced to produce a logic-1 output. The logic-1 level from output C is fed back to one input of gate 2; and this input, combined with the logic-1 level now being fed to the B input of gate 2, forced that NAND gate to yield a logic-0 output (applying all logic 1's to a NAND gate guarantees a logic-0 output). By the second line in the truth table, setting $A = 0$ and $B = 1$ produces outputs where $C = 1$ and $D = 0$.

The third line of the truth table in Fig. 6-1(a) can be analyzed in a similar way. Here $A = 1$ and $B = 0$; and the logic-0 state at the B input of gate 2 guarantees a logic-1 output from that gate. This logic-1 output is then NAND-ed with the logic-1 from input A to yield a logic-0 output at point C.

The problem now is to see what happens when the A and B inputs are both switched to logic 1 as indicated on the bottom line of the truth table. Note that the C and D outputs are not designated 1 or 0. In fact, the output of the circuit cannot be determined without knowing what its output state was *before* the transition to input $A = B = 1$ took place—and that is what distinguishes a sequential logic circuit from a combinatorial logic circuit.

To investigate the sequential properties of this circuit, suppose the inputs are $A = 0$ and $B = 1$. According to the truth table, this particular input condition should yield a $C = 1$, $D = 0$ output. And indeed it does. Figure 6-1(b) shows what happens as the inputs are then switched from $A = 0, B = 1$ to $A = B = 1$. Changing the A input from 0 to 1 does not affect the output of gate 1 at all because that gate is still seeing a logic 0 from output D; and as long as a NAND gate sees a logic-0 input, it generates a logic-1 output. Thus output C remains at logic 1, in spite of the transition from 0 to 1 at the A input. And since output C does not change, neither does output D—gate 2 still sees logic-1 inputs from B and C.

In short, switching from $A = 0, B = 1$ to $A = B = 1$ causes no change in the C and D outputs. They remain at $C = 1$, $D = 0$ as prescribed by the inputs prior to the transition to $A = B = 1$. Of course, switching back to inputs $A = 0, B = 1$ lets the outputs remain at $C = 1$, $D = 0$ according to the second line of the truth table in Fig. 6-1(a).

Now see what happens whenever the inputs are switched from $A = 1, B = 0$ to the input condition, $A = B = 1$, under consideration here. Figure 6-1(c) illustrates how the circuit responds to this particular transition. As long as $A = 1$ and $B = 0$, the third line of the truth table shows that the outputs are guaranteed to be $C = 0$, $D = 1$. What happens, though, as input B is switched to logic 1? Even after the B input changes to logic 1, gate 2 in Fig. 6-1(c) still sees a logic-0 state at output C, thus assuring a logic-1 output at D. Output D thus remains at logic 1 as B is switched from 0 to 1. And since the D output remains at logic 1, NAND gate 2 still sees all 1's, the logic condition necessary for producing a 0 state at output C. So output C remains fixed at logic 0.

A transition from $A = 1, B = 0$ to $A = B = 1$ does not cause any change in the circuit's output. The outputs remain at $C = 0$, $D = 1$ as prescribed by the input conditions that existed prior to the transition to $A = B = 1$.

The circuit in Fig. 6-1(a) thus has the ability to "remember" a given output condition whenever the inputs are switched so that A and B are both at logic 1. If the outputs happen to be $C = 0$, $D = 1$ prior to the transition to $A = B = 1$, the circuit "remembers" or *latches* that particular output until the A and B inputs are changed to some other condition. Or if the outputs happen to be $C = 1$, $D = 0$ prior to the transition to the memory mode ($A = B = 1$), the circuit latches that particular output.

The last line in the truth table in Fig. 6-1(a) shows that $C = C_{t-1}$ and $D = D_{t-1}$. The $t - 1$ subscript merely implies the notion "prior to the transition." In other words, setting $A = B = 1$ forces outputs C and D to hold the state that existed prior to the transition. There is no change in the outputs. The outputs are latched. The circuit "remembers" the output condition that existed prior to the transition to $A = B = 1$.

These are facts that have powerful implications for modern digital electronics.

This discussion has not yet considered what happens when the circuit's inputs are switched from $A = B = 0$ to $A = B = 1$. As indicated in Fig. 6-1(d), this is an invalid transition that actually produces a set of indeterminate outputs. An analysis of the circuit in Fig. 6-1(d) ought to show that the circuit becomes unstable if A and B are *simultaneously* switched from 0 to 1. The circuit actually oscillates until the two outputs somehow become complements of one another. The problem is that there is no rational way to determine which output will be at logic 1 and which one will be at logic 0 when the oscillation stops. Any circuit will tend to stop oscillating with the same output pattern, but the pattern can be different for the same circuit using different NAND gates.

The circuit in Fig. 6-1 is thus useful only as long as the two outputs remain in opposite logic states; and the best way to ensure that the circuit always operates in these three valid modes is to make provisions for preventing inputs A and B from going to 0 at the same time. Any other combination of inputs ensures a valid operating condition.

Study the operating features of the circuit in Fig. 6-1 thoroughly. The operation of every flip-flop depends on this circuit, and a complete understanding of the basic latching operation at this point will greatly simplify the analysis of more complex flip-flops described in this chapter.

6-2   The R-S Flip-Flop

An R-S flip-flop is so named because its inputs are traditionally labeled R and S: *RESET* and *SET*. Figure 6-2(a) shows the basic R-S flip-flop circuit, using NAND gates as the basic latch-action elements. Actually, this circuit is the same as the basic latch circuit in Fig. 6-1(a). The inverters at the input of the R-S flip-flop merely serve the purpose of making the truth table conform to standards specified for more complex R-S flip-flop devices. Without the inverters, the circuit is indeed identical to the one in Fig. 6-1(a).

The truth table for the R-S flip-flop in Fig. 6-2(a) is quite similar to that of the

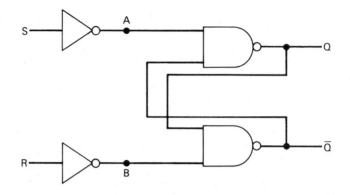

S	R	Q	$\overline{Q}$	Mode
0	0	Q_{t-1}	$\overline{Q}_{t-1}$	Memory
1	0	1	0	SET
0	1	0	1	RESET
1	1	1	1	Invalid

(a)

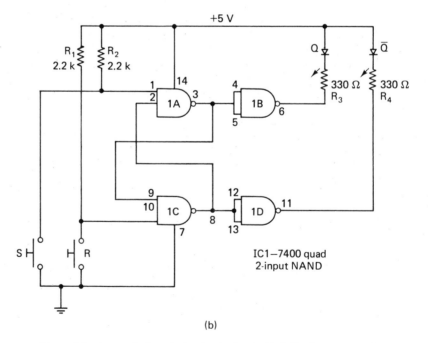

(b)

Figure 6-2 An equivalent logic circuit for an R-S flip-flop. (a) Logic diagram and truth table. (b) Practical implementation of the R-S flip-flop equivalent circuit.

basic latch circuit. In this instance, however, the invalid input condition is $A = B = 1$, while the memory or latch mode occurs when $S = R = 0$. The two inverters are wholly responsible for the differences between this R-S truth table and the table for the so-called $\bar{R}$-$\bar{S}$ flip-flop in Fig. 6-1(a).

The outputs of the R-S flip-flop are labeled Q and $\bar{Q}$. This notation conforms to modern standards for flip-flops of all kinds. Using outputs labeled Q and $\bar{Q}$ implies that the outputs are always complements of one another; which, indeed, they are as long as the user avoids the invalid input condition where $S = R = 1$.

The action of a basic R-S flip-flop can be summarized as follows:

1. The Q output follows the R-S input as long as the R and S inputs are different (complements).
2. The circuit "remembers" the last R-S condition as long as $R = S = 0$.
3. The invalid input condition is $R = S = 1$.

The circuit in Fig. 6-2(b) is a practical implementation of the basic R-S flip-flop using input and output interfacing techniques described in Chapter 5. The basic latching elements in this instance are NAND gates 1A and 1C. The NAND-equivalent inverters, 1B and 1D, merely invert the latch's outputs so that the output indicator LEDs (Q and $\bar{Q}$) are driven with the proper logic levels. LED Q, for instance, lights up whenever the circuit generates a $Q = 1$ output according to the truth table in Fig. 6-2(a). Similarly, the $\bar{Q}$ LED lights in response to a $\bar{Q} = 1$ output from the latch elements.

The inputs to this circuit are a pair of normally open push button switches S and R. Depressing these switches corresponds to a logic-1 input as designated in the truth table. Releasing the push buttons then corresponds to a logic-0 input. These two input switches, combined with their corresponding pull-up resistors (R_1 and R_2), actually invert the inputs as prescribed in the basic circuit diagram in Fig. 6-2(a). When neither push button is depressed—$S = R = 0$ input—the latch elements actually see a set of logic-1 inputs, and the system is in its memory mode.

Depressing the S push button sets up inputs $S = 1$, $R = 0$; then the circuit should respond by showing LED Q lighted and $\bar{Q}$ turned off. The lights should then remain in that state when the S push button is released. Depressing the R push button should then reverse the logic condition of the lamps, turning off Q and lighting $\bar{Q}$. Again, the outputs should remain in that particular condition after the R push button is released.

Depressing the S push button thus SETS the Q output to logic 1 and depressing the R push button RESETS the Q output to logic 0. That is the basic nature of the R-S flip-flop. Once the output is SET, it remains that way until it is RESET. And then it remains in the RESET condition until it is SET again.

The circuit works reliably unless the user places it into its invalid operating mode by depressing both input switches at the same time.

6-3 The Gated R-S Flip-flop

The circuit in Fig. 6-3(a) looks very much like the basic R-S flip-flop in Fig. 6-2(a). This circuit has a pair of NAND gates, gates 3 and 4, that uses the cross-coupled feedback connections to perform the basic latching operation. Gates 1 and 2 in Fig. 6-3(a) also invert the inputs S and R. But the common connection between the input gates to G lets the user gate off the S and R inputs.

Note from the truth table in Fig. 6-3(a) that setting the G input to logic 0 places the circuit in a memory mode and makes any inputs at S and R irrelevant. The circuit doesn't care what happens at the S and R inputs as long as $G = 0$—it remains in its memory mode.

Setting the G input to logic 1, however, lets it perform exactly like the basic R-S flip-flop. Compare the last four lines of the truth table in Fig. 6-3(a) with the table in Fig. 6-2(a).

Whenever the G input is set to logic 0, the outputs of NAND gates 1 and 2 are both forced to logic 1 (the output of any NAND gate goes to logic 1 whenever one or more inputs are set to logic 0); and no matter what states the R and S inputs might have, the outputs of gates 1 and 2 remain at 1 as long as $G = 0$. The signal inputs to gates 3 and 4 thus see logic-0 inputs that keep that output latch system in its memory mode.

Setting the G input to logic 1 transfers control to the R and S inputs.

The overall action of the gated R-S flip-flop can be summarized as follows:

1. The Q output follows the S input as long as $G = 1$ and the R and S inputs are complements of one another.
2. The circuit has two memory modes:
   (a) Whenever $G = 0$, regardless of the states of the R and S inputs
   (b) whenever $G = 1$ and $R = S = 0$
3. An invalid input occurs whenever $R = S = G = 1$.

The circuit in Fig. 6-3(b) is a practical implementation of this gated R-S flip-flop. Push-button switch G keeps the circuit normally in its gated-off memory mode. Depressing that push button allows R_3 to pull up the gate inputs to ICs 1A and 1B to logic 1, thus gating on the circuit and transferring control to the S and R inputs.

Switches S and R are toggle or slide switches. As long as these switches are set to their "1" positions, the flip-flop sees logic-1 inputs. Setting the switches to their "0" positions, however, grounds the inputs of gates 1A or 1B to provide logic-0 inputs.

The inverters, ICs 2A and 2B, at the output of the latching gates merely invert the output logic levels so that they drive LEDs Q and $\bar{Q}$ in the proper logic format, Whenever LED Q lights up, for instance, it indicates a $Q = 1$ output.

Operating this circuit is thus a matter of first setting the S and R switches for the desired input logic levels and then depressing push button G to see the circuit's

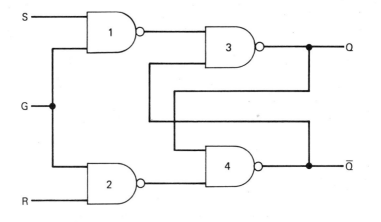

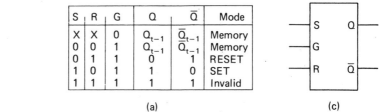

S	R	G	Q	$\overline{Q}$	Mode
X	X	0	Q_{t-1}	$\overline{Q}_{t-1}$	Memory
0	0	1	Q_{t-1}	$\overline{Q}_{t-1}$	Memory
0	1	1	0	1	RESET
1	0	1	1	0	SET
1	1	1	1	1	Invalid

(a)

(c)

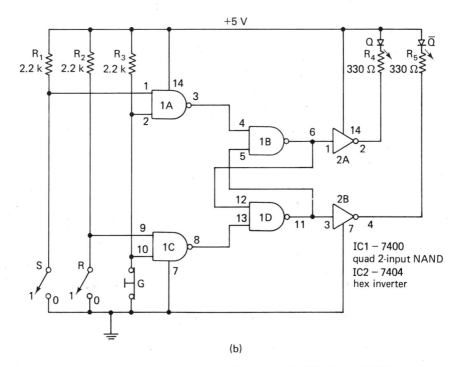

(b)

Figure 6-3   An equivalent logic circuit for a gated R-S flip-flop. (a) NAND gate equivalent and truth table. (b) Practical implementation of the NAND gate gated R-S flip-flop. (c) Logic symbol for a basic gated R-S flip-flop.

response at the LED outputs. After the push button is released, the inputs at S and R can be changed without affecting the outputs until G is depressed again.

The whole idea is to be able to set up the desired inputs, enter that set of inputs into the circuit by depressing switch G, and then store the response at the Q outputs after releasing the push button. This ability to store a desired 1 or 0 logic level is the heart of all digital IC memory circuits.

Suppose the user sets $S = 1$ and $R = 0$. He or she will not notice any change at the Q and $\bar{Q}$ outputs until the G push button is depressed. After the G push button is depressed the truth table indicates that the user will see $Q = 1$, $\bar{Q} = 0$ at the outputs. The circuit then stores this particular set of output logic levels after releasing push button G; and they will remain stored there until the user reverses the R and S logic levels and depresses the G push button once again.

If the user happens to gate in $S = R = 0$, the Q outputs do not change from their former state because this defines another memory mode. See the second line in the truth table in Fig. 6-3(a). The user should avoid gating in a $S = R = 1$ input, however, because that represents an invalid input condition that yields a $Q = \bar{Q} = 1$ output—an illogical state in which a pair of complemented outputs are equal.

6-4   The Level-D Flip-Flop

The R-S flip-flops described thus far in this chapter all have an invalid input condition that occurs whenever the R and S inputs are at logic 1 at the same time. It is possible to avoid this condition by modifying the gated R-S flip-flop so that one input is always the complement of the other. The result is a level-switched D flip-flop.

The circuit in Fig. 6-4(a) is a NAND-gate equivalent of a level-D flip-flop. It is similar to the gated R-S flip-flop, but it has an inverter at the "R" input that ensures that the two inputs are always complements of one another. This circuit is gated in exactly the same fashion as the gated R-S circuit.

Setting the G input to logic 0 gates off NAND gates 1 and 2, forcing them to yield logic-0 outputs, regardless of the state of the D input. And logic-0 outputs from gates 1 and 2 place the latch elements, NAND gates 3 and 4, into their memory mode. The circuit is thus in its memory mode as long as $G = 0$. See the first line in the truth table in Fig. 6-4(a).

Setting G to logic 1 transfers control of the circuit to the D input. According to the truth table, the Q output follows the D input as long as $G = 1$. The $\bar{Q}$ output is always the complement of the D input. In fact, setting $G = 1$ makes the circuit behave as though there were a straight-wire connection from D to Q and an inverter from D to $\bar{Q}$.

Note that the level-D circuit does not have an invalid input mode that typifies R-S flip-flops. As far as R-S flip-flops are concerned, the invalid mode occurs whenever the input elements both see a logic-1 level at the same time; but the inverter in Fig. 6-4(a) prevents this condition from occurring here.

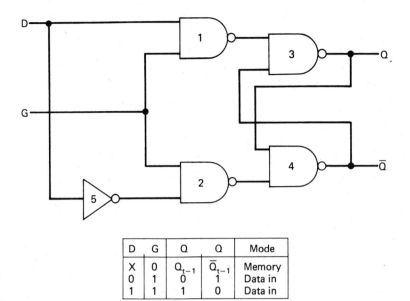

D	G	Q	Q̄	Mode
X	0	Q_{t-1}	$\bar{Q}_{t-1}$	Memory
0	1	0	1	Data in
1	1	1	0	Data in

(a)

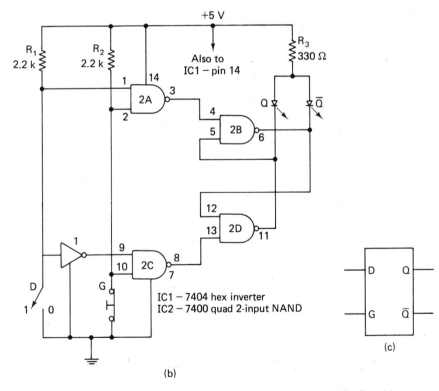

(b)

(c)

Figure 6-4 An equivalent logic circuit for a level-switched D flip-flop. (a) An equivalent logic circuit and truth table. (b) Practical implementation of the equivalent D flip-flop. (c) Logic symbol for a basic level-switched D flip-flop.

Using the inverter at the input of the D flip-flop also eliminates the $R = S = 0$ mode that characterizes a gated R-S flip-flop when $G = 1$. The D flip-flop can be placed into its memory mode only by setting $G = 0$.

The action of a level-D flip-flop can be summarized as follows:

1. The Q output follows the D input as long as $G = 1$.
2. The flip-flop is in its memory mode as long as $G = 0$.
3. The circuit has no invalid operating conditions.

The circuit in Fig. 6-4(b) is a practical implementation of a simple level-switched D flip-flop. The D input comes from a SPST toggle or slide switch, D. The circuit is gated by means of the normally closed push button switch, G.

This circuit is normally in its memory mode because switch G pulls down an input of both of the input logic NAND gates (2A and 2C) to ground and logic 0. In effect, these input gates are switched off, preventing any action at the D input from affecting the Q and $\bar{Q}$ outputs.

Depressing the G push button, however, opens its connection to ground and allows resistor R_2 to pull up the gate inputs to logic 1. The input gates are thus opened, letting the Q output follow whatever logic state is set at the D input.

The instant the G push button is released, the circuit returns to its memory mode, holding a Q output that was dictated by the D input just as the gate switched closed.

Notice that the interfacing to the output LEDs in Fig. 6-4(b) is somewhat different from that used for the R-S flip-flops described previously in this chapter. Since the circuit cannot generate an invalid $Q = \bar{Q} = 1$ output, it is possible to simplify the output interfacing scheme. For one thing, the circuit in Fig. 6-4(b) does not show inverters between the latch NAND gates (2B and 2D) and the LEDs. The cathodes of the LEDs are simply connected to NOT-ed versions of the output they are supposed to represent: The Q LED is connected to the latch circuit's $\bar{Q}$ output, and the $\bar{Q}$ LED is connected to the Q output. This inversion procedure totally eliminates the need for inverters between the latch elements and LED connections to $+5$ V.

The output circuit in Fig. 6-4(b) is also different because it uses a single ballast resistor for both LEDs. This is a part- and cost-saving technique that can be applied in instances in which only one LED can possibly light at a time. The invalid condition for the R-S flip-flops forces both LEDs to light up, making it necessary to use a separate ballast resistor for each of them.

The level-switched D flip-flop is the simplest flip-flop available in standard TTL IC packages today. R-S flip-flops are used as building blocks within certain kinds of more complex ICs, but they are no longer available as simple R-S flip-flop packages.

Figure 6-5 shows the internal arrangement of D flip-flops and pinout for the 7475 quad D latch. There are four D flip-flops in this package, each having a separate D input and a pair of Q and $\bar{Q}$ outputs. Note, however, that FF1

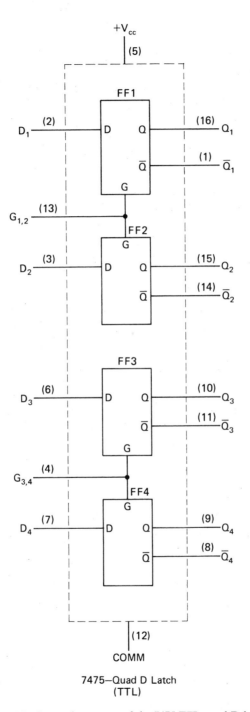

Figure 6-5 Internal structure of the 7475 TTL quad D latch.

and FF2 share the same gate input at pin 13, while F3 and F4 share a common gate input from pin 4.

This circuit is often used as a temporary storage place for 4-bit digital data. Used this way, the two gate inputs are tied together, making all four flip-flop sections respond to a G input at the same time. The user normally applies a pattern of 1's and 0's to the four D inputs and then sets the G inputs to logic 1. This action transfers the input data to the outputs where it is stored after the G input is returned to logic 0.

6-5 The Edge-Triggered D Flip-Flop

The gated R-S flip-flop and the level-switched D flip-flop described in the two previous sections of this chapter have gate inputs that allow the circuits' outputs to change in response to changes at the input as long as the gate input is in a logic-1 state. Switching the gate input to logic 0 places the circuits into their memory modes, effectively blocking any further changes from occurring at the outputs until the gate input is pulled up to logic 1 again. As long as the gate input is at logic 1, there can be a direct relationship between activity at the inputs and responses at the outputs.

An edge-triggered flip-flop, however, is normally in a memory mode: It is in a memory mode while the gate input is at logic 0 and it is in a memory mode while the gate input is at logic 1. *The only time the outputs can change state is during the brief interval of time it takes the gate signal to make a transition from 0 to 1, or in some circuits, from 1 to 0.* An edge-triggered flip-flop responds only to a rising or falling edge of the input gate waveform. Otherwise, the circuit is in its memory mode.

While gated flip-flops do indeed have some important applications, their popularity in modern digital systems is eclipsed by that of their edge-triggered counterparts.

6-5.1 The Basic Edge-Triggered D Flip-Flop

Figure 6-6 shows the logic symbol, truth table and NAND-gate equivalent of a basic edge-triggered D flip-flop. The Q output takes on the state of the D input, but only during the time the CLK (clock) input makes a transition from 0 to 1. At any other time, including the time the CLK falls from logic 1 to 0, the circuit is in its memory mode. The circuit, in other words, is a positive edge-triggered D flip-flop.

A complete analysis of an edge-triggered flip-flop is very involved, and few teachers and textbook writers feel that an in-depth analysis is justified. It is sufficient to say that the circuit is made up of three individual NAND-gate latches. See latches 1, 2 and 3 in Fig. 6-6. The inputs to latch 1 are CLK and the output of gate 4 in latch 2, the inputs to latch 2 are CLK, D and the output of gate 2 in latch 1, and the inputs to latch 3 are the output of gate 2 in latch 1 and gate 3 in

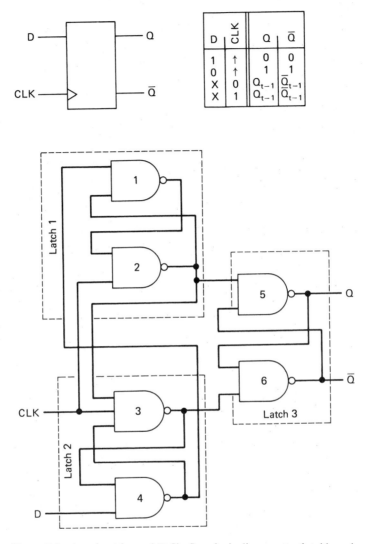

D	CLK	Q	Q̄
1	↑	0	0
0	↑	1	1
X	0	Q_{t-1}	$\bar{Q}_{t-1}$
X	1	Q_{t-1}	$\bar{Q}_{t-1}$

Figure 6-6   An edge-triggered D flip-flop: logic diagram, truth table and equivalent NAND gate circuit.

latch 2. Latches 1 and 2 have a cross-coupled feedback scheme connecting them together, and the overall effect is a circuit that is almost always in a memory mode— in a state that is determined by the previous CLK and *D* input conditions. The only operating condition that allows the *Q* and Q̄ outputs to change is one in which the CLK input rises from 0 to 1; and when that happens, the *Q* output takes on the logic state of the *D* input that prevails at the time. Otherwise, the circuit is insensitive to changes in the inputs.

The truth table in Fig. 6-6 shows that the Q̄ output is always the complement of the Q output. Also note that there are no invalid operating conditions for this particular flip-flop.

The truth table shows two lines with up-pointing arrows in the CLK column. These arrows denote the fact that the Q output equals the D input only after the CLK makes a positive-going level transition. According to the last two lines of the truth table, then, the circuit is in its memory mode while the CLK is fixed at 1 or 0, regardless of the logic level at the D input.

The logic symbol for the edge-triggered D flip-flop in Fig. 6-6 is a rectangle with input and output lines drawn for the D, CLK, Q and $\bar{Q}$ connections. The triangle at the CLK input indicates an edge-triggered input.

6-5.2 A Triggered D Flip-Flop with PRESET and CLEAR

The D flip-flop in Fig. 6-7 is very similar to the one just described. The circuit in Fig. 6-7, however, shows another set of inputs called PRESET and CLEAR. These two inputs influence the outputs without regard to any clocking operations, and their main purpose is to let the user set the outputs to any desired logic level before or after a positive-edge CLK operation occurs.

Setting the PRE (PRESET) input to logic 0, for example, guarantees that NAND gate 5 will show a logic 1 output, no matter what the CLK and D inputs are doing at the time. Similarly, the CLR (CLEAR) input will set the $\bar{Q}$ output of the circuit to logic 1 by directly gating off NAND gate 6. The truth table in Fig. 6-7 summarizes the operation of this circuit.

The truth table shows that the Q output directly follows the CLR input as long as the PRE and CLR inputs are complements of one another. The D and CLK inputs have no relevance at any time during these *asynchronous* (not clock-related) operating modes. Setting PRE and CLR to logic 1 at the same time, however, transfers control of the circuit to the D and CLK inputs, making the system behave exactly as the basic edge-triggered flip-flop in Fig. 6-6. The clocked operating modes are properly called the *synchronous* (clock-related) modes.

Note that the truth table indicates an invalid operating mode, one in which PRE = CLR = 0. Under this particular set of circumstances, the Q and $\bar{Q}$ outputs both go to logic 1, thus destroying the complemented relationship that is supposed to exist between the two.

The "bubbles" at the CLR and PRE inputs of the circuit's logic symbol in Fig. 6-7 imply an inverse logic relationship between these inputs and the response of the Q output. One would suppose that setting CLR to logic 1 would clear the Q output to logic 0; and, similarly, setting the PRE input to logic 1 would preset the Q output to logic 1. But the opposite is true in this case; so the logic diagram indicates an inverse relationship between the asynchronous inputs and the Q output.

The two D flip-flops described thus far are only representative of the triggered D flip-flop family now available in both TTL and CMOS IC packages. The following section is a brief summary of available D flip-flop ICs.

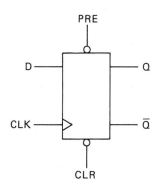

PRE	CLR	D	CLK	Q	Q̄
0	1	X	X	1	0
1	0	X	X	0	1
1	1	1	↑	1	0
1	1	0	↑	0	1

PRE = CLR = 0 is invalid

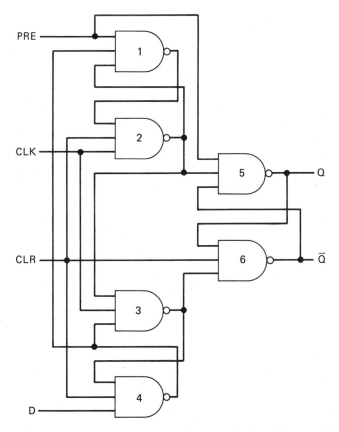

Figure 6-7 An edge-triggered D flip-flop with PRESET and CLEAR: logic diagram, truth table and equivalent NAND gate circuit.

6-5.3 Available Triggered D Flip-Flop ICs

Table 6-1 summarizes the triggered D flip-flops currently available in the TTL and CMOS families. Figure 6-8 shows their corresponding logic symbols.

All of the ICs, with the notable exception of the 4042, are positive edge triggered. The 4042 has a special *P* (POLARITY) input that lets the user select whether the circuit clocks on the positive-going or negative-going edge of the CLK pulse.

Table 6-1 AVAILABLE EDGE-TRIGGERED D FLIP-FLOP ICs.

TTL Family	
74174	Hex D flip-flop with positive-edge triggering, asynchronous CLR and Q output only [Fig. 6-8(a)].
74175	Quad D flip-flop with positive-edge triggering, asynchronous CLR and both Q and $\bar{Q}$ outputs [Fig. 6-8(b)].
7474	Dual D flip-flop with positive-edge triggering, asynchronous CLR and PRE and both Q and $\bar{Q}$ outputs [Fig. 6-8(c)].

CMOS Family	
4013	Dual D flip-flop with positive-edge triggering, positive-logic asynchronous PRE and CLR and both Q and $\bar{Q}$ outputs [Fig. 6-8(d)].
4042	Quad D flip-flop with programmable positive- or negative-edge triggering and both Q and $\bar{Q}$ outputs [Fig. 6-8(e)].

3-State Devices	
74173	3-state quad D flip-flop with positive-edge triggering, asynchronous CLR and both Q and $\bar{Q}$ outputs—TTL.
4076	3-state quad D flip-flop with positive-edge triggering, positive-logic asynchronous CLR and Q output only—CMOS.

With the exception of the 4042, all of these flip-flops have at least an asynchronous CLR input, and two of them have both CLR and PRE inputs. Note that the asynchronous inputs on the 4013 do not show "bubbles" at their input connections. The significance of this fact is that there is a positive-logic relationship between the CLR and PRE inputs and the response of the Q output: Setting CLR = 1 clears the Q output to logic 0, for instance. This is just the opposite of the negative-logic PRE and CLR operations for all the other D-type flip-flops.

The 74173 and 4076 CMOS D flip-flops are tri-state devices that are described in fuller detail in a later chapter.

6-6 T Flip-Flops

Figure 6-9(a) shows a logic diagram and truth table for a typical T, or *toggled*, flip-flop. According to the truth table, the Q output merely changes state each time a T input waveform makes a transition from 0 to 1. If the Q output happens to be

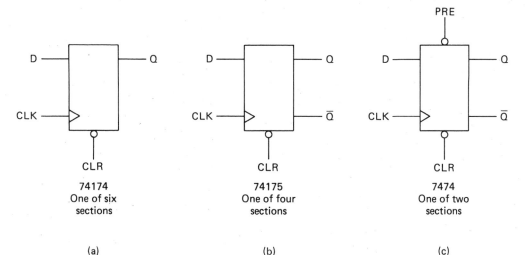

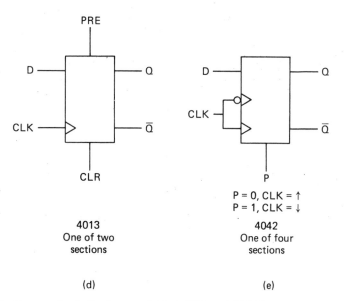

Figure 6-8   Logic diagrams for D flip-flops available in TTL and CMOS
IC packages. Compare with Table 6-1.

at logic 1 when a positive-going edge occurs at T, for example, the output switches
to logic 0. The next positive-going edge at T then switches the Q output back to
logic 1. This toggling action—switching from one output state to the other—continues as long as pulses are applied to the T input. The waveforms accompanying
Fig. 6-9(a) illustrate this positive-edge toggling effect.

The $\overline{Q_{t-1}}$ designation in the Q column indicates that the output switches to the

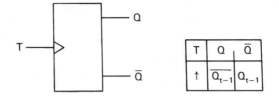

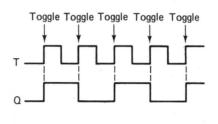

(a)

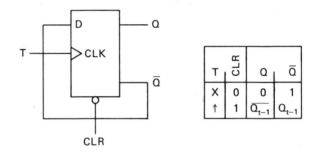

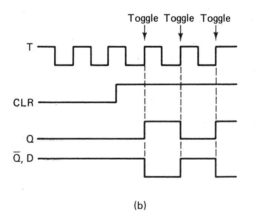

(b)

Figure 6-9 T flip-flops. (a) Logic diagram, truth table and waveforms. (b) Building a T flip-flop with asynchronous CLEAR from an edge-triggered D flip-flop: logic diagram, truth table and waveforms.

state the $\bar{Q}$ output had just prior to the toggle pulse. This circuit is not sensitive to the falling, or negative-going edge, of the toggling pulse waveform.

T flip-flops thus change output states each time a certain edge occurs at the toggling waveform at its T input. Some T flip-flops are positive edge-triggered as shown here, but others are negative edge-triggered.

6-6.1 Building a T Flip-Flop from a D Flip-Flop

T flip-flops are no longer very popular as IC devices. Many applications of switching circuits call for a T-type toggling action, however; and Fig. 6-9(b) shows how it is possible to build a T flip-flop from an edge-triggered D flip-flop IC. Note that the $\bar{Q}$ output is tied directly back to the D input, ensuring that the D input always sees a logic state that is opposite that of the Q output. Whenever the circuit is toggled, then, the Q output is always set to its opposite state.

This particular T flip-flop also includes a CLR input that lets the user asynchronously clear the Q output to zero whenever the CLR input is pulled down to logic 0; and since the asynchronous inputs on D flip-flops override any clocking action, it is possible to hold the circuit's output in its cleared state until the CLR input is set to logic 1 and the first positive-going edge of the toggle waveform appears at the T input. See the waveforms in Fig. 6-9(b).

6-6.2 Switch Debouncing for Toggle Inputs

Any toggled flip-flop must have a clean toggling input waveform for reliable operation. If the T input waveform happens to have any jitter or electrical noise riding on it, the T flip-flop will respond to that noise, switching output states back and forth until the noise stops; and there is no way to determine exactly how many times the toggle waveform will "bounce," and hence no way to determine exactly what state the circuit's output will have when the bouncing does finally stop.

Bouncing effects are especially critical when attempting to toggle a flip-flop that has a mechanical switch input. Anytime two metals slam in contact with one another, they are bound to bounce open and closed for a short period of time. Most switches settle down within 10 ms after their contacts close, but that is certainly long enough to play havoc with a toggled flip-flop.

Figure 6-10(a) shows a toggled flip-flop having a simple push button switch input. Whenever the user depresses the switch, presumably to switch the output states, contact bounce occurs at the flip-flop's T input. See the T waveform in Fig. 6-10(a). The flip-flop responds to this bouncing effect by toggling back and forth as shown in the Q waveform. In this particular example, the user gets the impression that the Q output doesn't change at all. The switch contact bounced an even number of times (four times in this example), finally leaving the Q output in its original logic-0 state. If the contact had bounced an odd number of times, Q would have settled down at logic 1, and the user would have the impression that the circuit worked properly. But switch bouncing is not reliable; it can bounce an even number of times at one closure, then an odd number of times when it is closed again.

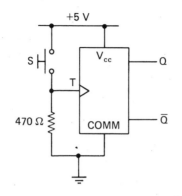
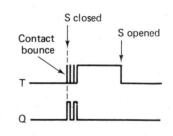

(a)

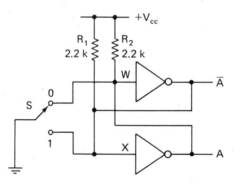

S	W	X	A	$\bar{A}$
0	0	1	0	1
1	1	0	1	0
↑↓	1	1	A_{t-1}	$\bar{A}_{t-1}$

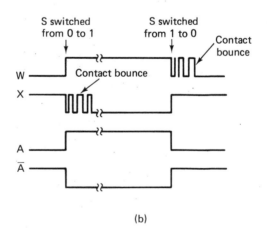

(b)

Figure 6-10 Debouncing mechanical switch inputs to toggled flip-flops. (a) The switch-bouncing problem. (b) A simple switch debouncing circuit: circuit diagram, truth table and waveforms.

The idea of toggling flip-flops directly from a switch interface is thus totally unacceptable. The switch contact must be debounced before its logic levels are applied to the toggling input. The circuit in Fig. 6-10(b) shows a simple contact debouncing scheme that calls for using a SPDT switch and a pair of inverters.

The logic inverters in this circuit have a cross-coupled feedback configuration that is typical of basic latch circuits. The circuit is, in fact, a latch circuit.

As long as switch S is resting in the 0 position, point W is pulled down to logic 0. Point X, however, is pulled up close to $+V_{cc}$ (logic 1) through resistor R_1. The output of the circuit thus shows $A = 0$, $\bar{A} = 1$. See the first line of the truth table in Fig. 6-10(b).

As the user switches S down toward the "1" position, the contacts make a clean break with the "0" position as shown in waveform W. (Switch contacts do not bounce when they are opened.) While the switch contact is moving toward its "1" position, points W and X are both at logic 1, putting the latch circuit into its memory mode, keeping the outputs at a clean $A = 0, \bar{A} = 1$ condition. The moment the contacts make the "1" position, they bounce several times as shown in waveform X.

At first contact, however, the output of the latch circuit is changed to $A = 1$, $\bar{A} = 0$; and as the contact bounces open for a moment, the latch circuit is placed into its memory mode again, holding the $A = 1, \bar{A} = 0$. No matter how many times the contacts bounce in the $S = 1$ position, the outputs do not change state. Compare the waveforms A and $\bar{A}$ with W and X just after S is switched from 0 to 1.

The same sort of process takes place whenever the switch is moved from its "1" to "0" position. The contacts bounce as the switch makes the "0" position, but the outputs respond only to the first contact; the bouncing is "memoried" out.

The up and down arrows in the S column of the truth table indicate the transition states of the switch: The up arrow signifies a transition from 0 to 1, and the down arrow signifies a transition from 1 to 0. Anytime the common terminal of the switch is not making contact with its "0" or "1" terminals, the latch circuit is holding the last solid switch connection it saw.

The switch debouncing circuit is necessary only when operating the toggle inputs of a flip-flop from a mechanical switch assembly, including relay contacts. A debouncing circuit is not necessary for any of the other kinds of flip-flop inputs such as asynchronous CLR and PRE, D inputs or R and S inputs.

6-6.3 A Toggled Flip-Flop with a Debounced T Input

Figure 6-11 shows a practical implementation of a T flip-flop built up from a positive edge-triggered D flip-flop. The T input switch is debounced by inverters I1 and I2 before its logic levels are fed to the flip-flop's CLK connection. As illustrated in this circuit, the D and CLR inputs do not have to be debounced.

The circuit is initially cleared by depressing the CLR push button. This action pulls the CLR input of the flip-flop down to logic 0, and it asynchronously pulls the Q output down to logic 0 while setting the $\bar{Q}$ output of the flip-flop to logic

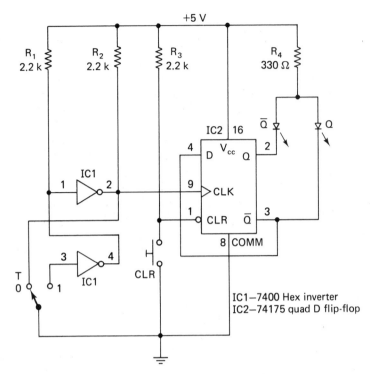

Figure 6-11 Practical implementation of a T flip-flop using a D flip-flop IC and switch debouncing features.

1. This particular output pattern allows LED Q to light up, and it extinguishes the $\bar{Q}$ LED. The output then remains in this state after the CLR push button is released.

If the T input switch is in the "0" position as shown in the diagram, setting it to its "1" position causes a clean transition from 0 to 1 at the CLK input of the flip-flop; and that is precisely the input required for toggling the circuit. The outputs thus switch state and remain fixed until the user returns the T switch to its "0" position *and back to "1" again*. The circuit toggles only when the switch is moved from 0 to 1 because this particular flip-flop is positive edge-triggered. The transition from 1 to 0 has no effect on the output.

In short, this circuit changes output states each time the user sets the T switch to "1" and then returns it to "0." It can be asynchronously cleared at any time by depressing the CLR push button.

6-7 The Master–Slave J-K Flip-Flop

Any complete discussion of flip-flops must culminate in a presentation of master–slave J-K flip-flops. According to the thinking of many circuit designers, the J-K flip-flop is *the* flip-flop to use at any time. This opinion is rather well justified on the grounds that it is possible to program a J-K flop-flop to work like any of the edge-triggered flip-flops described thus far in this chapter. Although certain opinions about J-K flip-flops might be questioned to some extent, one fact remains true: J-K flip-flops are the most versatile of all.

6-7.1 The Master–Slave Principle

Much of the usefulness of master–slave J-K flip-flops rests on the master–slave feature that is built into most of them. Any kind of flip-flop can be designed around the master–slave principle, but in actual practice it is only used in J-K flip-flops. Since we have already discussed R-S flip-flops in some detail in this chapter, it is more appropriate to introduce the notion of a master–slave flip-flop in the context of an R-S flip-flop. After that, we can extend the ideas to J-K flip-flops.

The NAND gate circuit in Fig. 6-12 illustrates the basic master–slave principle for an R-S flip-flop. Note that the circuit is made up of two complete R-S flip-flops: The master section is made up of an R-S flip-flop that consists of NAND gates 1 through 4; the slave section is made up of another R-S flip-flop using gates 5 through 8.

The critical difference between these two R-S flip-flop sections is the logic inverter that stands between the circuit's T input and the gate input the slave section. The two R-S flip-flop sections always see opposite logic levels at their gate inputs. While the T input is at logic 1, for instance, the master section is responding to the S and R inputs, reproducing those input logic levels at MQ and $M\overline{Q}$. At the same time, however, the slave section is in its memory mode because its gate input ($\overline{T}$) is at logic 0 at the time.

Whenever the T input is then changed from logic 1 to logic 0, the master section is placed into its memory mode, holding the S and R input levels it saw just prior to the T transition from 1 to 0. At that same time, then, the slave section is "opened up" to receive the logic levels stored at MQ and $M\overline{Q}$. Whatever logic levels are stored at the output of the master section thus appear at the slave's Q and $\overline{Q}$ outputs.

The operation of a master–slave flip-flop is a two-step operation. The T input must make a positive-going transition in order to get data from the R and S inputs to the output of the master section, and then the T input must make a negative-going transition to get the data from the master section to the outputs of the slave. The truth table in Fig. 6-12 implies the need for both the positive-going and negative-going edges by showing a pulse waveform in the T column. In other words, the T input must experience a *complete* pulse waveform before the Q and $\overline{Q}$ outputs show the states indicated on the truth table.

R and S data are fed into the master section while the T input is at logic 1, but then that same set of R and S data is fed to the Q and $\overline{Q}$ outputs of the slave section only as the T input falls from 1 to 0. In effect, this is a negative edge-triggered device; the user can see a change in the Q outputs only after the T input makes a negative-going transition.

Looking at this master–slave R-S flip-flop from a slightly different point of

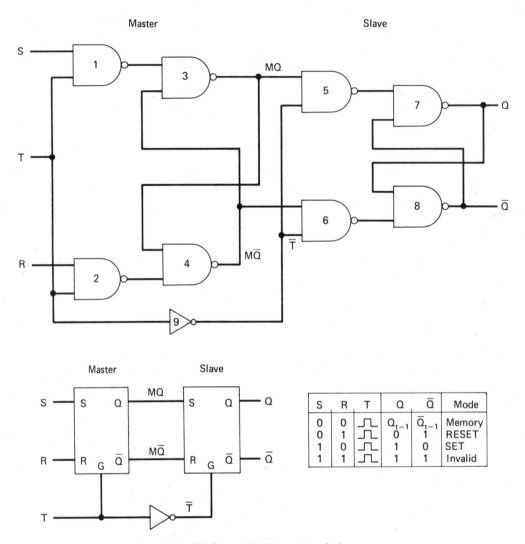

Figure 6-12 A master–slave RST flip-flop: NAND gate-equivalent circuit, RST-equivalent circuit and truth table.

view, consider how the MQ and $M\bar{Q}$ outputs of the master section respond while $T = 1$. While $T = 1$, the MQ and $M\bar{Q}$ connections respond directly to any changes at the R and S inputs. In a sense, there is a direct electrical connection between S and MQ and between R and $M\bar{Q}$ while $T = 1$. At that same time, however, the slave section is seeing $\bar{T} = 0$, and its outputs are totally immune to changes at the S and R inputs: The slave section is "remembering" the MQ and $M\bar{Q}$ inputs it saw at some earlier point in time, when the T input rose from 0 to 1.

The master and slave sections then change roles when $T = 0$. While $T = 0$, the master section is latching the S and R inputs it saw just as T dropped from 1 to 0. Thus, any changes at the R and S input do not cause any change at the MQ and $M\bar{Q}$ connections. And although the slave section is now able to respond directly to MQ and $M\bar{Q}$, the fact that these connections cannot change (the master is in its memory mode) fixes the Q and $\bar{Q}$ outputs at the logic states presented to the slave's inputs just as T dropped to 0. Again, the output of the slave section is immune to changes at the R and S inputs.

The only time the Q outputs can change state is during the negative-going edge of the T waveform. And at that time, the outputs take on the logic states determined by the master section just prior to the negative-going edge of the T waveform. The Q outputs of a master–slave flip-flop are thus totally isolated from the device's data inputs at any given instant. *The outputs respond only to input logic levels that prevailed at some earlier time—when the* T *waveform last made a transition from 1 to 0.*

The circuit in Fig. 6-12 can be properly called a master–slave R-S flip-flop. It can be reclassified as a simple kind of J-K flip-flop, however, by merely relabeling the inputs: Call the S input J and the R input K.

Figure 6-13 is a demonstration circuit that clearly shows the master–slave action of a simple J-K master–slave flip-flop. The J and K inputs are SPST slide or toggle switches. The CLK waveform comes from a debounced SPDT push button switch.

The master section is made up of the four sections of IC2. A pair of LEDs at the output of the master section shows the output states of the master. While these lamps are not necessary to the actual operation of the circuit, being able to keep track of the flow of data through the master and slave sections can be very instructive at this point.

The slave section includes the four sections of IC2 and the output LEDs, Q and $\bar{Q}$.

According to the truth table, setting $J = 1$, $K = 0$ ought to light the Q output LED and extinguish the $\bar{Q}$ LED—*after* one complete clock pulse occurs. Suppose the CLK switch is setting at logic 0 when the user sets switches J and K to 1 and 0, respectively. There should be no changes in the status of any of the four LEDs. While CLK $= 0$, the master section is in its memory mode; and as long as the output of the master section does not change, the slave section cannot change either.

Now suppose the user depresses the CLK push button. The instant the button is depressed, the MQ output of the master section should light and the $M\bar{Q}$ lamp should go out. This is the response of an R-S flip-flop that is being gated on with $S = 1$ and $R = 0$. There should be no change at the Q outputs, however, because the slave section is now in its memory mode.

The moment the user releases the CLK push button, the slave section is gated on and responds by lighting Q and turning off $\bar{Q}$. The cycle is complete. The state of the J and K inputs are finally seen at the Q outputs. The master section is again returned to its memory state, making it immune to any further changes at the J and K inputs until the CLK is set to logic 1 again, thus beginning another master–slave operating cycle.

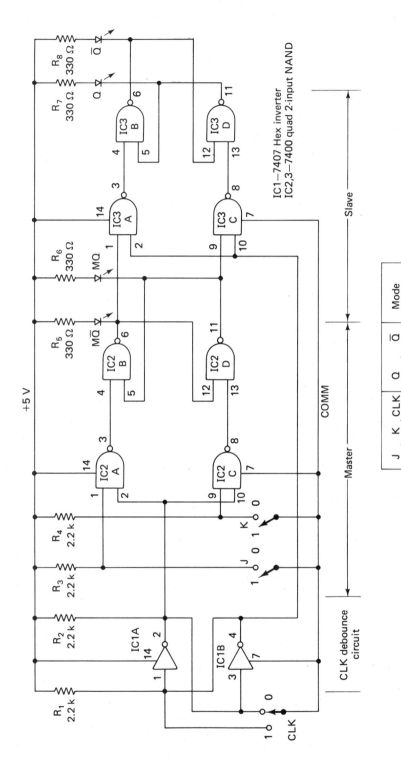

J	K	CLK	Q	Q̄	Mode
0	0	⊓	Q_{t-1}	$\overline{Q_{t-1}}$	Memor
1	0	⊓	1	0	J SET
0	1	⊓	0	1	K RESET
1	1	⊓	1	1	Invalid

Figure 6-13 Practical implementation of a master–slave J-K flip-flop.

The master section responds to the J-K inputs only while the CLK input is at logic 1. The slave section then responds only when the CLK input then drops to logic 0 again. The circuit in Fig. 6-13 is unique in that the user does not usually have access to the output connections of the master section. This arrangement allows the user to observe the master–slave action firsthand.

6-7.2 The Basic J–K Master–Slave Flip-Flop

Figure 6-14 shows the basic logic diagram and truth table for the most complete type of J-K flip-flop. Aside from the J, K and CLK inputs already described, the basic J-K flip-flop also includes a set of asynchronous PRE and CLR inputs.

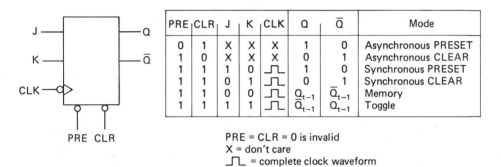

PRE	CLR	J	K	CLK	Q	Q̄	Mode
0	1	X	X	X	1	0	Asynchronous PRESET
1	0	X	X	X	0	1	Asynchronous CLEAR
1	1	1	0	⎍	1	0	Synchronous PRESET
1	1	0	1	⎍	0	1	Synchronous CLEAR
1	1	0	0	⎍	Q_{t-1}	$\bar{Q}_{t-1}$	Memory
1	1	1	1	⎍	$\bar{Q}_{t-1}$	Q_{t-1}	Toggle

PRE = CLR = 0 is invalid
X = don't care
⎍ = complete clock waveform

Figure 6-14 The basic J-K master–slave flip-flop: logic symbol and truth table.

The truth table clearly shows that the flip-flop is in one of its two asynchronous operating modes as long as PRE and CLR are complements of one another. In these asynchronous modes, the PRE and CLR inputs override any ongoing activity at the J, K and CLK inputs—J, K and CLK are not relevant. Also note that the Q output follows the CLR input while the flip-flop is in one of its asynchronous operating modes.

Setting PRE and CLR to logic 1 at the same time, however, transfers control to the J, K and CLK inputs. While J and K are complements of one another, the Q output takes on the logic state of the J input after one complete clock pulse occurs. In a sense, the CLR and J inputs influence the Q output in the same manner. When using the J input, however, the flip-flop must be clocked.

The circuit can be put into a memory mode by setting PRE = CLR = 1 and J = K = 0. Once a complete clock pulse occurs, the circuit "remembers" the status it held at the end of the previous clock pulse.

And, finally, the truth table in Fig. 6-14 shows that the flip-flop can operate in a toggle mode as long as PRE = CLR = J = K = 1. The negative-going edge of each clock pulse switches the output state.

The master section of this J-K flip-flop "reads" the J and K inputs only on the positive-going edge of the CLK waveform. The master's data is then directed to the Q outputs during the negative-going edge of the CLK pulse. The synchronous modes of the circuit are thus completely edge controlled.

PRE = CLR = 0 is an invalid operating mode that should be avoided. Most flip-flop manufacturers also recommend that the J and K inputs be changed only while the CLK pulse is at logic 0. Since the master section is positive edge operated, any change at the J and K inputs while the CLK is high will not appear at the Q outputs until the CLK input drops to 0 and is cycled one more time.

6-7.3 Available J–K Flip-Flop ICs

Table 6-2 and Fig. 6-15 summarize the J-K flip-flops available in the TTL and CMOS families.

The 7476 is the basic TTL J-K master–slave flip-flop. Its truth table is identical to the one in Fig. 6-14. Figure 6-16 shows a demonstration circuit that exhibits all the characteristics of this valuable flip-flop circuit.

Table 6-2 AVAILABLE J-K FLIP-FLOP ICs

TTL Family	
7476	Dual J-K flip-flop with asynchronous PRE and CLR and both Q and $\bar{Q}$ outputs [Fig. 6-15(a)].
7473	Dual J-K flip-flop with asynchronous CLR and both Q and $\bar{Q}$ outputs [Fig. 6-15(b)].
74109	Dual J-$\bar{K}$ flip-flop with positive edge triggering, asynchronous PRE and CLR and both Q and $\bar{Q}$ outputs [Fig. 6-15(c)].
7472	AND-gated J-K flip-flop with asynchronous PRE and CLR both Q and $\bar{Q}$ outputs [Fig. 6-15(d)].
7470	AND-gated J-K flip-flop with positive edge triggering, asynchronous PRE and CLR and both Q and $\bar{Q}$ outputs [Fig. 6-15(e)].
CMOS Family	
4027	Dual J-K master–slave flip-flop with positive edge triggering, asynchronous PRE and CLR, and both Q and $\bar{Q}$ outputs [Fig. 6-15(f)].

The 7473 in Fig. 6-15(b) is very similar to the 7476, except the 7473 lacks a PRE (PRESET) input. The 74109, however, is very different in several respects. For thing, the 74109 is not really a master–slave flip-flop; its output responds directly to the synchronous inputs the instant the CLK input waveform rises from 0 to 1. Also take note of the fact that the K input is inverted. The PRE and CLR inputs perform their usual function, however.

The 7472 and 7470 both feature AND-ed J and K inputs. In the case of the 7472, all three J inputs and the CLK must be at logic 1 before the internal J' connection sees a logic 1. The same requirements apply to the K inputs as well. After determining the outputs of the AND gates in this circuit, the J' and K' input logic levels can then be considered in light of the usual J-K master–slave truth table in Fig. 6-14.

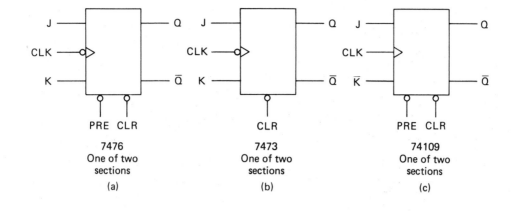

7476
One of two
sections

(a)

7473
One of two
sections

(b)

74109
One of two
sections

(c)

$J' = CLK \cdot J1 \cdot J2 \cdot J3$
$K' = CLK \cdot K1 \cdot K2 \cdot K3$

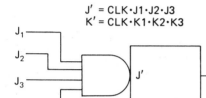

PRE   CLR

7472

(d)

$J' = J1 \cdot J2 \cdot \overline{J3}$
$K' = K1 \cdot K2 \cdot \overline{K3}$

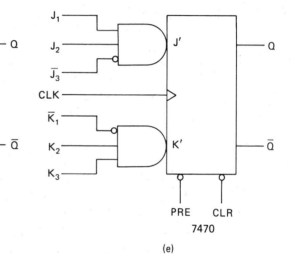

PRE   CLR

7470

(e)

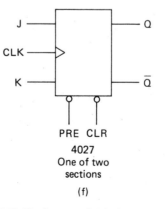

4027
One of two
sections

(f)

Figure 6-15 Logic diagrams for J-K flip-flops available in TTL and CMOS IC packages. Compare with Table 6-2.

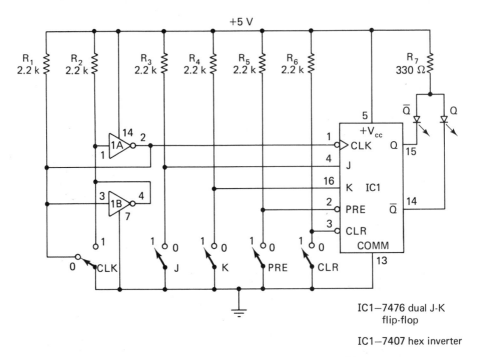

Figure 6-16 A circuit for demonstrating the essential qualities of a master–slave J-K flip-flop.

The 7470 features three separate J and K inputs, too, but in this instance, one of the J and K inputs is inverted, and the CLK input is not included in the AND operations.

The purpose of the multiple J and K inputs on the 7472 and 7470 will become apparent in later chapters dealing with synchronous counters and shift registers.

The 4027 shown in Fig. 6-15(f) is the basic CMOS J-K master–slave flip-flop. Its truth table is identical to the one in Fig. 6-14, except that it is positive edge-triggered. Data from the J and K inputs are fed to the master section on the negative-going edge of the CLK waveform.

6-7.4 Building Other Flip-Flop Functions from a J-K Flip-Flop

Generally speaking, the master–slave J-K flip-flop is an all-purpose flip-flop. With the notable exception of a level-triggered D flip-flop, a master–slave J-K can be "programmed" to operate as any of the other basic flip-flop circuits. Figure 6-17 summarizes the techniques for wiring a J-K flip-flop to work as a simple R-S, RST, edge-triggered D and T flip-flop.

The labels inside the box symbols indicate the actual J-K flip-flop connections; the labels outside the box indicate the pertinent inputs for each type of flip-flop. The circuit in Fig. 6-17(a), for instance, shows a J-K flip-flop wired to function as

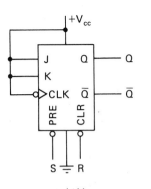

R	S	Q	Q̄
0	0	Invalid	
0	1	1	0
1	0	0	1
1	1	Memory	

(a) R-S flip-flop

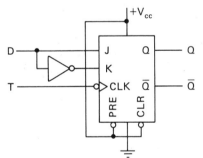

R	S	T	Q	Q̄
0	0	⊓	Memory	
0	1	⊓	1	0
1	0	⊓	0	1
1	1	⊓	Toggle	

(b) RST flip-flop

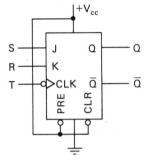

D	T	Q	Q̄
1	⊓	1	0
0	⊓	0	1

Edge-triggered D flip-flop

(c)

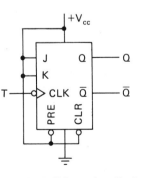

$$Q = \bar{Q}_{t-1}$$

$$\bar{Q} = Q_{t-1}$$

T flip-flop

(d)

Figure 6-17 Building other flip-flop functions from a master–slave J-K flip-flop. (a) An R-S flip-flop. (b) An RST flip-flop. (c) An edge-triggered D flip-flop. (d) A T flip-flop.

a simple R-S flip-flop. The J-K's *J*, *K* and CLK inputs are all disabled by connecting them together at $+V_{cc}$. Labeling the PRE input *S* and the CLR input *R* yields the RS truth table in Fig. 6-17(a). Compare this truth table with the standard J-K truth table in Fig. 6-14 and the basic RS truth table in Fig. 6-2. The invalid and memory input conditions are reversed, but the *Q* output does, indeed, follow the *S* input as prescribed for R-S flip-flop action.

The RST flip-flop in Fig. 6-17(b) uses the synchronous J-K inputs. Compare this truth table with the J-K table in Fig. 6-14 and the one for the basic RST master–slave flip-flop in Fig. 6-12. Note that the J-K version does not have an invalid state that characterizes the RST flip-flop; the J-K version has the advantage of a toggle mode that exists whenever $R = S = 1$.

The J-K circuit in Fig. 6-17(c) serves the purposes of an edge-triggered D flip-flop. The inverter connected between the *J* and *K* inputs ensures that the *J* and *K* inputs are always complements of one another. The *Q* output thus follows the *D* input upon completing one input trigger or CLK pulse, and the circuit is in its memory mode at all times except when the *T* input is making a transition from logic 1 to logic 0.

Figure 6-17(d) shows a J-K flip-flop that is wired to operate in its toggle mode only. Used in this fashion, it works as a simple T flip-flop, toggling to the opposite output condition each time the negative-going edge of the *T* waveform occurs.

Exercises

1. Draw up a truth table for the alternate latch circuit in Fig. E-6.1. Compare your results with that of the NAND gate version in Fig. 6-1. What input transition is invalid in this case?

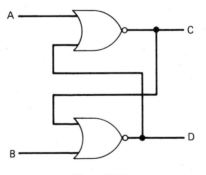

Figure E-6.1

2. Describe how a simple R-S flip-flop (Fig. 6-2) is different from a gated R-S flip-flop (Fig. 6-3).

3. Why is the input $S = R = G = 1$ considered an invalid input for the gated R-S flip-flop in Fig. 6-3?

4. What is the essential difference between a level-switched D flip-flop and an edge-triggered D flip-flop?

5. Explain why the Q output of the D flip-flop in Fig. 6-11 is connected to the $\bar{Q}$ lamp and why the flip-flop's $\bar{Q}$ output is connected to the Q lamp.

6. Suppose you want to use the J-K flip-flop in Fig. 6-15(e) in its toggle mode. Describe the logic levels that must be applied to the six J and K inputs to accomplish this.

7

MONOSTABLE AND ASTABLE MULTIVIBRATORS

Although it is possible to construct some highly sophisticated control circuits around the various varieties of triggered flip-flops, they are entirely useless without a good, clean source of trigger pulses. Pulse waveforms and the circuits that generate them are important to digital electronics, and this chapter describes some of the most common and useful monostable and astable waveform generators.

7-1 Monostable and Astable Effects from Logic Gates

This section deals with some common procedures for generating pulses using common logic gates. These circuits often lack the precision of their more sophisticated counterparts described in later sections of this chapter, but they are certainly easy to design and analyze.

7-1.1 Simple Monostable Trigger Generators

The circuit in Fig. 7-1(a) illustrates a simple technique for generating a relatively short trigger pulse at the leading edge of a relatively long gate pulse from any sort of TTL or CMOS device. The duration of the output pulse, t_o, is approximately RC, and it is inverted compared to the input pulse, t_i.

Resistor R keeps the input of the inverter normally pulled down to logic 0; and since the t_i waveform is also normally at logic 0, it follows that the capacitor is normally discharged. The instant the t_i pulse rises toward $+V_{cc}$, however, the

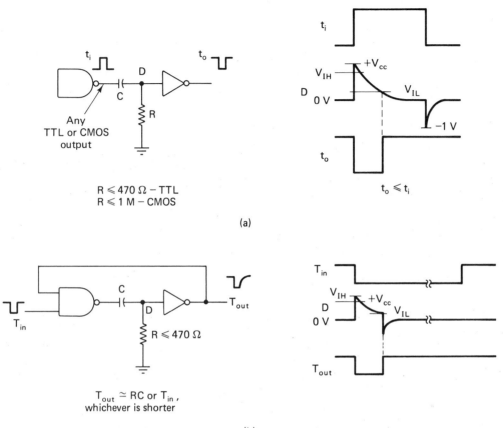

Figure 7-1   Two simple trigger-pulse generators.

capacitor begins charging through *R*, generating the sharp rise to V_{cc} shown in waveform *D*. The output of the inverter responds immediately by switching to its logic-0 state.

The t_o waveform remains at logic 0 until the capacitor charges to a point where the voltage across *R* is less than the inverter's definition of V_{IL}. At that moment, the output of the inverter switches to logic 1 again, thus completing the output cycle.

The circuit is reset only after the input waveform returns to logic 1 to discharge the capacitor. The capacitor's discharge current produces a negative potential at the input of the inverter, but of course this voltage has no effect on the output since the inverter is already generating a logic 1 level.

The circuit in Fig. 7-1(a) is often used as a trigger-pulse generator, and it should not be considered any more useful than that. Some of the disadvantages of the circuit include the fact that the output pulse cannot be any longer than the input pulse. Another problem is that the width of the output pulse changes with any variation in supply voltage; and, finally, it must be noted that the inverter is highly susceptible to any external noise occurring during the interval its input is falling between V_{IH} and V_{IL}: The inverter's input is actually in an undefined state between V_{IH} and V_{IL}, and any external switching or power supply noise is likely to create some output "bouncing" effects.

The pulse circuit in Fig. 7-1(b) is a slightly improved version of the first circuit. The operating principles are basically the same, but the feedback loop eliminates any output bouncing that might otherwise create more than one output pulse. The occurrence of some external noise during the capacitor's charge time might cut the output waveform short, but the circuit is noise-free in the sense that it is impossible to get more than one pulse out for each gate pulse in.

Incidentally, the circuit in Fig. 7-1(b) cannot generate an output pulse that is longer than the input pulse. This might appear possible at first sight, but if the inverter's input is in the noise region between V_{IH} and V_{IL} when the input pulse returns to logic 1, the resulting switching transient is sufficient to trigger the output waveform high again.

While the circuits in Fig. 7-1 do, indeed, suffer from a lack of precision, they are useful in situations calling for an inexpensive source of relatively short trigger pulses. Resistor R can be replaced with a variable resistor to give the user some measure of control over the output pulse width.

The circuits described thus for assume a clean, noise-free source of input pulses. Some triggering situations, however, call for generating a clean trigger pulse from the closure of a mechanical switch. The circuit in Fig. 7-2 shows a switch debouncing circuit (Sec. 6-6.2) serving as the input to a simple monostable trigger generator.

Whenever the user switches the input switch to its TRIG position, the

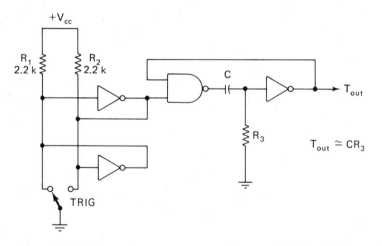

Figure 7-2   A trigger-pulse generator operated by a mechanical switch.

debouncing circuit guarantees a clean, negative-going transition at the input of the NAND gate. This action then causes the trigger generator to produce a relatively short, negative-going pulse at T_{out}. If the RC time constant of the trigger generator is kept fairly short (less than 100 μs), this circuit provides a simple and reliable means for generating short trigger pulses from a mechanical switch input.

7-1.2 Simple Free-Running Multivibrators

The circuits in Fig. 7-3 illustrate a simple procedure for generating relatively high frequency pulses. The "flat-out" oscillator in Fig. 7-3(a) takes advantage of the fact that every logic gate has a certain propagation delay interval. In the case of TTL inverters, the propagation delay for each gate can be on the order of 10 ns.

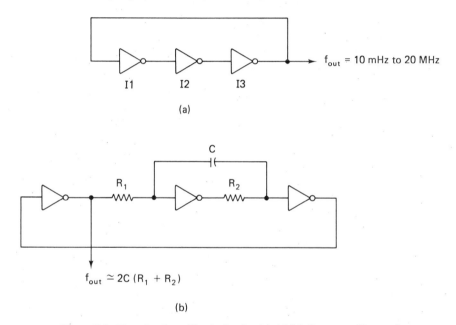

Figure 7-3   Two simple oscillator circuits. (a) A high-frequency "flat-out" oscillator. (b) An oscillator with controlled output frequency.

Suppose, for example, the input to I1 is making a transition from logic 0 to logic 1. About 10 ns later, its output responds by falling from 1 to 0. Inverter I2 responds to this situation by making a 1-to-0 transition about 10 ns later; and, finally, I3 responds about 10 ns later by switching from 1 to 0. The output of I3 is fed back to the input of I1, setting the input of that inverter back to logic 0 to begin the cycle all over again.

The total propagation delay time is equal to the sum of the individual delay times, about 30 ns in this example. The circuit completes one half-cycle of its output in that period, making it possible to express the circuit's free-running frequency as $1/2\, d_t$, where d_t is the total propagation delay time of the inverters. If each inverter happens to have a delay time of 10 ns, the circuit's operating frequency is on the order of 16 MHz.

Actually, this circuit works on the same basic principles as an ordinary *RC* phase-shift oscillator. The phase shift in this case is provided by the inverters' propagation delay time.

The circuit is most useful in clocking operations that do not call for any particular frequency. Since it is difficult to determine the exact propagation delay time of any logic IC, the user really has no control over what the frequency might be. Given the upper and lower figures for propagation delay, all that can be said is that the frequency will be between 10 MHz and 20 MHz.

The circuit designer can get more control over the final output frequency by inserting some timing elements into the circuit as shown in Fig. 7-3(b). The capacitor greatly reduces the operating frequency to about $2C(R_1 + R_2)$ and makes the overall operation independent of the inverters' propagation delay times.

Neither of the circuits in Fig. 7-3 should be used in systems calling for precise and stable operating frequencies. They are very useful, however, whenever it is necessary to get some high-frequency trigger pulses at a very low cost.

The circuits described in the remainder of this chapter do have the precision required for most digital applications.

7-2 The 555 and 556 Timers

The 555 timer is one of the most versatile sequential logic devices on the market today. Its control inputs and output are directly compatible with both TTL and CMOS logic circuits, and it can be connected to operate in a wide variety of monostable and astable modes. Some of the more popular circuits built around the 555 timer include trigger-pulse generators, astable multivibrators with fixed or variable duty cycles, voltage-controlled oscillators, missing-pulse detectors, sequential pulse timers, frequency dividers, ramp voltage generators and switch contact debouncers. Actually, the variety of possible applications is only limited by the user's imagination and understanding of the device's basic operating principles.

Figure 7-4 is a simplified diagram of the internal structure of the 555 timer. First note the three comparator circuits at the inputs and the resistor voltage divider made up of R_1, R_2 and R_3. Comparator B compares voltage levels from the TRIGGER input and a reference voltage at R_3 that is equal to one-third the $+V_{cc}$ level. Comparator B thus changes output state as the TRIGGER voltage passes through $1/3V_{cc}$.

Comparator A compares voltages from the THRESHOLD input and $2/3V_{cc}$ taken from the common point between R_1 and R_2. This comparator changes output state whenever the THRESHOLD voltage passes through $2/3V_{cc}$.

The RESET comparator, comparator C, is internally referenced at a maximum of $+1$ V, regardless of the value of $+V_{cc}$. The output of comparator C changes state any time the RESET input passes through its fixed reference potential.

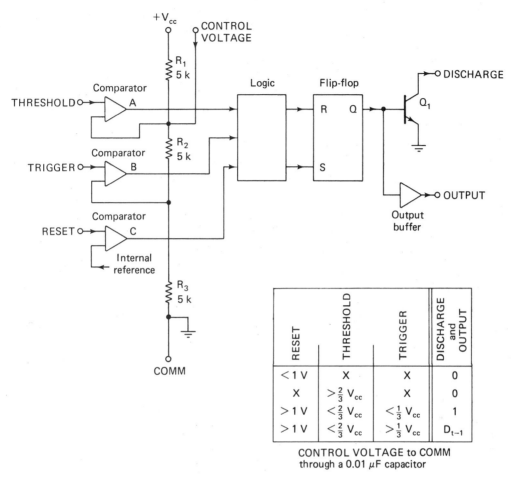

RESET	THRESHOLD	TRIGGER	DISCHARGE and OUTPUT
< 1 V	X	X	0
X	$> \frac{2}{3} V_{cc}$	X	0
> 1 V	$< \frac{2}{3} V_{cc}$	$< \frac{1}{3} V_{cc}$	1
> 1 V	$< \frac{2}{3} V_{cc}$	$> \frac{1}{3} V_{cc}$	D_{t-1}

CONTROL VOLTAGE to COMM
through a 0.01 μF capacitor

Figure 7-4   Basic internal structure of a 555 IC.

The outputs of these three comparator circuits control the status of a simple R-S flip-flop; and that flip-flop controls the conduction of the DISCHARGE output transistor Q_1 and determines the logic state of the OUTPUT buffer amplifier circuit. Transistor Q_1 and the output buffer are arranged so that OUTPUT is at logic 1 (near $+V_{cc}$) whenever Q_1 is switched off, and OUTPUT is at logic 0 (near COMM potential) when Q_1 is on.

The function truth table accompanying the diagram in Fig. 7-4 summarizes the basic action of the 555 timer. Note that pulling the RESET input below 1 V overrides any other ongoing activity, switching on Q_1 to set the DISCHARGE output to logic 0 and pulling OUTPUT down to logic 0.

The THRESHOLD input has a similar effect on the DISCHARGE and OUTPUT terminals. Whenever THRESHOLD rises above $2/3V_{cc}$, it pulls both DISCHARGE and OUTPUT down to their logic-0 states.

The TRIGGER input is relevant only when RESET is above 1 V and THRESHOLD is below $2/3V_{cc}$. Under these conditions at RESET and THRESHOLD, pulling TRIGGER below $1/3V_{cc}$ guarantees that Q_1 will turn off and that the OUTPUT will be pulled up close to $+V_{cc}$.

The D_{t-1} notation in the last line of the truth table indicates the circuit's memory mode. The implication is that the outputs will hold the logic states they had prior to setting up the conditions specified on that line of the truth table.

All of the discussions thus far assume that the CONTROL VOLTAGE input is uncommitted—not connected to any sort of voltage source. Whenever the CONTROL VOLTAGE terminal is not committed to any outboard circuitry, it shows the $2/3V_{cc}$ potential that serves as a reference input for comparator A. Applying a potential between COMM and V_{cc} to this terminal alters the reference voltages for both comparators A and B. Pulling the CONTROL VOLTAGE input below $2/3V_{cc}$, for instance, lowers the circuit's threshold and trigger voltages; pulling CONTROL VOLTAGE above $2/3V_{cc}$ raises the threshold and trigger references above their normal levels. Unless specified otherwise, however, the CONTROL VOLTAGE input is left uncommitted, or at least tied to COMM through a 0.01-μF capacitor.

Figure 7-5 shows the pinouts and principal electrical characteristics of the 555 timer and its dual-555 counterpart, the 556. Note the wide range of supply voltages

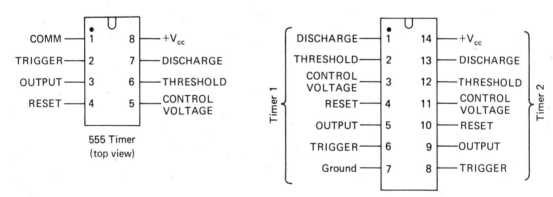

Supply voltage $(+V_{cc})$ 18 V max., +5 V to +15 V typical

Output Voltage, Logic 0   0.35 V max., V_{cc} = 5 V @ 10 mA
                                0.75 V max., V_{cc} = 15 V @ 50 mA
                                2.5 V max., V_{cc} = 15 V @ 200 mA

Output Voltage, Logic 1   12.5 V typical, V_{cc} = 15 V @ 200 mA
                                3.3 V typical, V_{cc} = 5 V @ 100 mA

Minimum TRIGGER pulse duration is 1 μs

See manufacturers' data sheets for more complete and detailed information

Figure 7-5   Pinouts and general specifications for the 555 and 556 timers.

(normally between $+5$ V and $+15$ V) and the fact that the OUTPUT connection can both sink and source up to 200 mA.

7-2.1 555 Monostable Multivibrators

Figure 7-6 shows two basic monostable configurations for the 555 timer. The only difference between the two circuits as far as external electrical connections are concerned is that the RESET terminal in Fig. 7-6(a) is tied to $+V_{cc}$, but it is connected to the TRIGGER input in Fig. 7-6(b). The output load in either instance can be connected either to $+V_{cc}$ for an inverted response or to COMM for a true, or non-inverted, response. Under normal operating conditions, the period of the logic-1 output is equal to 1.1 RC.

First consider the details in Fig. 7-6(a). The trigger input is normally high, while the DISCHARGE and OUTPUT are low. Holding DISCHARGE low in this fashion keeps C completely discharged.

The moment the TRIGGER input drops below its $1/3V_{cc}$ reference level, DISCHARGE and OUTPUT switch to their logic-1 states, pulling the OUTPUT connection near $+V_{cc}$ and allowing C to begin charging toward $+V_{cc}$ through R. The circuit remains in this condition—OUTPUT high and the capacitor charging— even after the TRIGGER input is returned to logic 1. When the capacitor voltage finally reaches $2/3V_{cc}$, as sensed by the THRESHOLD connection, the output returns to logic 0 and the DISCHARGE connection is returned to COMM potential to discharge the capacitor rather rapidly. Only then is the circuit ready for triggering again.

Note on the waveforms in Fig. 7-6(a) that the timing operation cannot be interrupted by dropping the TRIGGER input to logic 0. One complete output timing operation must be completed before the circuit can be retriggered.

Let as summarize the operation of the circuit in Fig. 7-6(a): The output timing interval is initiated when the TRIGGER input falls below $1/3V_{cc}$, and it is terminated only when the capacitor charges to $2/3V_{cc}$. The output timing interval is equal to 1.1 RC. For proper operation, the logic-0 TRIGGER interval must be shorter than the circuit's monostable timing interval.

While the RESET function is totally disabled by connecting it to $+V_{cc}$ in Fig. 7-6(a), it plays a vital role in the operation of the circuit in Fig. 7-6(b).

The monostable circuit in Fig. 7-6(b) is interruptible. The timing cannot begin when the TRIGGER input drops to logic 0 because the RESET input also goes to zero to override the timing operation. Timing begins the moment the TRIGGER and RESET inputs return to logic 1. In effect, the monostable shown here is positive edge-triggered.

Once the timing operation begins, OUTPUT is set to logic 1 and C begins charging toward $+V_{cc}$ through R. Unless the circuit is retriggered with a negative-going pulse waveform at the TRIGGER/RESET inputs, the output remains at logic 1 until the capacitor voltage reaches $2/3V_{cc}$ as sensed by the THRESHOLD terminal. The uninterrupted timing interval is equal to 1.1 RC.

If the circuit is retriggered during the timing interval, the OUTPUT and

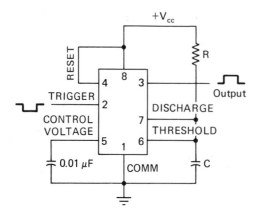

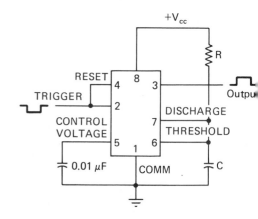

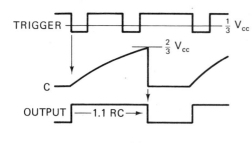

(a)

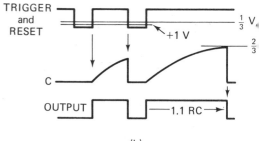

(b)

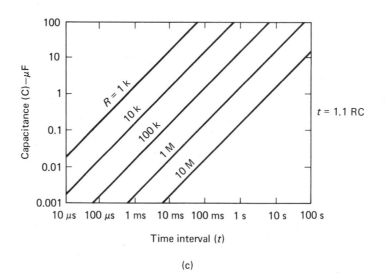

(c)

Figure 7-6 Basic circuit configurations for 555-type monostables. (a) RESET connected to $+V_{cc}$ for non-interruptible operation. (b) RESET connected to TRIGGER for interruptible operation. (c) Basic monostable design chart.

DISCHARGE connections respond immediately, setting the output to logic 0 and discharging C. The timing is then restarted when the TRIGGER/RESET terminals are pulled up to logic 1 again.

Figure 7-6(c) is a 555/556 monostable design chart that is very useful for estimating the values of R and C for any output timing interval, t, between 10 μs and 100 s. The usual design procedure is to select a standard value of capacitance first and then use the chart to *estimate* the necessary value of R. If it turns out that the value of R is less than 1 k or greater than 10 M, select a different value for C and then use the chart for estimating a more suitable value for R. Finally, use the equation $R = t/1.1\ C$ to determine the exact value for R.

The internal structures of the 555/556 IC devices limit the shortest possible monostable timing interval to 10 μs, but there is no theoretical limit on how long the output timing interval can be. The accuracy and reliability of timing intervals longer than 100 s are strictly determined by the quality of the external timing components, especially the capacitor. The longest possible output timing interval is actually determined by the capacitor's leakage current. Very large electrolytic capacitors used for very long output time constants might have leakage currents that exceed the circuit's charge current; and the result is that the capacitor can never charge to the $2/3V_{cc}$ point required for setting OUTPUT low again.

7-2.1.1 Some practical monostable timing circuits.
Figure 7-7 shows two practical monostable timing circuits. Either circuit can be built around a single 555 timer or one section of a 556.

Suppose that the START switch in Fig. 7-7(a) is not depressed and that the circuit is not executing an output timing interval. Under these circumstances, resistor R_1 is pulling up the TRIGGER and RESET inputs to logic 1, and LED $\bar{Q}$ is lighted because the OUTPUT terminal is resting at logic 0.

Nothing happens when the user depresses the START push button. Although this action pulls the TRIGGER input down to COMM potential, it also pulls down the RESET input to keep the output set at logic 0. The timing interval in this case begins the instant the user releases the push button. LED $\bar{Q}$ switches off and Q turns on; and the outputs remain in that condition until the timing interval is over.

The monostable timing interval for the circuit in Fig. 7-7(a) is equal to $1.1C_2(R_2 + R_3)$, providing an adjustable interval between about 6.5 s and 13 s. By adjusting the value of R_2, this circuit can be used for a 10-s timer. Using an adjustable resistor in this fashion lets the user compensate for the tolerance of the timing capacitor.

Aside from making the timing interval begin on the positive-going edge of the input START pulse, the fact that TRIGGER and RESET are tied together in Fig. 7-7(a) makes the timer interruptible. See the waveforms in Fig. 7-6(b).

The monostable circuit in Fig. 7-7(b) is connected for negative-going edge-triggering, and it has an output timing interval of about 1.1 s. The load for the 555 device is a relay coil having a rating of 12 V at about 100 mA. Whenever the circuit

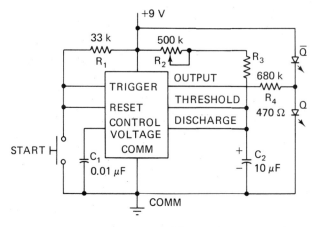

(a)

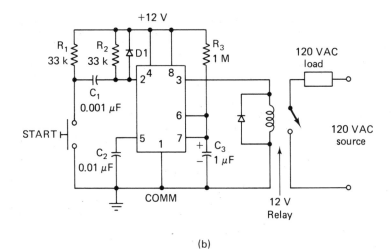

(b)

Figure 7-7 555 monostable demonstration circuits. (a) A timer with adjustable output duration and LED indicators. (b) A fixed-interval timer driving a relay coil.

is not executing a timing interval, the relay is de-energized and its contacts are open to prevent 120 VAC power from reaching the 120-VAC load (a motor, lamp, heating element, etc.).

The TRIGGER connection is normally pulled up to $+V_{cc}$ through resistor R_4, but the instant the user depresses the START push button, the TRIGGER input is momentarily pulled down close to COMM potential through the charge action of C_1. As long as this pulse interval exceeds about 2 μs, the 555 can be switched to its output timing mode—and it cannot be restarted or interrupted until the 1.1-s timing interval is over. *Using the coupling capacitor,* C_1, *ensures that the circuit times 1.1s, even if the user happens to hold down the START push button for a longer period of time.*

Diode D1 prevents the pin-2 TRIGGER input voltage from rising to twice the supply voltage ($2 \times 12 = 24$ V in this case) whenever the user depresses the START push button. This clamping diode is required in any situation where the timer is triggered through a series-connected capacitor such as C1.

As far as the user is concerned, depressing the START push button initiates one complete timing cycle; and the button must be released and depressed again to run another cycle.

7-2.1.2 *Practical time delay circuits.*

The output timing intervals for the circuits in Fig. 7-7 are initiated the moment the START push button is released or depressed. It is more often desirable to construct a time delay circuit, a circuit that is ener-gized only after some predetermined time interval passes. The circuits in Fig. 7-8 illustrate two kinds of time delay circuits built around 555/556 timers.

The time delay interval in Fig. 7-8(a) is initiated the instant the user depresses the START push button. Momentarily grounding the TRIGGER input through C_1 begins an output timing interval that is equal to $1.1C_3(R_3 + R_4)$; and when the output of the 555 device drops low at the end of that timing interval, coupling capacitor C_4 delivers a negative-going spike to the CLK input of a J-K flip-flop. Since the circuit is arranged so that $J = 1$ and $K = 0$ at all times, the delayed trigger pulse from the timer always sets the $\bar{Q}$ output of the flip-flop to logic 0, thereby lighting the LED. The flip-flop then remains in this ON condition until the user depresses the RESET push button, grounding the CLR input and asynchronously clearing the output. Another time delay cycle can then be ini-tiated by releasing the RESET button and depressing the START push button once again.

The circuit in Fig. 7-8(b) uses a 556 dual timer. Both units are connected as monostable multivibrators, with the output of IC1-A initiating the timing interval for IC1-B. This is a two-step timer, a time delay circuit that applies power to the load for a predetermined period of time.

The negative-going edge of the input TRIG pulse initiates the timing interval of IC1-A. The timing interval in this instance is equal to $1.1C_3(R_2 + R_3)$. At the end of this timing interval (the delay interval) the negative-going edge of the wave-form from IC1-A triggers on monostable IC1-B. The output of that circuit then remains at logic 1 for a period of time determined by $1.1C_6(R_5 + R_6)$. See the waveforms accomanying the diagram in Fig. 7-8(b).

The circuit in Fig. 7-8(b) can be considered a two-step sequence timer. It is possible to cascade any number of timers in this fashion to produce some rather elaborate and lengthy timing cycles.

7-2.1.3 *Switch debouncing circuits.*

The circuits in Fig. 7-9 illustrate three types of switch debouncing circuits. These three circuits are actually practical applications of some of the monostable timers already described in this section. Their main job

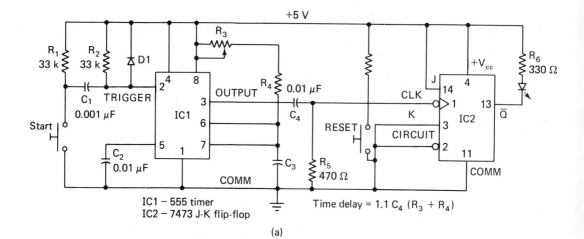

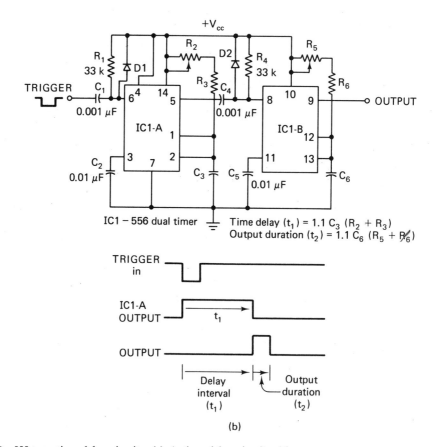

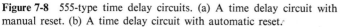

Figure 7-8 555-type time delay circuits. (a) A time delay circuit with manual reset. (b) A time delay circuit with automatic reset.

144

here is to mask contact bounce noise that occurs whenever a mechanical switch contact closes.

The debouncing circuit in Fig. 7-9(a) is an application of the basic 555 mono-stable circuit. As long as switch S is in its normally open state, pull-up resistor R_2 fixes the TRIGGER input of the 555 timer at logic 1; and as long as this condition exists, the output of the timer sets at logic 0. This output is inverted by IC2-A and applied as a logic-1 input to the output NAND gate, IC2-B. The second input to this NAND gate is normally connected to a logic-1 source through R_1. The S' output, in other words, is normally resting at logic 0.

Whenever the user depresses switch S, the TRIGGER input of the timer is pulled down to COMM potential through the charging action of C_1, thus ini-tiating the circuit's 11-ms output timing interval.

While the monostable is timing out, the inverted output reaching the output NAND gate is at logic 0, ensuring a logic-1 output at S', regardless of any contact bouncing from the switch. IC2-B actually serves the purpose of an OR circuit; so its output remains at logic 1 as long as S is depressed, even after the monostable timing interval is over. See the waveforms accompanying the circuit in Fig. 7-9(a).

The timer in Fig. 7-9(a) actually masks the switch's bouncing effect, and the output NAND gate lets switch S take over control of the S' output once the timer does its job. In short, this circuit generates a noise-free output that follows the closure of switch S.

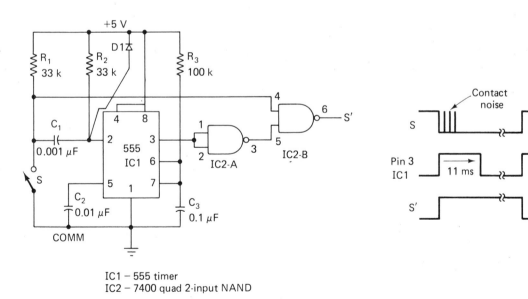

(a)

Figure 7-9   Switch contact debouncing circuits. (a) Single switch de-bouncing.

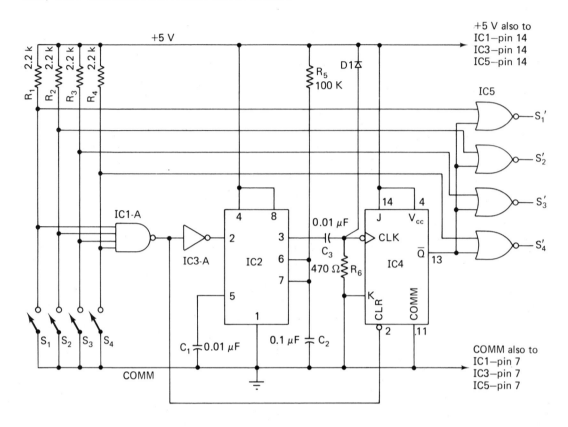

IC1—7420 dual 4-input NAND
IC2—555 timer
IC3—7404 hex inverter
IC4—7473 J-K flip-flop
IC5—7402 quad 2-input NOR

(b)

Figure 7-9 (*cont.*) (b) A 4-switch debouncing circuit in which the output remains energized as long as the corresponding key is energized.

The switch debouncing circuit in Fig. 7-9(b) is an application of the time delay circuit already described in connection with Fig. 7-8(a). The use of two ICs in the time delay portion of the circuit is justified by the circuit's ability to debounce any number of switch contacts. The example given here only shows four input switches, but an additional 4-input NAND gate and another set of four NOR gates could expand the number of possible inputs to eight.

As long as all four input switches, S1 through S4, are open, the output of IC1-A is at logic 0; and this particular logic state keeps the J-K flip-flop in an asynchronous condition in which $\bar{Q}$ equals logic 1. As long as this logic-1 level is

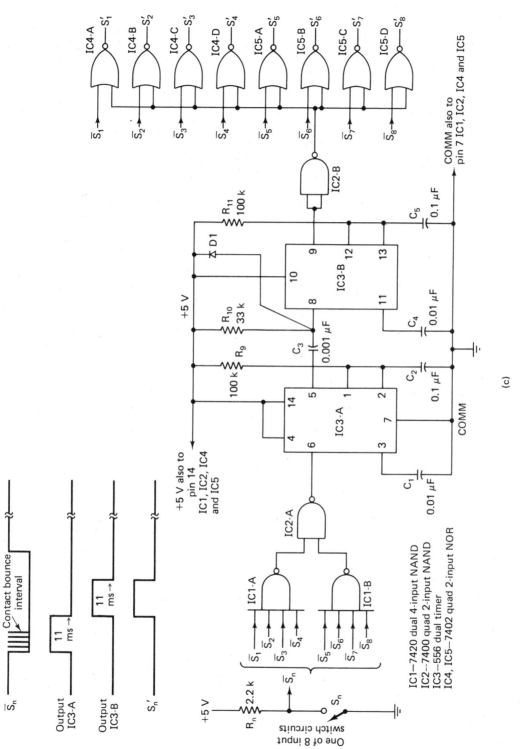

Figure 7-9 (*cont.*) (c) An 8-key debouncing circuit that generates a pulse output.

147

being fed to one input of each of the output NOR gates, those gates will output a logic 0.

The same logic-0 output from IC1-A that keeps the flip-flop cleared is inverted and connected to the TRIGGER input of the monostable circuit, IC2. That circuit is thus ready to accept a negative-going trigger pulse that will initiate its timing cycle.

Closing any one of the input switch contacts immediately begins the timing cycle for the monostable circuit and, at the same time, releases the asynchronous clearing at the J-K flip-flop. The flip-flop does not change output state, however, until the monostable times out 11 ms later. When the flip-flop is finally clocked, its $\bar{Q}$ output switches to logic 0, effectively opening the output NOR gates. The input switch contact that is closed provides a logic 0 input to its corresponding NOR gate, and the output of that one gate goes to logic 1. If the user depresses S2, for instance, output S2′ will go to logic 1 11 ms later—after the contact has had time to stop bouncing.

The output of the selected switch remains at logic 1 until the user releases the push button. With all four switches open again, the flip-flop is immediately cleared so that its $\bar{Q}$ output is at logic 1; and as a result, the outputs of all the NOR gates return to zero.

The overall action of the debouncing circuit in Fig. 7-9(b) is identical to that of the simpler circuit in Fig. 7-9(a); the appropriate output remains at logic 1 as long as the input switch is depressed. The more complicated circuit has provisions for expanding the number of input switch contacts indefinitely, however.

The contact debouncing circuit in Fig. 7-9(c) has been expanded to accommodate eight input switch contacts; but of course it could be modified to handle even more inputs if necessary. The same expansion technique used here can be applied equally well to the circuit in Fig. 7-9(b).

The circuit in Fig. 7-9(c) is quite different from the other two in one important respect: Instead of generating a logic 1 output as long as the user depresses one of the input switches, this circuit generates a pulse of a fixed interval, no matter how long the user depresses an input switch.

The circuit here is actually an application of the two-step sequence timer described in connection with the circuit in Fig. 7-8(b). Closing any one of the eight input switch contacts initiates the two-step timing cycle, masking any switch bounce during the first phase and holding the corresponding S' output high for a period of time determined by the time constant of the second phase. See the waveforms associated with the circuit in Fig. 7-9(c).

7-2.1.4 *Frequency dividers and missing-pulse detectors.* All of the monostable circuits described thus far in this section have inputs that can occur at irregular intervals as typified user-operated input switches. The two circuits in this section assume a source of clean square-wave or trigger pulses from another type of multivibrator device, such as a fixed-frequency astable multivibrator or a voltage-controlled oscillator.

The circuit in Fig. 7-10(a) looks very much like a basic 555-type monostable.

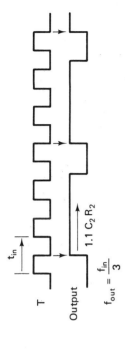

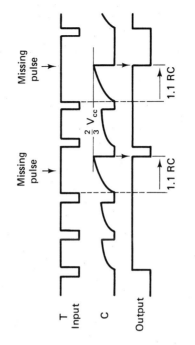

(a)

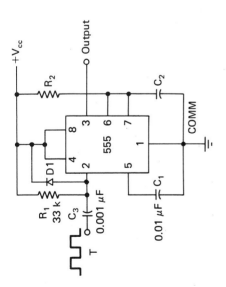

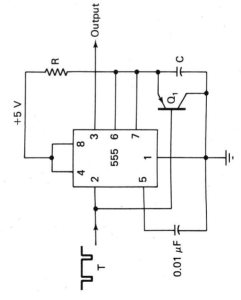

(b)

Figure 7-10 Special 555-type circuits. (a) Frequency divider. (b) Missing-pulse detector.

The trigger source in this instance, however, is a set of fixed-frequency pulses. Since the monostable is connected for non-interruptible operation, it follows that the circuit will respond only to those trigger input pulses that catch it in its resting state.

Suppose, for example, that the input trigger pulses are arriving at a 10-kHz rate. This means that the period of that waveform is 100 μs. Now suppose that the monostable is set for an output timing interval of 250 μs. The overall effect is that the monostable responds only to every third input pulse; once the monostable's output is set high, it is impossible to start another timing interval until the timer completes its cycle and another input pulse occurs. See the waveforms in Fig. 7-10(a).

This frequency-divider circuit is especially useful when it is necessary to divide a given frequency by some number that is not an integer power of two. Flip-flop frequency dividers described in the next chapter are better for dividing a given frequency by 2, 4, 8, 16, 32 and so on. The circuit shown here is better for frequency division by 3, 5, 7, 9 and so on.

The missing-pulse detector circuit in Fig. 7-10(b) works much like the frequency divider, except that the transistor in the missing-pulse detector discharges the timing capacitor each time an input pulse occurs. The effect is that the timing interval is restarted with each incoming pulse, but the output is not allowed to drop to logic 0.

As long as the period of the incoming waveform is shorter than the charge time of the timing capacitor, C_3, the output of the circuit remains set at logic 1. If the period of the incoming waveform ever extends beyond the monostable's timing interval, the monostable is allowed to time out and its output drops to logic 0. See the waveforms accompanying the diagram in Fig. 7-10(b).

Setting the monostable's output timing interval just a bit beyond the normal period of the incoming waveform makes it operate as a missing-pulse detector. And by the same token it can be used as a low-frequency detector. If the input frequency drops below a point where its period exceeds the monostable's output timing interval, the monostable begins generating negative-going pulses that can be used for energizing an alarm LED or setting a flip-flop to a particular output state.

7-2.1.5 A pulse-width modulator. A 555-type monostable circuit can serve as a pulse-width modulator by applying a varying voltage level to the CONTROL VOLTAGE input. Referring back to Fig. 7-4, note that the 555's CONTROL VOLTAGE input is connected directly to the portion of the resistor voltage divider that normally sets the device's THRESHOLD trigger level at $2/3V_{cc}$. Applying a voltage greater or less than $2/3V_{cc}$ to the CONTROL VOLTAGE input thus alters the THRESHOLD trigger level.

Pulling the CONTROL VOLTAGE potential above $2/3V_{cc}$, for instance, raises the device's THRESHOLD level and tends to increase the output timing interval. Pulling the CONTROL VOLTAGE potential below $2/3V_{cc}$, however, lowers the THRESHOLD trigger level and decreases the monostable's timing interval.

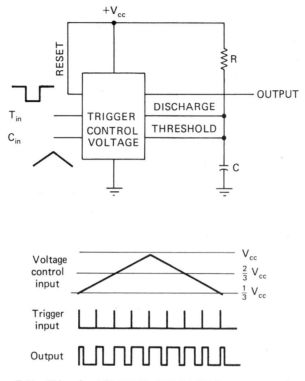

Figure 7-11   Using the CONTROL VOLTAGE input to control output pulse width.

The control voltage can be applied to the 555 directly as shown in Fig. 7-11. The circuit designer must bear in mind, however, that the CONTROL VOLTAGE terminal shows a rather constant 5-k load to $+V_{cc}$ and a 10-k load to COMM. It is often easier to achieve satisfactory results by coupling the control voltage to the CONTROL VOLTAGE terminal through a capacitor.

Figure 7-11 shows how the basic 555 monostable responds to a triangular waveform at the CONTROL VOLTAGE input. The values of the timing resistor and capacitor fix the duration of the monostable's output whenever V_i is equal to $2/3V_{cc}$. The output timing then varies in proportion to the change in the V_i potential around that $2/3V_{cc}$ reference point.

7-2.2 555 Astable Multivibrators

The circuit in Fig. 7-12 shows the basic configuration for an astable multivibrator built around a 555 timer. Notice that THRESHOLD and TRIGGER are connected together at the junction between R_B and C and that DISCHARGE connects with the junction of R_A and R_B. These special connections, plus the use of two timing resistors, make up the primary differences between 555-type monostable and astable multivibrators.

To see how this circuit works, suppose that the circuit is in an operating mode

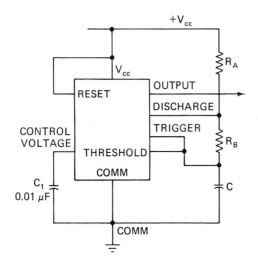

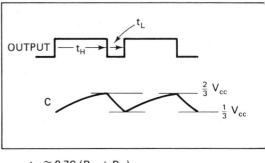

$$t_H \simeq 0.7C\,(R_A + R_B)$$
$$t_L \simeq 0.7CR_B$$
$$T = t_H + t_L \simeq 0.7C\,(R_A + 2R_B)$$
$$f = \frac{1}{T} \simeq \frac{1.4}{C(R_A + 2R_B)}$$
$$\text{Duty cycle} = \frac{t_H}{T} = 1 - \frac{R_B}{R_A + 2R_B}$$

(a)

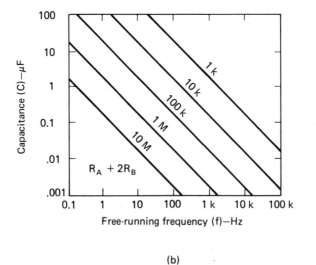

(b)

Figure 7-12 The basic 555-type astable multivibrator. (a) Circuit configuration, waveforms and relevant design equations. (b) Basic astable design chart.

where C is charging and the output is set high. See the t_H interval in the waveforms accompanying Fig. 7-12(a). The capacitor is charging toward $+V_{cc}$ through timing resistors R_A and R_B; but as soon as the THRESHOLD connection senses $2/3V_{cc}$, the 555's internal circuitry sets the OUTPUT to zero and begins the DISCHARGE operation. The capacitor cannot discharge immediately as it does in a monostable configuration, however. In this case, the capacitor must discharge through R_B, thus prolonging the discharge interval until the TRIGGER input senses $1/3V_{cc}$. See interval t_L on the waveforms.

The instant the voltage across C drops below the $1/3V_{cc}$ trip point for THRESHOLD, the 555 sets output high once again, removes the path to ground through DISCHARGE and allows C to begin charging toward $+V_{cc}$ potential once again. The cycle continues as long as power is applied to the circuit, with the voltage across C alternately charging up to $2/3V_{cc}$ and discharging down to $1/3V_{cc}$.

All of the equations for the 555-type astable multivibrator are based on the fact that C *charges* through both R_A and R_B and *discharges* through R_B only. The charge interval is equal to $0.7C(R_A + R_B)$ and the discharge interval is very close to $0.7CR_B$; and since the OUTPUT terminal is near V_{cc} while C is charging, it follows that $t_H = 0.7C(R_A + R_B)$. By the same token, the OUTPUT terminal is near COMM potential while C is discharging; so $t_L = 0.7CR_B$.

The other equations accompanying the schematic and waveforms in Fig. 7-12(a) use these two basic equations to determine the period of the output cycle (T), the frequency (f) and the duty cycle.

The design chart in Fig. 7-12(b) provides a convenient means for selecting values of R_A, R_B and C for any desired operating frequency between 0.1 Hz at 100 kHz. Carefully note that the resistance values on this chart show $R_A + 2R_B$, and not the value of either resistor alone. Of course, the same diagram can be used for analyzing the operation of an existing 555-type astable circuit.

Before describing some specific procedures for designing a 555-type astable multivibrator, it is important to note that the t_H interval must be longer than t_L. The equation for t_H contains both R_A and R_B, while the expression for t_L contains only R_B; t_H is always $0.7CR_A$ s longer than t_L. This fact can pose something of a problem at times, but some discussions later in this section describe ways to get around it.

Designing a 555-type astable multivibrator is generally a matter of picking a frequency and then some value for C. The designer then uses the design chart or rearranges the equation for f to determine $R_A + 2R_B$. The next step is to select a suitable duty cycle, keeping in mind the fact that it must be greater than 0.5. Rearranging the equation for duty cycle and substituting known values yields a value for R_B; and since the expression $R_A + 2R_B$ is already known, substituting the value for R_B leads to the necessary value of R_A.

Now this might seem to be a rather complicated design procedure, but it is really easier to carry out than to explain in words. Follow this specific example: A designer wants a 1-kHz oscillator that has a duty cycle of 90 percent, and the designer would like to use a $0.1\text{-}\mu\text{F}$ capacitor that is handy at the time.

Step 1.   Rearrange the equation for f to solve for $R_A + 2R_B$.

$$R_A + 2R_B = \frac{1.4}{fC}$$

$$= \frac{1.4}{(1 \times 10^3)(0.1 \times 10^{-6})}$$

$$= 14 \times 10^3 \ \Omega$$

Note that 14 k corresponds to the resistance value shown at the intersection of 1 kHz and 0.1 μF on the design chart.

Step 2.   Rearrange the equation for duty cycle to solve for R_B.

$$R_B = (R_A + 2R_B)(1 - D)$$

$$= (14 \times 10^3)(1 - 0.9)$$

$$= (14 \times 10^3)(0.1)$$

$$= 1.4 \ \text{k}$$

Step 3.   Use the expression $R_A \div 2R_B = 14$ k to solve for R_A.

$$R_A = 14 \ \text{k} - 2R_B$$

$$= 14 \ \text{k} - 2.8 \ \text{k}$$

$$= 11.2 \ \text{k}$$

The circuit will oscillate at 1 kHz with a duty cycle of 90 percent if $C = 0.1 \ \mu$F, $R_A = 11.2$ k and $R_B = 1.4$ k.

Determining the operating frequency and duty cycle of an existing astable circuit is simply a matter of solving the equations for f and duty cycle as they appear in Fig. 7-12(a).

The circuit in Fig. 7-13 is a practical demonstration circuit. The operating frequency in this instance can be varied between approximately 1 Hz to 10 Hz. The Q lamp is lighted during the t_H output intervals, while $\bar{Q}$ lights during the t_L intervals.

7-2.2.1 *Achieving duty cycles of 50 percent or less.* While the 555-type astable multivibrator is, indeed, a versatile and relatively simple device, it is not possible to force it to generate duty cycles of 50 percent or less. The two circuits in Fig. 7-14 show how it is possible to achieve lower duty cycles with the aid of some extra circuitry.

The circuit in Fig. 7-14(a) is an example of a fixed-frequency trigger-pulse generator. The 555 naturally generates negative-going pulses; but whenever it is necessary to get short, positive-going pulses, the output can be connected to a simple logic inverter. The result is the desired positive-going pulse waveform.

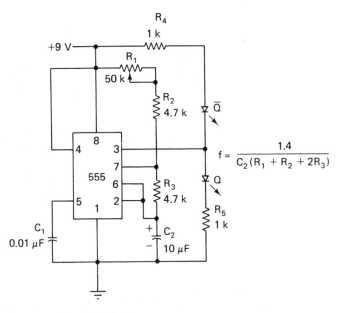

1 Hz to 10 Hz frequency range

Figure 7-13   A practical demonstration circuit for 555-type astable multi-vibrators.

Manufacturers' literature for the 555 often shows a special external arrangement of timing components to achieve a 50 percent duty cycle directly from the output of the IC. There are some special limitations on the selected values of the timing resistors, however; so the best all-around approach for getting a precise 50 percent duty cycle is to follow the output of the 555 with a T flip-flop. See the circuit in Fig. 7-14(b).

Recall that a toggled flip-flop responds only to the negative-going edge of pulses applied to its T input. Since the waveform from the 555 astable multivibrator has equal time periods between successive negative-going edges (regardless of the duty cycle), it follows that the Q output of the flip-flop will show a duty cycle of exactly 50 percent *and a frequency equal to one-half that of the oscillator*. Of course, the T flip-flop can be built from a J-K master–slave flip-flop as described in Sec. 6-7.4.

One of the special features of the circuit in Fig. 7-14(b) is that the frequency can be varied by means of R_1 without affecting the precise 50 percent duty cycle from the flip-flop. The duty cycle from the 555 device changes, but the duty cycle from the flip-flop does not.

7-2.2.2 A 556 oscillator with variable frequency and fixed output pulse width. The circuit in Fig. 7-15 takes advantage of the two separate timer circuits in the 556 IC to produce an astable multivibrator that has a variable output frequency but

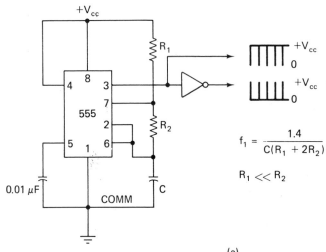

$$f_1 = \frac{1.4}{C(R_1 + 2R_2)}$$

$$R_1 \ll R_2$$

(a)

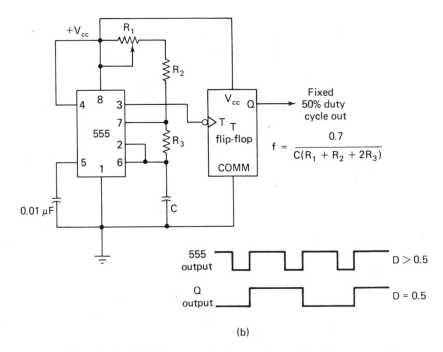

$$f = \frac{0.7}{C(R_1 + R_2 + 2R_3)}$$

(b)

Figure 7-14 Achieving operating duty cycles equal to 50 percent or less. (a) Short duty cycles by inverting the oscillators output. (b) Fifty percent duty cycle by triggering a T flip-flop from the output of a 555-type oscillator.

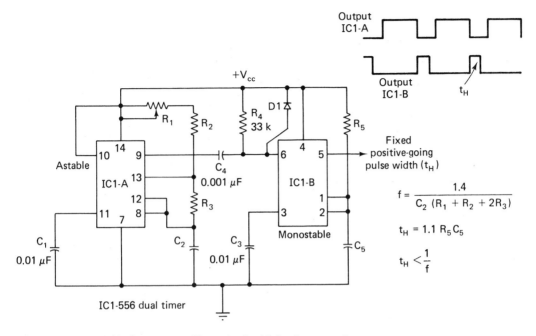

Figure 7-15   A variable-frequency oscillator circuit with fixed output pulse intervals.

a fixed positive-going pulse width. The basic idea is to follow a standard astable multivibrator with a monostable circuit. The values of R_1, R_2, R_3 and C_2 fix the circuit's operating frequency, while the values of R_5 and C_5 fix the duration of the positive-going output pulse. See the relevant equations and waveforms accompanying the diagram.

The coupling capacitor, C_4, permits output pulse times that are shorter than the low time of the oscillator. Of course, the monostable's output timing can be increased to a point where it is nearly equal to the period of the waveform from the free-running oscillator.

7-2.2.3 A simple tone burst generator. The circuit in Fig. 7-16 uses a pair of astable-connected timer circuits to produce a tone burst generator effect or "beeper." IC1-B is adjusted to produce the frequency of the tone, 1 kHz in this example.

This higher-frequency oscillator, however, can be effectively turned off whenever its RESET input at pin 4 is pulled down to logic 0.

IC1-A is also connected as an astable multivibrator, but this one is fixed for a much lower operating frequency, about 2 Hz in this example. As the output of the lower-frequency oscillator switches between logic 1 and logic 0, it alternately switches the higher-frequency oscillator on and off.

The overall effect is a string of tone bursts having a repetition rate equal to the frequency of IC1-A and a tone frequency equal to the operating frequency of

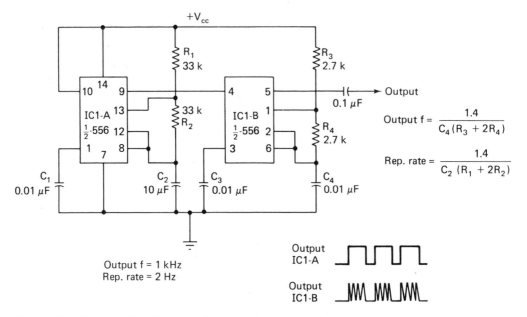

Figure 7-16 Basic circuit configuration for a tone burst generator.

IC1-B. Connecting the output of this circuit to an audio amplifier and a loudspeaker makes it an effective "beep" tone alarm.

7-2.2.4 Controlling frequency via the CONTROL VOLTAGE input. It is possible to vary the output frequency of a 555-type astable multivibrator by applying a changing voltage level to the device's CONTROL VOLTAGE input. The mechanism is identical to that described in connection with a variable pulse-width monostable circuit described in Sec. 7-2.1.5.

Increasing the CONTROL VOLTAGE potential above $2/3 V_{cc}$ increases the time-high interval of the astable circuit, thereby decreasing the operating frequency. Pulling the CONTROL VOLTAGE input below $2/3 V_{cc}$, however, decreases the time-high output interval and increases the operating frequency. The circuit's center frequency, as determined by the values of the external timing components, occurs whenever the CONTROL VOLTAGE potential is exactly $2/3 V_{cc}$.

Review the discussion in Sec. 7-2.1.5 to appreciate the limitations of the CONTROL VOLTAGE input techniques.

7-3 The 74123 Dual Monostable

As versatile and simple as the 555/556 timers might be, they are limited to pulse widths no shorter than 10 μs or an operating frequency no greater than 100 kHz. Many digital applications call for much shorter monostable intervals and higher operating frequencies; and that is where the advantages of the 74123 IC become apparent.

In a practical sense, the 74123 takes up where the basic 555 device leaves off.

Figure 7-17 completely summarizes the operating features of the 74123 dual monostable IC. The circuit in Fig. 7-17(a), for instance, shows the basic circuit arrangement for monostable operation. The output timing interval is determined by the values of R and C, and using either input A or B determines whether the output waveform will be initiated on the positive- or negative-going edge of the input waveform.

The circuit has both Q and $\bar{Q}$ outputs, thereby giving the user the option of selecting positive-going or negative-going output waveforms. There is also a CLR input that allows the circuit to be asynchronously cleared, or held inoperative, as long as the CLR terminal is fixed at logic 0.

The truth table in Fig. 7-17(b) shows how the outputs respond to different conditions at the A, B and CLR inputs. The circuit generates an output pulse, for example, whenever input A is fixed at logic 0 and input B sees a positive-going

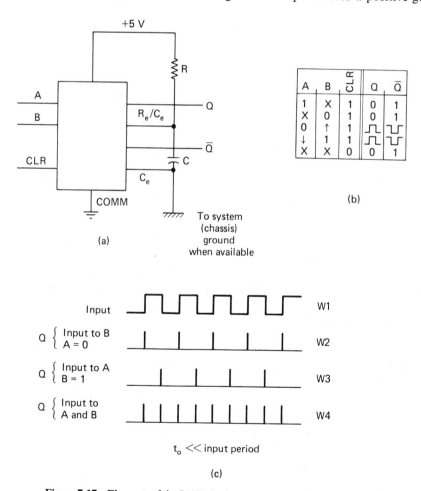

(b)

(a)

(c)

$t_o \ll$ input period

Figure 7-17 Elements of the 74123 dual monostable. (a) Basic monostable circuit configuration. (b) Truth table. (c) Waveforms.

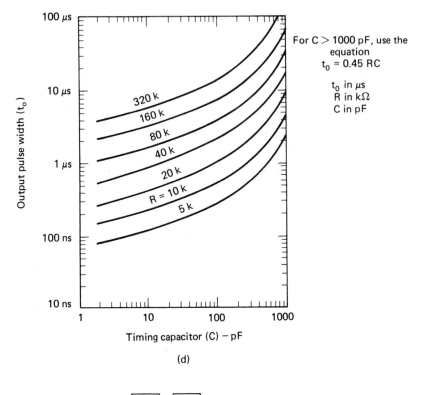

For C > 1000 pF, use the equation
$$t_0 = 0.45 \, RC$$

t_0 in μs
R in kΩ
C in pF

Output pulse width (t_0)

320 k
160 k
80 k
40 k
20 k
R = 10 k
5 k

Timing capacitor (C) – pF

(d)

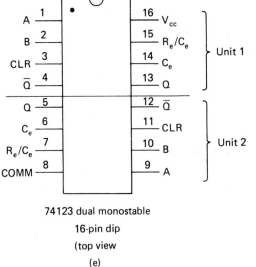

74123 dual monostable

16-pin dip

(top view

(e)

Figure 7-17 (*cont.*) (d) Design chart for circuits using capacitor values less than 1000 pF and equations for using capacitors larger than 1000 pF. (e) Pinout for the 74123.

trigger pulse. Fixing input B at logic 1, however, makes it possible to trigger an output pulse on the negative-going edge of a pulse at input A. At any other time, Q is fixed at logic 0 and $\bar{Q}$ is at logic 1.

The waveforms in Fig. 7-17(c) demonstrate the triggering properties of the 74123 monostable. Waveform $W1$ represents the triggering input in any of the three possible operating modes. If this waveform is applied to input B while $A = 0$, waveform $W3$ shows that Q yields monostable output pulses each time the input trigger pulse goes positive. Waveform $W3$, however, shows output pulses occurring on the negative-going or trailing edge of the input waveform. This action takes place whenever the trigger input is applied to terminal A while B is fixed at logic 1.

Applying the trigger input to both A and B produces monostable output pulses on both the positive-going and negative-going edges as shown in waveform $W4$. The implication here is that the 74123 can be used as a frequency doubler circuit.

The curves in Fig. 7-17(d) can be used for determining the output pulse time of an existing 74123 monostable or finding the values of R and C for a given output pulse time. Note that the circuit is good for pulses as short as 2 ns. The curves become linear for capacitor values larger than 1000 pF; and whenever such a condition arises, the circuit can be analyzed or designed according to the simple equation $t_o = 0.45RC$. Note, however, that t_o is expressed in units of nanoseconds, while R and C are in units of kilohms and picofarads, respectively.

Unfortunately, the 74123 becomes rather unstable for very long output pulse widths; so most digital designers prefer to use the 74123 for pulse widths less than 10 μs and resort to a 555-type monostable for pulse widths greater than 10 μs. In a practical sense, the 74123 and 555 complement one another very nicely.

Figure 7-17(e) shows the pinout of the 74123 dual monostable IC chip.

Since the 74123 is a true TTL device, its power supply voltage, input and output parameters conform to the standard TTL format.

7-3.1 Pulse-Sequence Operations

It is possible to cascade 74123 units to produce a variety of pulse-sequence operations. The example in Fig. 7-18 is especially useful for tachometer functions described later in this book. In this particular instance, the leading monostable, IC1-A, is triggered on the positive-going edge of a TRIG IN control waveform. The Q output of IC1-A is then connected to the A input of IC1-B, making that monostable generate its output pulse at the end—on the trailing edge—of the first monostable pulse. The monostable intervals are about 1 μs in both cases, but of course they can be altered by changing the values of the timing components.

7-3.2 74123 Astable Multivibrators

The pulse-sequence circuit just described can be modified to transform a pair of 74123 monostable units into a free-running multivibrator. The idea is to return the output of the second monostable back to the A input of the first one. See the circuit and waveforms in Fig. 7-19.

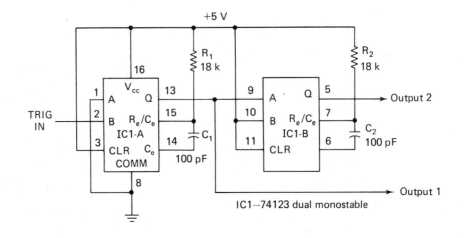

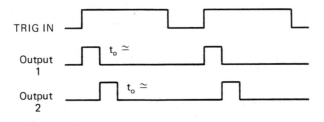

Figure 7-18 Sequence timing with the 74123 dual monostable IC.

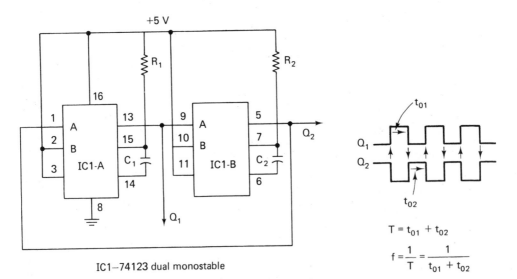

IC1–74123 dual monostable

$$T = t_{01} + t_{02}$$

$$f = \frac{1}{T} = \frac{1}{t_{01} + t_{02}}$$

Figure 7-19 An astable multivibrator built around the 74123.

Monostable IC1-A generates an output pulse having a duration t_{o1} determined by the values of R_1 and C_1. The negative-going edge of this waveform triggers IC1-B, making its Q output generate an output pulse determined by the values of R_2 and C_2. Returning this pulse to the A input of IC1-A then sustains oscillation by forcing the first monostable to produce yet another output pulse.

The output can be taken from the Q (or $\bar{Q}$) outputs of either monostable section. The duty cycle of the chosen output waveform depends on the relative timing periods of the two circuits, and the maximum operating frequency is about 250 MHz, a figure determined by the 2-ns minimum timing interval of the 74123.

Exercises

1. What is the primary advantage of the trigger circuit in Fig. 7-1(b) compared to that in Fig. 7-1(a)?

2. If one of the inverters in a "flat-out" oscillator is replaced with a 2-input NAND gate as shown in Fig. E-7.2, what effect does the C input have on the operation of the circuit?

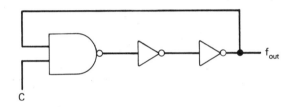

Figure E-7.2

3. Without referring to the text, draw the basic circuit diagram, label the inputs and outputs and sketch the relevant waveforms for the following circuits:
   (a) a non-interruptible 555-type monostable multivibrator
   (b) an interruptible 555-type monostable multivibrator
   (c) a 555-type astable multivibrator

4. What is the essential difference between the circuits in Figs. 7-8(a) and 7-8(b)?

5. How would increasing the value of R_2 in Fig. 7-8(b) affect the operation of the circuit?

6. What is the duty cycle of the astable multivibrator in Fig. 7-12 if $R_A = R_B$?

7. Referring to the circuit in Fig. 7-18, suppose that the CLR input at pin 3 is disconnected from $+5$ V and connected to COMM or logic 0. How would this action influence the operation of the circuit? How could you use this feature to build a tone burst generator?

8

PRINCIPLES OF DIGITAL COUNTERS

The subject of digital counters and counter circuits can be an especially exciting one because it brings together the essential features of combinatorial logic, flip-flops and astable and monostable multivibrators. Digital counting is a big subject, however; and it is difficult to deal with it properly in a single lesson or chapter. This chapter thus introduces the most basic digital counting principles and circuits, leaving some of the finer practical details and applications to Ch. 9.

8-1 Basic Digital Counter Principles

Figure 8-1(a) shows a J-K master–slave flip-flop connected to operate in its toggle mode. The CLK input receives trigger pulses from T, and the Q output changes state each time the T input switches from logic 1 to logic 0. This behavior is typical of any toggled flip-flop.

Figure 8-1(b) shows a pair of toggle-mode flip-flops that are cascaded in such a way that the Q output of FF-A serves as the trigger input for FF-B. The output of FF-B thus changes state each time the Q output of FF-A drops from logic 1 to logic 0.

A careful comparison of the circuits in Figs. 8-1(a) and 8-1(b) shows that the circuit in Fig. 8-1(a) has only two output states: 0 and 1; The two-flip-flop circuit has four distinctly different output states: $Q_A = AB = 0$; $Q_A = 1$ and $Q_B = 0$; $Q_A = 0$ and $Q_B = 1$; and $Q_A + Q_B = 1$. The circuit in Fig. 8-1(b) cycles through this pattern of four output states continuously, or at least as long as trigger pulses are applied to the first flip-flop.

Technically speaking, the circuit in Fig. 8-1(a) is said to have a *modulus* of 2 and the circuit in Fig. 8-1(b) is said to have a modulus of 4. *The modulus of a counter is the number of different output states it has.*

The truth table accompanying the diagram and waveforms in Fig. 8-1(b) shows the four possible output conditions available from Q_A and Q_B. These four states indicate four different binary numbers, and the DECIMAL column shows their decimal equivalents. Binary output 00, for instance, is equal to decimal 0, binary 01 is equal to decimal 1, binary 10 is equal to decimal 2 and so on. This is actually a standard 2-bit binary counting sequence that is capable of generating decimal equivalents of 0 through 3.

The circuit in Fig. 8-1(c) is expanded to include three cascaded flip-flops. Each time the CLK input of one of these flip-flops sees a transition from 1 to 0, its output changes state; and using three flip-flops in this fashion expands the counting mod-

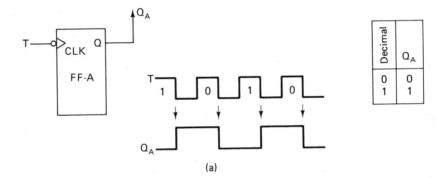

Decimal	Q_A
0	0
1	1

(a)

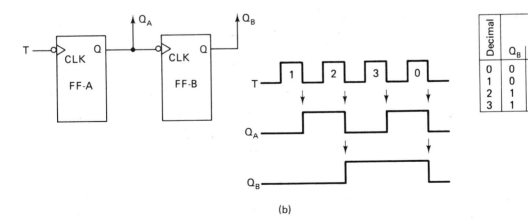

Decimal	Q_B	Q_A
0	0	0
1	0	1
2	1	0
3	1	1

(b)

Figure 8-1   Basic flip-flop counter circuits, waveforms and counting tables.
(a) Modulo-2. (b) Modulo-4.

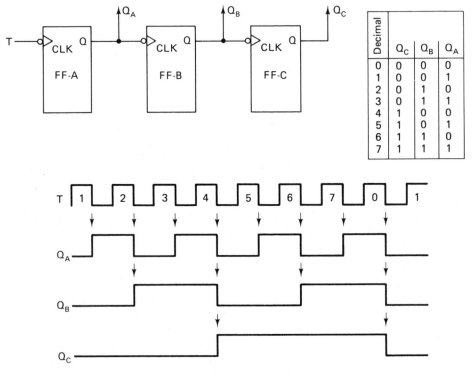

(c)

Figure 8-1 *(cont.)* (c) Modulo-8.

ulus to 8. As shown in the truth table in Fig. 8-1(c) there are eight different output states that represent decimal numbers 0 through 7. This is an example of a 3-bit binary counter.

Adding a fourth flip-flop to the sequence as shown in Fig. 8-1(d) produces a modulo-16 counter. There are 16 different output conditions representing decimal numbers 0 through 15. The circuit can be properly called a 4-bit binary counter.

Of course, it is possible to expand the number of outputs indefinitely, cascading another flip-flop's CLK input to the Q output of the preceding one, but most of the discussions in this chapter will deal with 4-bit counters.

While the circuits in Fig. 8-1 are properly considered binary counting circuits, it is also possible to view them as frequency dividers. Consider, for instance, the waveforms in Fig. 8-1(d). Recall that any toggled flip-flop very naturally divides its CLK frequency by 2; so it follows that the output of FF-A has exactly one-half the frequency of the T source: Q_A completes one cycle for every two CLK pulses.

FF-B then divides the output of FF-A by two, yielding an output frequency that is exactly one-fourth that of the T input. Similarly, FF-C divides the signal frequency at Q_B by two, generating a square-wave signal at Q_C that has one-eighth the input frequency at T. And, finally, FF-D divides the signal frequency

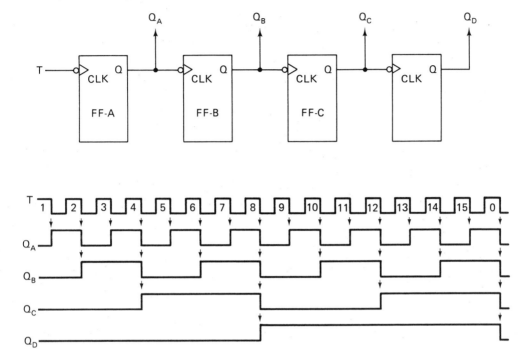

Decimal	Q_D	Q_C	Q_B	Q_A
0	0	0	0	0
1	0	0	0	1
2	0	0	1	0
3	0	0	1	1
4	0	1	0	0
5	0	1	0	1
6	0	1	1	0
7	0	1	1	1
8	1	0	0	0
9	1	0	0	1
10	1	0	1	0
11	1	0	1	1
12	1	1	0	0
13	1	1	0	1
14	1	1	1	0
15	1	1	1	1

(d)

Figure 8-1 *(cont.)* (d) Modulo-16.

by two again, producing an output that is one-sixteenth that of the T input.

Suppose, for example, the trigger frequency at input T is 160 Hz. The output at Q_A is then 80 Hz, the frequency at Q_B is 40 Hz, the frequency at Q_C is 20 Hz and the frequency at Q_D is 10 Hz. A fifth flip-flop added to this circuit would reduce the frequency by one-half again, yielding an output of 5 Hz.

The amount of frequency division is actually equal to the modulus of the counter's output. Taking the output from Q_D in Fig. 8-1(d) (a point where the counter's modulus is 16) produces a frequency equal to the input frequency divided by 16.

As useful as the 4-bit binary counter in Fig. 8-1(d) might be, it is limited by the fact that it only counts a single sequence from binary 0 through 15, and it is capable of frequency divisions of 2, 4, 8 or 16, depending on which Q output is used. The bulk of this chapter deals with procedures for altering the counting sequence and making it possible to divide the input frequency by other numbers such as 3, 5, 6, 7 and so on.

8-1.1 A 4-Bit Binary Counter Demonstration Circuit

Figure 8-2 is a practical implementation of the basic 4-bit binary counter configuration shown in Fig. 8-1(d). The circuit in this case is triggered at about 0.7 Hz from a 555 astable multivibrator, IC1. Each time the output of the oscillator goes to its logic-1 state, LED indicator T lights up.

All four J-K flip-flops in this circuit are connected for their toggle mode of operation. Each negative-going edge of the oscillator's waveform thus toggles the output of IC2-A; and then each time the Q output of IC2-A drops from 1 to 0, it toggles the output of IC2-B. All of the flip-flops are cascaded in such a way that a negative-going transition from one Q output toggles the next stage down the line.

Carefully note that the display outputs, Q_A through Q_D, are connected to the flip-flops' $\overline{Q}$ outputs. The signals for the LEDs must be taken from the inverted outputs so that the lights respond in the proper phase; lamp Q_A lights up, for instance, whenever the Q output of IC2-A is at logic 1.

The lamps thus respond as the outputs of a conventional 4-bit binary counter. See the truth table in Fig. 8-1(d). Note that it generates binary equivalents of decimal numbers 0 through 15; and if the circuit is looked at as a frequency divider, the output of each flip-flop stage has a frequency that is exactly one-half that of the previous stage.

Of course, it is possible to speed up the counting by decreasing the value of R_1 in the oscillator. Increasing the value of that resistor then lowers the circuit's operating frequency.

8-1.2 Asynchronous Clearing for Binary Counters

Figure 8-3(a) shows a 4-bit binary counter that has the CLR inputs of each flip-flop connected to a common CLEAR bus. As long as the logic level on the CLEAR bus is high, the counter works in its usual fashion, generating binary numbers 0 through 15.

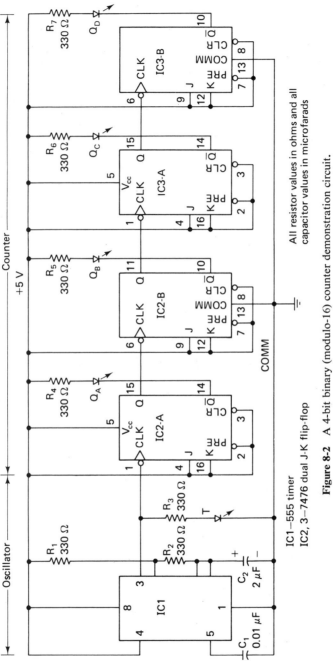

Figure 8-2 A 4-bit binary (modulo-16) counter demonstration circuit.

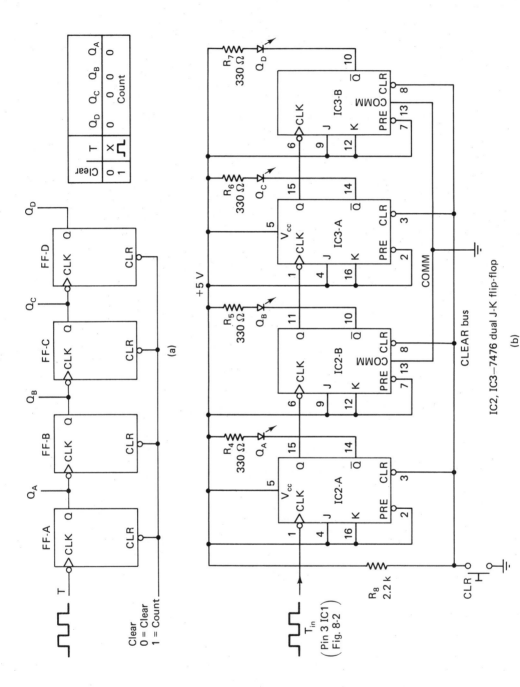

Figure 8-3 Four-bit binary counters with asynchronous clear. (a) Basic logic diagram and function table. (b) A demonstration circuit.

Setting the CLEAR bus to logic 0, however, asynchronously sets the outputs of all the flip-flops to logic 0, effectively resetting the counter to 0000. And since the CLR input overrides any clocking activity, *setting the CLEAR bus to 0 both resets the count to 0000 and stops the counting action.* Counting begins from 0000 only after the CLEAR bus is returned to logic 1. The truth table in Fig. 8-3(a) shows that the outputs are all set to 0 whenever CLR = 0, regardless of any toggle pulses appearing at the T input.

The circuit in Fig. 8-3(b) is a practical implementation of the resettable 4-bit binary counter in Fig. 8-3(a). It is actually a modified version of the counter circuit in Fig. 8-2, and the input trigger pulses can come from a 555-type astable multi-vibrator used in that circuit.

The circuit shown here can be manually stopped and reset to 0000 by depressing the CLR push button. Releasing that push button then allows the counting action to begin from 0000.

8-1.3 Count Enable/Disable

While clearing all the flip-flops in a binary counter both stops the counting action and resets the outputs to 0000, it is possible to modify the basic circuit such that it is possible to stop the count *without* clearing all the outputs. The latter feature is often called a count enable/disable function.

The circuit in Fig. 8-4(a) looks very much like the basic 4-bit binary counter in Fig. 8-1(d). In fact, the only difference is an input connection to the J-K inputs of FF-A.

Recall that it is possible to place a J-K master–slave flip-flop into a memory mode by setting its J and K inputs to logic 0 at the same time. If the J and K inputs are both set to logic 0 while the CLK input is at logic 0, the next CLK pulse has no effect on the circuit's output status; if it is at logic 1, it remains at logic 1 as long as $J = K = 0$. Returning the J and K inputs to logic 1 again allows the toggling action to resume.

FF-A in Fig. 8-4(a) responds in this fashion to its COUNT CONTROL input. Setting COUNT CONTROL to logic 1 allows it to toggle in response to trigger pulses at its CLK input. Setting COUNT CONTROL to logic 0, however, stops the toggling action, but it allows the Q_A output to retain the logic level it had *before* COUNT CONTROL was set to 0.

The COUNT CONTROL connection in Fig. 8-4(a) thus directly influences the operation of FF-A. But what about the other three flip-flops? What effect does the COUNT CONTROL input at FF-A have on the operation of FF-B, FF-C and FF-D? It turns out that stopping the toggling action of the first flip-flop stops all the other flip-flops. If the first flip-flop stops toggling by setting its J and K inputs to logic 0, it cannot possibly generate a negative-going output for toggling FF-B; and if FF-B never toggles, it cannot generate a negative-going output for toggling FF-C. Each flip-flop is toggled by a negative-going edge from the output of the previous one; so if the first flip-flop stops toggling for any reason, it follows that the others also stop toggling.

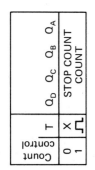

Count control		Q_D Q_C Q_B Q_A
0	X	STOP COUNT
1	⎍	COUNT

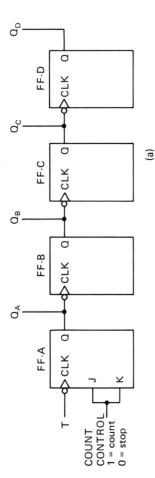

(a)

COUNT
CONTROL
1 = count
0 = stop

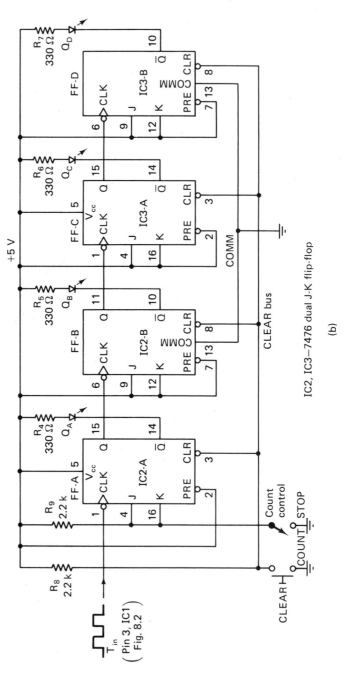

IC2, IC3—7476 dual J-K flip-flop

(b)

Figure 8-4 Four-bit binary counters with count control. (a) Basic logic diagram and function table. (b) A demonstration circuit.

Setting the COUNT CONTROL input to logic 0 thus stops the counting action, but it lets the counter retain the number it was showing just before the COUNT CONTROL was set to 0. Counting then resumes from that stored number after returning COUNT CONTROL to logic 1. See the truth table in Fig. 8-4(a).

This COUNT CONTROL action is notably different from the asynchronous clearing action described in the previous section of this chapter. Clearing the counter stops the count, but it also resets the outputs to 0.

The circuit in Fig. 8-4(b) is a modified version of the 4-bit binary counter demonstration circuit. The circuit now includes the COUNT CONTROL feature as well as CLR, and the truth table in Fig. 8-4(b) summarizes its main operating states. Using trigger pulses from an oscillator such as the one in Fig. 8-2, the circuit counts normally as long as the CLR push button is not depressed and COUNT CONTROL is set to its COUNT position.

Depressing the CLR push button both stops the counting action and resets the outputs to 0000. The counting then advances the moment the CLR button is released and a trigger pulse arrives from the oscillator.

Setting the COUNT CONTROL switch to STOP stops the counting action, but it lets the counter retain the number showing just before setting that switch to the STOP position. Counting then resumes from that stored number the moment the COUNT CONTROL switch is returned to COUNT and a trigger pulse arrives from the oscillator.

Incidentally, the COUNT CONTROL must be set to STOP only while the trigger waveform is at logic 0—the T lamp is out. Setting COUNT CONTROL to STOP while the trigger waveform is at logic 1 will allow the count to advance one more number before it stops.

8-1.4 Up/Down Counting

All of the counters described thus far in this section have been *up counters*; their binary counting sequence advances from the binary equivalent of decimal 0 through 15. It is possible, and frequently desirable, to construct a binary *down counter*: a counter that counts backward from 15 to 0, for example.

The circuit in Fig. 8-5 shows a basic 4-bit binary down counter. The circuit is very much like the 4-bit binary up counter in Fig. 8-1(d), but there is one important difference: The down counter has the CLK inputs of FF-B, FF-C and FF-D connected to the $\bar{Q}$ outputs of the previous stage. The observed outputs, however, are still taken from the flip-flops' Q outputs.

The flip-flops respond to the negative-going edge of logic levels from the previous stage, $\bar{Q}$ logic levels in this instance; but the waveforms in Fig. 8-5 show that clocking from the $\bar{Q}$ outputs while reading the count from the Q outputs yields a down-counting effect.

Whenever the observed outputs and clock outputs to the next stage are complements of one another, the overall result is a down-counting action. Observing and clocking the next stage from the same point produce an up-counting action.

Note from the waveforms, for example, that a 0000 output from the Q connections appears as a 1111 at the $\bar{Q}$ connections; and by the same token, a 1101 at the Q connections appears as 0010 at the $\bar{Q}$ positions. The flip-flops still toggle on the negative-going edge of their respective CLK input waveforms, but the observer gets the impression that the counter is running backward.

Now notice that the down counter in Fig. 8-5 has a CLR bus that is connected to the PRE inputs instead of to the CLR inputs of each flip-flop. Since a down counter counts backward, it is usually desirable to clear it to its highest count instead of to its lowest count.

Whenever the CLR bus in this circuit is set to logic 1, the counter is able to count down as described; but setting the CLR bus to logic 0 both stops the count and resets the outputs to 1111. Returning the CLR bus to logic 1 lets the counting begin from 1111 and count down toward 0000.

Many digital counters are constructed for either up or down counting, but it is possible to include some simple logic elements that give the user the option of changing the direction of counting by means of a single UP/DOWN control. The block diagram in Fig. 8-6(a) represents such a circuit. The inputs to this circuit are the Q and $\bar{Q}$ outputs of one particular flip-flop stage and a pair of control inputs, UP and DOWN. The output of this little logic block is a CLK waveform for the next flip-flop down the line.

The logic block's main job is to direct either the Q or $\bar{Q}$ output of one stage to the CLK input of the next. Presumably, the logic levels at the UP and DOWN control inputs determine whether the next stage sees Q or $\bar{Q}$ waveforms. Since the counter's observed outputs are taken from the Q connections, it follows that sending Q waveforms to the CLK input of the next stage will produce an up-

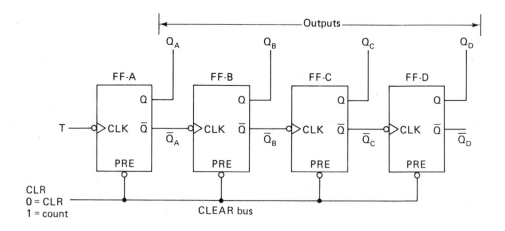

(a)

Figure 8-5 A 4-bit binary down counter. (a) Basic logic diagram.

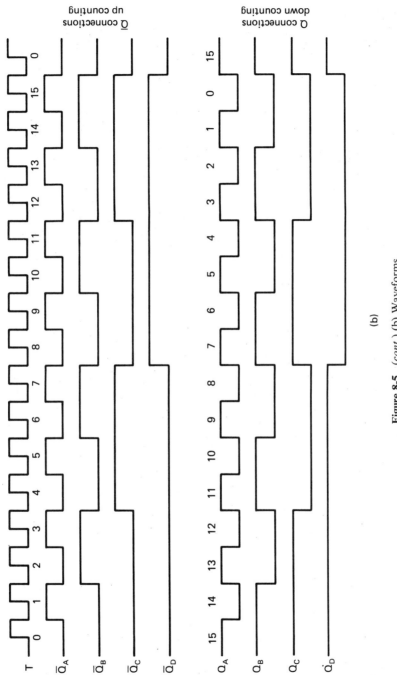

Figure 8-5 *(cont.)* (b) Waveforms.

175

CLR	Decimal	Q̄ Connections				Decimal	Q Connections (outputs)			
		$\bar{Q}_D$	$\bar{Q}_C$	$\bar{Q}_B$	$\bar{Q}_A$		Q_D	Q_C	Q_B	Q_A
0	0	0	0	0	0	15	1	1	1	1
1	1	0	0	0	1	14	1	1	1	0
1	2	0	0	1	0	13	1	1	0	1
1	3	0	0	1	1	12	1	1	0	0
1	4	0	1	0	0	11	1	0	1	1
1	5	0	1	0	1	10	1	0	1	0
1	6	0	1	1	0	9	1	0	0	1
1	7	0	1	1	1	8	1	0	0	0
1	8	1	0	0	0	7	0	1	1	1
1	9	1	0	0	1	6	0	1	1	0
1	10	1	0	1	0	5	0	1	0	1
1	11	1	0	1	1	4	0	1	0	0
1	12	1	1	0	0	3	0	0	1	1
1	13	1	1	0	1	2	0	0	1	0
1	14	1	1	1	0	1	0	0	0	1
1	15	1	1	1	1	0	0	0	0	0
1	0	0	0	0	0	15	1	1	1	1

(c)

Figure 8-5 *(cont.)* (c) Operating truth table.

counting effect and that sending $\bar{Q}$ waveforms to CLK of the next stage will yield a down-count action.

The logic circuit in Fig. 8-6(b) shows the internal structure of the UP/DOWN logic block. The circuit uses one portion of a dual AOI (AND-OR-INVERT) IC package. The logic equation analysis and truth table show that $\bar{Q}$ pulses appear at the CLK connection whenever control input U is at logic 1 and that Q pulses appear at CLK whenever the control input is at logic 0. The overall effect, summarized in the simple truth table in Fig. 8-6(b), is that the circuit will down count whenever $U = 1$ and that it will up count whenever $U = 0$.

Figure 8-6(c) shows how this UP/DOWN control logic block can be fitted into a standard 4-bit binary counter.

The circuit in Fig. 8-7 is the 4-bit binary counter demonstration circuit that has been gradually evolving through this discussion of basic digital counters. As shown here, the circuit now includes an UP/DOWN counting control switch. The truth table accompanying that diagram summarizes all of its counting modes.

In a typical operating situation, the operator might decide to set up the circuit for down counting. The operator first depresses the CLR_{15} push button to stop the counting action and set the outputs to 1111. After making certain the U/D switch is set for DOWN counting, the operator then releases the CLR_{15} push button. Counting will then begin from 1111, counting backward continuously.

Setting the COUNT CONTROL switch to STOP will always stop the counting action and force the flip-flops to hold the last-triggered number. The counting then

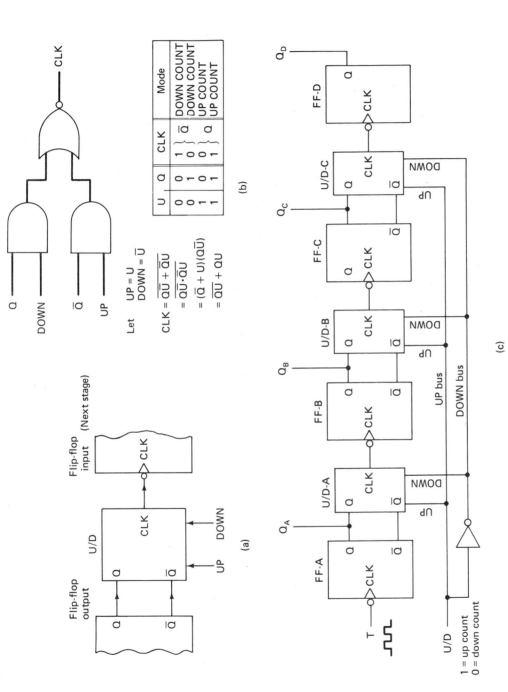

Let $UP = U$
$DOWN = \overline{U}$

$CLK = \overline{Q\overline{U} + \overline{Q}U}$
$\quad\quad = \overline{Q\overline{U}} \cdot \overline{\overline{Q}U}$
$\quad\quad = (\overline{Q} + U)(Q\overline{U})$
$\quad\quad = \overline{Q}\overline{U} + QU$

U	Q	CLK		Mode
0	0	1	} $\overline{Q}$	DOWN COUNT
0	1	0		DOWN COUNT
1	0	0	} Q	UP COUNT
1	1	1		UP COUNT

(b)

(a)

(c)

Figure 8-6 Up/down counter principles. (a) The up/down logic block. (b) Logic diagram of the up/down logic block. (c) Basic logic diagram for a 4-bit binary up/down counter.

177

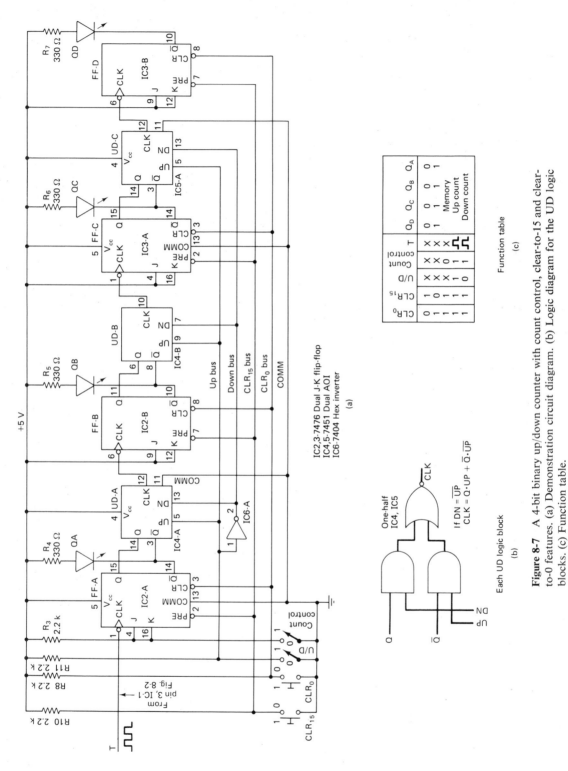

Figure 8-7 A 4-bit binary up/down counter with count control, clear-to-15 and clear-to-0 features. (a) Demonstration circuit diagram. (b) Logic diagram for the UD logic blocks. (c) Function table.

IC2,3-7476 Dual J-K flip-flop
IC4,5-7451 Dual AOI
IC6-7404 Hex inverter

(a)

One-half
IC4, IC5

If DN = $\overline{UP}$
CLK = Q·UP + $\overline{Q}$·$\overline{UP}$

Each UD logic block

(b)

CLR₀	CLR₁₅	U/D	Count control	T	Q_D	Q_C	Q_B	Q_A	
0	1	×	×	×	0	0	0	0	
1	0	×	×	×	1	1	1	1	
1	1	×	0	×					Memory
1	1	1	1	⌐⌐				0	Up count
1	1	0	1	⌐⌐				1	Down count

Function table

(c)

resumes from that point after returning the COUNT CONTROL switch to COUNT. Of course, whether the counting is up or down depends on the setting of the U/D switch.

8-1.5 Presettable Binary Counters

Although the counter circuits evolving through this discussion seem to be becoming rather complex, there is yet one more feature that must be added: an asynchronous preset feature.

A presettable counter gives the user the option of asynchronously preloading the counter's output with any desired number and then letting the count resume from that point. Recall that the Q output of a J-K flip-flop can be asynchronously set to 1 or 0 by setting CLR = 1 and PRE = 0 or setting CLR = 0 and PRE = 1, respectively. A presettable counter uses some logic circuits to take advantage of these asynchronous preset and clear features.

The logic circuit in Fig. 8-8(a) shows how it is possible to load the Q output of a flip-flop to any desired logic level at input P. In this instance, the Q output will take on the same logic level as the P input as long as control input L is at logic 1. The function table accompanying the diagram in Fig. 8-8(a) completely summarizes the influence the L and P inputs have on the flip-flop's Q output, but it is very instructive to analyze the circuit by means of some basic logic equations.

According to the function table in Fig. 8-8(a), setting the L input to logic 0 sets the PRE and CLR inputs of the flip-flop to logic 1 (recall that a logic-0 input at the input of a NAND gate always establishes a logic-1 output). With PRE and CLR both at logic 1, the flip-flop is in its toggle mode; so every toggle pulse occurring at the T input causes the Q outputs to switch state. The P input has no influence on the flip-flop under these circumstances.

Setting L to logic 1, however, guarantees that the logic levels at the PRE and CLR inputs are complements of one another; and that is the condition prescribed for operating the flip-flop in its asynchronous PRESET or CLEAR modes. If $P = 1$ when $L = 1$, for instance, the flip-flops PRE input sees a logic 0 (both inputs to NAND gate 1 are at logic 1). Inverter 3 inverts the $P = 1$ input, thus providing NAND gate 2 with a 1, 0 input, the conditions necessary for setting the flip-flop's CLR input to logic 1. Whenever $L = 1$ and $P = 1$, PRE = 0 and CLR = 1; and the flip-flop's Q output is asynchronously set to logic 1.

Use the same line of reasoning to show that the Q output will be asynchronously set to logic 0 whenever $L = 1$ and $P = 0$.

Once the output of the flip-flop is loaded so that its Q output is equal to the P input, setting the L input to 0 allows normal toggling to take place from that point.

Figure 8-8(b) shows how the preset feature can be incorporated into a 4-bit binary up counter. Each of the P blocks represent the two NAND gates and inverter shown in Fig. 8-8(a). Normal counting takes place as long as $L = 0$, but setting it to logic 1 immediately stops the counting action and loads each flip-flop with the data appearing at the P inputs. Counting resumes only after the L input is returned to logic 0.

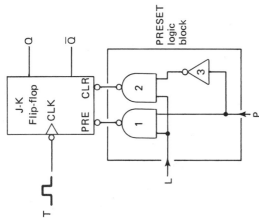

L	P	T	$\overline{PRE}$	$\overline{CLR}$	Q	$\overline{Q}$	Mode
0	X	⎍	1	1	Q_{t-1}	Q_{t-1}	Toggle
1	1	X	0	1	1	0	Load 1
1	0	X	1	0	0	1	Load 0

(a)

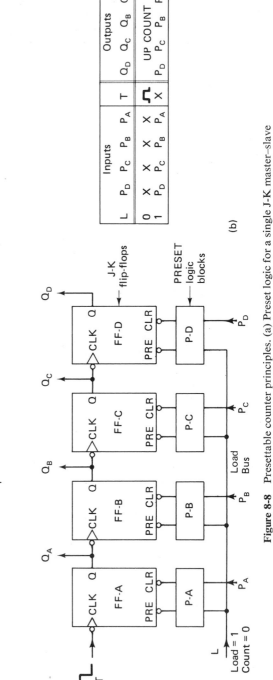

Inputs					T	Outputs			
L	P_D	P_C	P_B	P_A		Q_D	Q_C	Q_B	Q_A
0	X	X	X	X	⎍	UP COUNT			
1	P_D	P_C	P_B	P_A	X	P_D	P_C	P_B	P_A

(b)

Figure 8-8 Presettable counter principles. (a) Preset logic for a single J-K master–slave flip-flop. (b) Implementing the present logic in a 4-bit binary up counter.

180

Suppose $L = 0$ and the counter is running normally when the user sets in the binary number 1101 at the P inputs. No matter what the Q outputs might be showing at the time, they are set to 1101 the instant the user switches the L input to logic 1. The counter then retains that loaded number until the user sets the L input back to 0; and at that time, normal counting resumes from that number.

Figure 8-9 is the basic 4-bit binary counter circuit now expanded to include the presettable input feature. This counter can be described as a presettable 4-bit binary up/down counter with enable. The function table accompanying this drawing completely summarizes its operating features. At this point in the discussion of binary counters, the user ought to be able to use this table to determine how the circuit can be set to operate in its various modes.

Take special note of the fact that this circuit no longer has a separate CLEAR input. It is possible to clear the counter, however, by first setting the P inputs to 0000 and then operating the LOAD push button.

8-2   BCD and Other Fixed-Modulus Counters

The discussion in Sec. 8-1.2 shows how it is possible to reset a counter to 0000 by pulling the CLR inputs of all the flip-flops down to logic 0. This is an asynchronous operation that overrides any incoming clock pulses and holds the counter's output at 0000 as long as desired.

All of the examples cited thus far use a manually operated switch to do the clearing operation, but it is altogether possible to accomplish the same thing automatically by using some simple logic circuitry.

Figure 8-10, for example, shows a NAND gate having its inputs connected to counter outputs Q_B and Q_D. The output of the NAND gate is connected directly to the counter's CLEAR bus. Whenever $Q_B = Q_D = 1$, the output of the NAND gate drops to logic 0 and the counter is immediately reset to show an output 0000. The moment the counter is reset or cleared, the inputs to the NAND gate are no longer equal to logic 1; therefore, the output of the NAND gate returns to logic 1; And as long as the NAND gate shows an output of logic 1, the counter runs through its normal up-counting sequence. As soon as the counter reaches a point where $Q_B = Q_D = 1$ again, the output of the NAND gate switches to logic 0 and restarts the cycle.

This is an example of automatic asynchronous clearing as it applies to a *fixed-modulus counter*.

8-2.1 Basic BCD Counters

The waveforms and truth table in Fig. 8-10 show the counting sequence for a basic BCD (binary-coded decimal) counter. If the count starts at binary 0000, it progresses normally through decimal 9 or binary 1001. The next clock pulse sets $Q_B = Q_D = 1$, however, and the NAND gate responds by clearing the counter to 0000. See the waveforms in Fig. 8-10(b). In principle at least, the counter shows

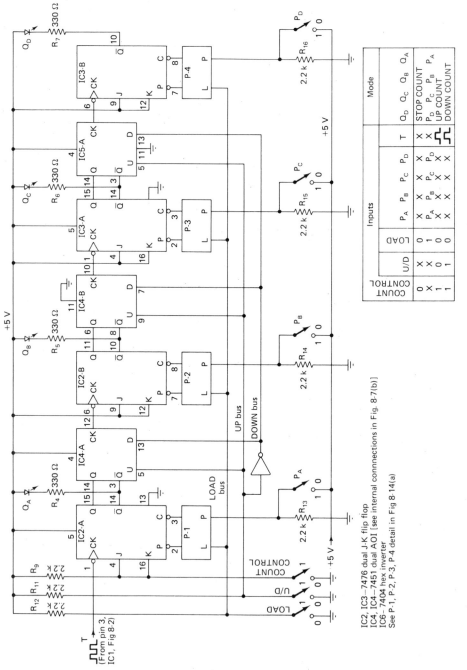

Figure 8-9 Demonstration version of a presettable 4-bit binary up/down counter with enable.

IC2, IC3—7476 dual J-K flip flop
IC4, IC4—7451 dual AOI [see internal connnections in Fig. 8-7(b)]
IC6—7404 hex inverter
See P-1, P-2, P-3, P-4 detail in Fig 8-14(a)

CONTROL			Inputs					Mode
COUNT	U/D	LOAD	P_A	P_B	P_C	P_D	T	
0	X	0	X	X	X	X	X	Q_D Q_C Q_B Q_A STOP COUNT
X	X	1	P_A	P_B	P_C	P_D	⊐⌐	P_D P_C P_B P_A
1	0	0	X	X	X	X	⊐⌐	UP COUNT
1	1	0	X	X	X	X	⊐⌐	DOWN COUNT

182

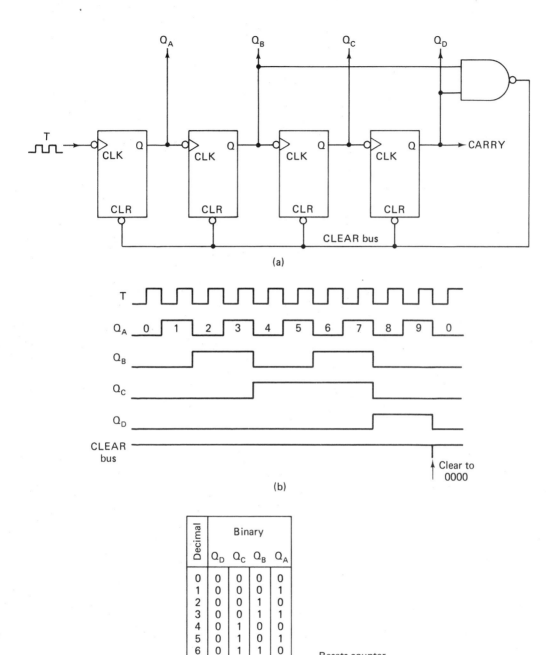

Figure 8-10 A basic BCD counter. (a) Logic diagram. (b) Waveforms. (c) Function table.

decimal 10 or binary 1010 only long enough to reset the counter to 0. For all practical purposes, however, the circuit never shows the decimal 10 output; it counts the sequence 0, 1, 2, 3, 4, 5, 6, 7, 8, 9, 0, 1, 2, ...

This particular count sequence represents normal decimal counting. The outputs, however, are binary coded. Thus the name *binary-coded decimal* counter.*

BCD counters are among the most popular counters in the digital business today because they generate the ten numbers required for ordinary decade (decimal 0 through 9) counting. And if the Q_D or CARRY output of one BCD counter is connected to the T input of another decade counter, the two circuits generate two full decades of decimal numbers, 00 through 99. See Fig. 8-11(a).

To see how this cascading principle works, note from waveform Q_D in Fig. 8-10(b) that the CARRY output shows a negative-going edge *only* when the counter is being asynchronously reset to 0. This is the exact moment the next counter down the line is to be advanced. Suppose, for example, a pair of cascaded decimal counters are showing the numeral 19. The next trigger pulse into the least-significant digit counter will reset it to 0; but as the Q_D output of that counter responds by dropping to logic 0, it triggers the ten's counter, changing its output count to decimal 2. The overall result is that the count changes from 19 to 20.

Of course, it is possible to cascade BCD counters in this fashion to count any number of decades. The 4-decade counter in Fig. 8-11(b), for instance, can count ordinary decimal numbers 0000 through 9999. The outputs are still in the BCD format, but they can be translated into a decimal equivalent by using a suitable code converter and display assembly. The CARRY output of the most-significant digit counter can serve as an overflow trigger, perhaps triggering a flip-flop that will indicate a count that exceeds the circuit's capacity. That is the function of FF-O in Fig. 8-11(b).

A BCD counter has a modulus of 10 when looking at the Q_D or most-significant bit output. This means that BCD counters can be used for dividing pulse frequencies by a factor of 10. Note from the waveforms in Fig. 8-10(b) that the Q_D output shows one complete pulse for every ten pulses delivered to the circuit's T input.

---

*By convention, binary numbers are usually written with the least-significant bit (the Q_A bit, the one that changes most frequently) showing at the right-hand end of the expression. See, for example, the BCD code in the truth table in Fig. 8-10(c). This convention, however, often conflicts with another convention whereby electronic circuits are drawn such that a sequence of operations runs from right to left; thus the flip-flop for the least-significant bit appears at the left-hand end of the drawing. See the drawing in Fig. 8-10(a).

Many texts attempt to clarify the situation by drawing binary counters backward, with the first flip-flop (FF-A) appearing at the right-hand end of the drawing. This text, however, follows the procedures used in the electronics industry: Unless clearly stated otherwise, all binary expressions are written in the conventional manner, with the least-significant bit at the right-hand end. Drawings for counter circuits, though, normally show the flip-flop for the least-significant bit at the left-hand end.

In any case, there is no cause for confusion as long as the reader is able to identify the position of the least-significant bit, whether it appears at the right-hand end of a binary expression or at the left-hand end of a counter circuit.

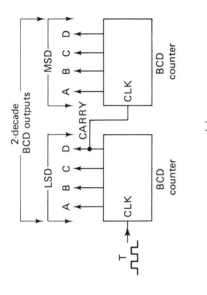

(a)

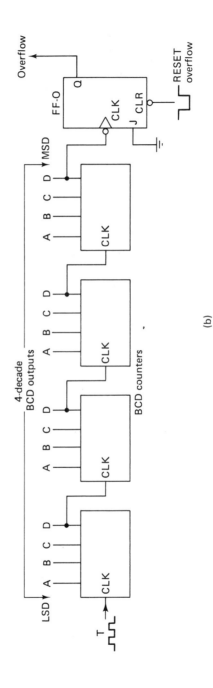

(b)

Figure 8-11 Cascading BCD counters. (a) A 2-decade counter capable of up counting decimal 00 through 99. (b) A 4-decade counter with overflow indicator. The useful counting range is between decimal 0000 and 9999.

Cascading two or more BCD counters permits frequency divisions in steps of 10. Suppose you are applying a 10-kHz waveform to the input of a 4-decade BCD counter chain. The first counter divides the frequency by 10, outputting a 1-kHz waveform. The second decade then divides by 10 again, yielding a 100-Hz output; and the third decade produces an output frequency of 10 Hz. The D output of the final stage shows a 1-Hz pulse waveform.

The overall frequency division of a BCD counter chain is equal to 10^n, where n equals the number of counters in the circuit. Such counter chains are often used in precision timing equipment in which the output of a crystal-controlled oscillator is divided down to intervals of 0.01 s 0.1 s and 1 s—a digital stopwatch circuit, for instance.

8-2.2 Other Fixed-Modulus Counters

The principle of automatic asynchronous clearing demonstrated for the BCD counter in Fig. 8-10 can be used for building counters having any desired fixed modulus. Suppose a standard 4-bit binary counter is modified so that its Q_D, Q_C and Q_A outputs are connected to the inputs of a 3-input NAND gate and suppose that the output of that gate is tied to the circuit's CLEAR bus. This counter would count normally between decimal-equivalent 0 through 12. The next clock pulse would set the counter's output to 13, but that 4-bit pattern (1101) would cause the NAND gate to clear the outputs to 0. The counter, in other words, would continuously cycle between 0 and 12, making it a modulo-13 counter that would count "dozens" of pulses.

The modulus of a counter is always equal to the number of binary numerals it generates. A BCD counter, for example, has a modulus of 10 because it counts 10 different numerals, 0 through 9. The "dozens" counter is a modulo-13 counter because it counts 13 different numerals, 0 through 12.

Now notice that the highest numeral a binary up counter shows is one less than its modulus: The highest numeral for a BCD up counter is 9; the highest numeral for a modulo-13 "dozens" counter is 12. The reason for this is that the numeral 0 is one of the valid output states that must be included in the modulus designation.

The NAND gate in the feedback circuit of a fixed-modulus counter must respond either to a numeral equal to the desired modulus or to one numeral larger than the largest one to be displayed. The NAND gate in a BCD counter, for instance, responds to binary ten (1010), while the gate in the "dozens" counter responds to the binary equivalent of the numeral 13 (1101).

Figure 8-12 summarizes the simplest procedures for transforming a basic 4-bit binary counter into counters having a fixed modulus between 2 and 16. Outputs Q_A, Q_B, Q_C, and Q_D generate counts having a modulus of 2, 4, 8 and 16, respectively; but the other fixed-modulus counters require a NAND-gate feedback to the circuit's CLEAR bus.

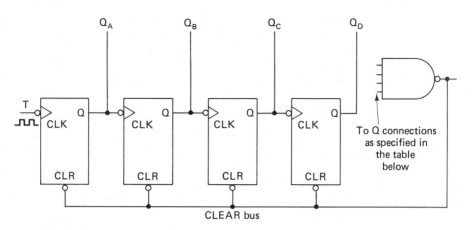

Counter modulus	NAND gate inputs	Counting range	Frequency division
2	None	$0 \rightarrow 1$	$f_{Q_A} = f_T/2$
3	Q_A, Q_B	$0 \rightarrow 2$	$f_{Q_B} = f_T/3$
4	None	$0 \rightarrow 3$	$f_{Q_B} = f_T/4$
5	Q_A, Q_C	$0 \rightarrow 4$	$f_{Q_C} = f_T/5$
6	Q_B, Q_C	$0 \rightarrow 5$	$f_{Q_C} = f_T/6$
7	Q_A, Q_B, Q_C	$0 \rightarrow 6$	$f_{Q_C} = f_T/7$
8	None	$0 \rightarrow 7$	$f_{Q_D} = f_T/8$
9	Q_A, Q_D	$0 \rightarrow 8$	$f_{Q_D} = f_T/9$
10	Q_B, Q_D	$0 \rightarrow 9$	$f_{Q_D} = f_T/10$
11	Q_A, Q_B, Q_D	$0 \rightarrow 10$	$f_{Q_D} = f_T/11$
12	Q_C, Q_D	$0 \rightarrow 11$	$f_{Q_D} = f_T/12$
13	Q_A, Q_C, Q_D	$0 \rightarrow 12$	$f_{Q_D} = f_T/13$
14	Q_B, Q_C, Q_D	$0 \rightarrow 13$	$f_{Q_D} = f_T/14$
15	Q_A, Q_B, Q_C, Q_D	$0 \rightarrow 14$	$f_{Q_D} = f_T/15$
16	None	$0 \rightarrow 15$	$f_{Q_D} = f_T/16$

Figure 8-12   Summary of feedback requirements for fixed-modulus counters.

If the fixed-modulus counters in Fig. 8-12 are viewed as frequency dividers, it turns out that the frequency of the waveform appearing at their most-significant bit (MSB) positions is exactly equal to the counter's input frequency divided by its modulus. All of these circuits can thus be considered counters, frequency dividers or both.

Cascading two or more fixed-modulus counters produces a larger counter circuit having a modulus equal to $n_1 \times n_2 \times n_3 \ldots$, where each n represents the modulus of an individual counter. Connecting the MSB output of a BCD counter to the trigger input of a modulo-12 counter, for instance, produces a counter circuit having all the characteristics of a modulo-120 counter.

8-3 Variable-Modulus Counters

Although fixed-modulus counters do, indeed, have some important functions in many different kinds of digital circuits (most notably BCD counter circuits), it is often necessary to change the modulus of a counter to suit the requirements of other types of digital systems.

There are several basic approaches to building variable-modulus counters, counters that give the user the option of selecting the counting range or amount of frequency division in the simplest possible way. The most popular approach to building a variable-modulus counter combines the features of a binary down counter with those of a presettable counter.

Figure 8-13 shows a presettable 4-bit binary down counter having a special NAND gate and inverter feedback connection to the LOAD bus. The inputs to the NAND gate are connected to the flip-flops' Q outputs, thereby making it responsive to output 1111 or binary 15. The circuit thus counts down normally until it reaches 1111. At that point in the operation, the output of the NAND gate goes to logic 0, and the inverter switches the logic level to 1. Applying a logic 1 to the LOAD input in this fashion then loads the counter with any number appearing at the P inputs.

Suppose, for instance, the user sets the number 12 into the P inputs. Counting begins from the binary equivalent of 12 and then decrements (counts down) toward 0. The counter will eventually show output 0000; but on the next trigger pulse, when the count attempts to go to 1111, the feedback circuit responds almost immediately, resetting the counter's output to binary 12. The counting cycle thus runs continuously from 12, down to 0, and back to 12 again. The counter in this case has a modulus of 13.

Changing the P inputs to binary 10 changes the counting cycle. Instead of cycling between 12 and 0, the circuit cycles between 10 and 0, giving it a modulus of 11. The modulus of this circuit can be set anywhere between 1 and 15 by setting the P inputs to any binary number between 0000 and 1110, respectively. The table accompanying the drawing in Fig. 8-13 summarizes the preset inputs required for operating the circuit in all of its possible counting patterns.

The NAND gate feedback circuit is sometimes called a *15's detector*, *underflow detector* or *borrow* output. A similar kind of circuit taking its inputs from the flip-flops' $\bar{Q}$ connections would respond to the 0000 output pattern of the Q outputs, making it serve the function of a *zero detector*, *overflow detector* or *carry* output.

If a zero detector, rather than a 15's detector, is used for loading the preset, the counter never shows a 0 output and the modulus is equal to the number set on the P inputs.

Figure 8-14 is a demonstrator version of a "universal" counter circuit. It can be described as a presettable 4-bit binary up/down counter with enable input and with carry and borrow outputs. The table in Fig. 8-14(b) represents this circuit's basic logic truth table, and the table in Fig. 8-14(c) shows how the circuit can be used in a wide variety of different ways.

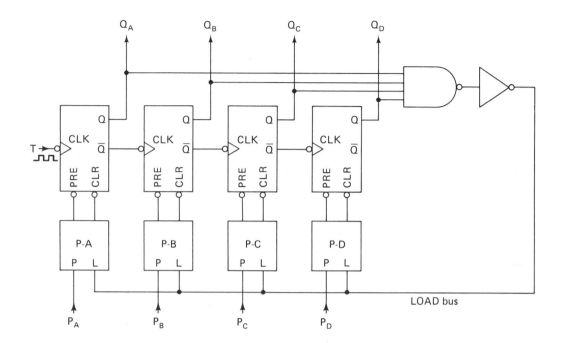

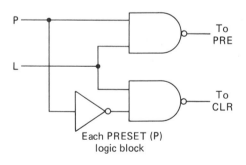

Each PRESET (P)
logic block

Dec.	PRESET inputs				Count sequence	Modulus
	Binary					
	P_D	P_C	P_B	P_A		
0	0	0	0	0	Always zero	1
1	0	0	0	1	1 → 0	2
2	0	0	1	0	2 → 0	3
3	0	0	1	1	3 → 0	4
4	0	1	0	0	4 → 0	5
5	0	1	0	1	5 → 0	6
6	0	1	1	0	6 → 0	7
7	0	1	1	1	7 → 0	8
8	1	0	0	0	8 → 0	9
9	1	0	0	1	9 → 0	10
10	1	0	1	0	10 → 0	11
11	1	0	1	1	11 → 0	12
12	1	1	0	0	12 → 0	13
13	1	1	0	1	13 → 0	14
14	1	1	1	0	14 → 0	15
15	1	1	1	1	Always zero	1

Figure 8-13 A 4-bit binary counter with variable modulus.

189

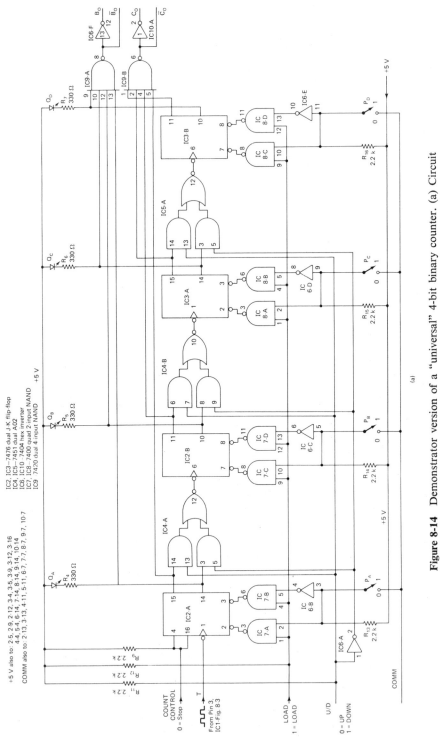

Figure 8-14 Demonstrator version of a "universal" 4-bit binary counter. (a) Circuit diagram.

(a)

190

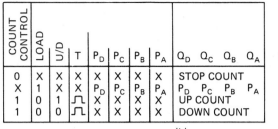

COUNT CONTROL	LOAD	U/D	T	P_D	P_C	P_B	P_A	Q_D Q_C Q_B Q_A
0	X	X	X	X	X	X	X	STOP COUNT
X	1	X	X	P_D	P_C	P_B	P_A	P_D P_C P_B P_A
1	0	1	⎍	X	X	X	X	UP COUNT
1	0	0	⎍	X	X	X	X	DOWN COUNT

$$C_O = Q_D \cdot Q_C \cdot Q_B \cdot Q_A$$
$$B_O = \overline{Q}_D \cdot \overline{Q}_C \cdot \overline{Q}_B \cdot \overline{Q}_A$$

(b)

CONTINUOUS MODULO-16 UP OR DOWN COUNTING

Counter runs at the T clock rate, continuously cycling through all binary numbers between 0 and 15.

COUNT CONTROL = 1
LOAD = 0
U/D = 0 for up counting
1 for down counting

CONTINUOUS MODULO-10 (BCD) DOWN COUNTING

Counter runs at the T clock rate, continuously cycling through all binary numbers, downward from 9 through 0.

COUNT CONTROL = 1   PRESET INPUTS = binary 9
LOAD = C_O
U/D = 1

CONTINUOUS MODULO-n DOWN COUNTING

Counter runs at the T clock rate, continuously cycling through all binary numbers, downward from n through 0.

COUNT CONTROL = 1
LOAD = C_O
PRESET INPUTS = binary n
U/D = 1

STOP-AT-15 MODULO-n UP COUNTING

Counter runs at the T clock rate, beginning from 16-n and incrementing to binary 15 where it stops until the Load input is momentarily set to logic 1.

COUNT CONTROL = C_O
LOAD = 0 for counting operation
= 1 for restarting
PRESET INPUTS = binary 16-n
U/D = 0

STOP-AT-ZERO MODULO-n DOWN COUNTING

Counter runs at the T clock rate, beginning from n-1 and decrementing to 0 where it stops until the Load input is momentarily set to logic 1.

COUNT CONTROL = B_O
LOAD = 0 for counting
= 1 for restarting count
U/D = 1

(c)

Figure 8-14 *(cont.)* (b) General function table. (c) Summary of possible operating modes.

8-4 Synchronous Counters

All of the counters described thus far in this chapter have been asynchronous or *ripple counters*. The first flip-flop (the one generating the least-significant bit) is triggered directly from an outside source of clock or trigger pulses; every flip-flop after that, however, gets its trigger signal from an output of the preceding stage. All of the flip-flops, except for the first one, depend on a change in the output of the previous flip-flop.

Flip-flops cannot change state instantaneously; there is always some amount of propagation delay time from the instant the waveform at the CLK input changes until the Q outputs respond. This propagation delay time is on the order of 25 ns for TTL J-K flip-flops and is about 100 ns for CMOS versions. Propagation delay time slows down the operation of a ripple counter, adding another delay interval for each stage that changes state.

Changing the output of a ripple counter from 1110 (binary 14) to 1111 (binary 15) calls for changing only the output of the first flip-flop; therefore, the overall process requires only about 25 ns. A change from 1011 (binary 11) to 1100 (binary 14), however, calls for changing the outputs of three of the flip-flops in succession; and that means a total propagation delay or *settling time* of about 75 ns for TTL devices.

The worst-case condition for a 4-bit binary ripple counter is the change from 1111 to 0000. In this instance, all four flip-flops must change state. The effect of the input clock pulse ripples down the line, setting the output of each flip-flop to 0 in a fashion similar to knocking over a row of four dominoes. The total propagation delay time is the sum of the delay times for each flip-flop—on the order of 100 ns for four TTL flip-flops or about 400 ns for CMOS flip-flops.

The practical significance of total propagation delay time or settling time of a ripple counter is that it puts a tight ceiling on the counter's maximum operating frequency. If the worst-case settling time of a 4-bit counter is 100 ns, for example, the counter cannot possibly work reliably at frequencies above 10 MHz; and if the counter is built from CMOS circuits, the upper operating frequency is limited to $\frac{1}{400}$ ns, or 2.5 MHz. Although these upper operating frequencies might suit many kinds of low-frequency digital circuits quite nicely, they are wholly inadequate for high-speed computer systems operating at basic clock frequencies on the order of 25 MHz or more.

The answer to the problem of getting faster counters is not one of developing better and faster flip-flops. The most suitable solution to the problem is using any entirely different counter technique known as *synchronous counting*. As shown in Chap. 9, most counter ICs are synchronous counters, counters that have a total settling time equal to the propagation delay time of the slowest flip-flop in the circuit. And that means synchronous 4-bit counters can run at frequencies between 10 MHz (CMOS) and 40 MHz (TTL).

8-4.1 Synchronous 4-Bit Binary Counters

Figure 8-15(a) is a diagram of a 4-bit binary counter that uses synchronous clocking. Note that this circuit differs from a 4-bit ripple counter in two important respects: The synchronous counter has direct input clocking connections to each flip-flop, and the Q outputs are coupled to their following stages via multiple J and K inputs.

The fact that all the CLK inputs are connected to a common bus line means that all four flip-flops are clocked at precisely the same time; there is no propagation delay of the type that accumulates through the switching action of a ripple counter. And by using the flip-flops' multiple J-K inputs, each unit "knows" what output state it is supposed to take on *before* the negative-going edge of the clock pulse occurs. Synchronous counters are built around a *look-ahead* principle. That

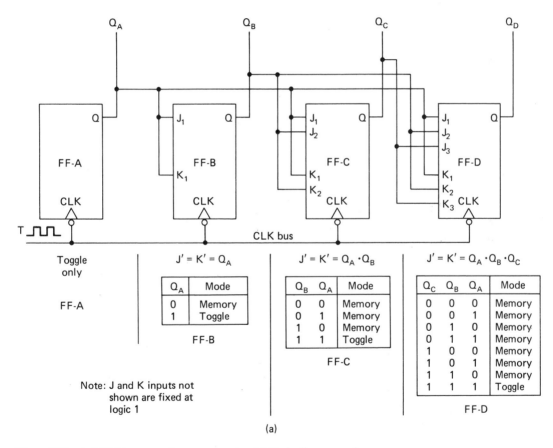

(a)

Figure 8-15 A 4-bit binary synchronous counter. (a) Basic diagram and truth tables for each flip-flop.

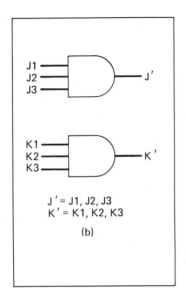

J' = J1, J2, J3
K' = K1, K2, K3

(b)

Dec.	Binary				Clock transition	Operating mode			
	Q_D	Q_C	Q_B	Q_A	T	FF-A	FF-B	FF-C	FF-D
	0	0	0	0	↑	T	M	M	M
1	0	0	0	1	↓				
	0	0	0	1	↑	T	T	M	M
2	0	0	1	0	↓				
	0	0	1	0	↑	T	M	M	M
3	0	0	1	1	↓				
	0	0	1	1	↑	T	T	T	M
4	0	1	0	0	↓				
	0	1	0	0	↑	T	M	M	M
5	0	1	0	1	↓				
	0	1	0	1	↑	T	T	M	M
6	0	1	1	0	↓				
	0	1	1	0	↑	T	M	M	M
7	0	1	1	1	↓				
	0	1	1	1	↑	T	T	T	T
8	1	0	0	0	↓				
	1	0	0	0	↑	T	M	M	M
9	1	0	0	1	↓				
	1	0	0	1	↑	T	T	M	M
10	1	0	1	0	↓				
	1	0	1	0	↑	T	M	M	M
11	1	0	1	1	↓				
	1	0	1	1	↑	T	T	T	M
12	1	1	0	0	↓				
	1	1	0	0	↑	T	M	M	M
13	1	1	0	1	↓				
	1	1	0	1	↑	T	T	M	M
14	1	1	1	0	↓				
	1	1	1	0	↑	T	M	M	M
15	1	1	1	1	↓				
	1	1	1	1	↑	T	T	T	T
0	0	0	0	0	↓				

↑ Positive-going clock transition
↓ Negative-going clock transition

T Toggle mode
M Memory mode

(c)

Figure 8-15 (*cont.*) (b) J-K input logic for each flip-flop. (c) Master function table.

is, the master section of each flip-flop "reads" its instructions on the positive-going edge of each clock pulse and then executes those instructions on the negative-going edge of the input T waveform.

To see exactly how this look-ahead or synchronous counting takes place, first recall that J-K master–slave flip-flops having multiple J and K inputs have those inputs ANDed together as shown in Fig. 8-15(b). See the detailed drawing of the 7472 master–slave J-K flip-flop in Fig. 6-15 for example. The small function tables appearing below FF-B, FF-C and FF-D in Fig. 8-15(a) show how these multiple J and K inputs work to set the operating modes of the flip-flops under all possible combinations of inputs. Since the J and K inputs to FF-A are uncommitted, that particular flip-flop is always in its basic toggle mode of operation.

FF-B, however, can be in either a memory or toggle mode, depending on the output of FF-A while the common CLK bus shows a positive-going transition. If Q_A happens to be at logic 0 during the positive-going edge of the T waveform, the master section of FF-B is loaded with 0's to set it up for a memory (no change) operation when the negative-going edge occurs. But if the output of FF-A is at logic 1 when the positive-going edge of the input clock pulse occurs, the master section of FF-B is loaded with 1's, thus setting up that flip-flop for a toggling operation when the CLK bus falls to logic 0 again.

FF-B and FF-C respond in a similar fashion to the outputs of all preceding stages. In short, FF-A and FF-B are normally in their memory modes, being programmed for a toggling operation only when the outputs of all preceding flip-flops are at logic 1 when the positive-going edge of the T clock pulse occurs.

Figure 8-15(c) is a master function table for any synchronous 4-bit binary counter. The arrows in the T column indicate the transitions of the input clock pulse; an up-arrow indicates a positive-going CLK pulse that loads data into the flip-flops' master sections and a down-arrow indicates the negative-going edge that forces each flip-flop to execute its stored instructions.

Suppose, for example, the counter is showing binary 5 (0101) when the T pulse goes to logic 1. The outputs of the flip-flops do not change at that moment, but according to Fig. 8-15(c), FF-A and FF-B are set for toggling, while FF-C and FF-D are set for a memory operation. When the T pulse then falls back to logic 1, Q_A is toggled to 0, Q_B is toggled to 1, and Q_C and Q_D remain at logic 1 and logic 0, respectively. The counter's output is thus incremented to binary 6 (0110).

While synchronous counters are, indeed, somewhat more complicated than their ripple-counter counterparts, the fact that the flip-flops in synchronous counters are all clocked at the same instant gives them operating frequencies that are limited only by the slowest flip-flop in the circuit. If the slowest flip-flop happens to have a propagation delay of 25 ns, for example, the counter can operate at frequencies as high as 40 MHz, and that is a 4:1 increase in speed over a comparable 4-bit ripple counter. Most of the counter ICs described in Chap. 9 are, in fact, synchronous counters.

The counting pattern of any synchronous pattern is exactly the same as that of a corresponding ripple counter. The only difference as far as an outside observer is concerned is that synchronous counters can be run much, much faster.

8-4.2 Synchronous BCD Counters

Any synchronous counter can be modified to count any fixed modulus by using the same feedback technique summarized in Fig. 8-12. The feedback mechanism in these instances, however, represents an asynchronous operation that slows down the resetting operation by as much as three propagation delay times. A basic synchronous counter using asynchronous clearing circuits is actually a *semi-synchronous counter*.

It is possible to incorporate a fixed-modulus counter of any number without destroying the synchronous feature by modifying the patterns of connections to the multiple *J* and *K* inputs. Figure 8-16 summarizes the operating features of a purely synchronous BCD counter. Note especially the synchronous feedback connection between the $\bar{Q}$ output of FF-D and the J_2 input of FF-B.

The small function tables below each flip-flop show the operating modes under all possible input conditions at the *J* and *K* connections. The master function table in Fig. 8-16(b) shows the overall operation of the circuit.

One of the critical points in the operation of a BCD synchronous counter takes place as the outputs are showing binary 9 (1001) while the *T* clock pulse makes a

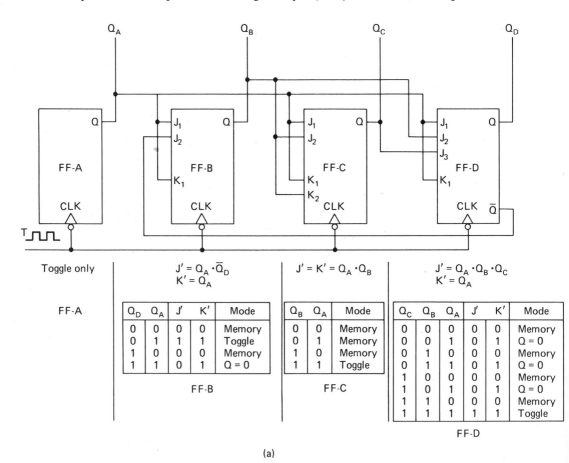

(a)

Figure 8-16 A synchronous BCD counter. (a) Basic diagram and input truth tables for each flip-flop.

Outputs					Clock transition	Operating mode			
	Binary					FF-A	FF-B	FF-C	FF-D
Dec.	Q_D	Q_C	Q_B	Q_A	T				
	0	0	0	0	↑	T	M	M	M
1	0	0	0	1	↓				
	0	0	0	1	↑	T	T	M	Q = 0
2	0	0	1	0	↓				
	0	0	1	0	↑	T	M	M	M
3	0	0	1	1	↓				
	0	0	1	1	↑	T	T	T	Q = 0
4	0	1	0	0	↓				
	0	1	0	0	↑	T	M	M	M
5	0	1	0	1	↓				
	0	1	0	1	↑	T	T	M	Q = 0
6	0	1	1	0	↓				
	0	1	1	0	↑	T	M	M	M
7	0	1	1	1	↓				
	0	1	1	1	↑	T	T	T	T
8	1	0	0	0	↓				
	1	0	0	0	↑	T	M	M	M
9	1	0	0	1	↓				
	1	0	0	1	↑	T	Q = 0	M	Q = 0
0	0	0	0	0	↓				

↑ Positive-going clock transition
↓ Negative-going clock transition

T Toggle mode
M Memory mode
Q = 0, J = 0, K = 1 synchronous mode

(b)

Figure 8-16 (*cont.*) (b) Master function table.

positive-going transition. At that moment, FF-A is programmed for a toggling operation; and FF-C, seeing 0's at J_2 and K_2 from Q_B, is set for a memory (no change) operation. FF-B and FF-D, however, are seeing at least one 0 at their J inputs and all 1's at their K inputs. Whenever $J = 0$, $K = 1$ in this kind of flip-flop, it is being programmed to set its Q output at logic 0.* So when the T pulse returns to logic 0, FF-A toggles to 0, FF-C remains unchanged at 0 and the outputs of FF-B and FF-D are set to 0. A count output of binary 9 thus changes to 0 the moment the T pulse makes its return to logic 0.

Exercises

1. What is meant by the *modulus* of a counter? What is the modulus of a single flip-flop? A BCD counter? A 5-bit binary counter? How many flip-flop stages are required for building a modulo-256 counter?

2. Why are the LED indicator lamps Q_A, Q_B, Q_C and Q_D connected to the $\bar{Q}$ outputs of the counter circuit in Fig. 8-2 when it is the Q outputs that are generating active-high or true counting outputs?

*See the standard J-K master–slave truth table in Fig. 6-14.

3. In what way is the asynchronous clearing feature different from the enable/disable feature of a counter circuit?

4. State the principle regarding the electrical connections that are different for up counting and down counting.

5. What is the largest *decimal* number that can be counted with three cascaded BCD counters?

6. What is the modulus of an up counter that automatically clears to 0 when the feedback circuit "sees" the binary numeral 101101?

7. What is the difference between a *carry* and a *borrow* output available from a binary ripple counter?

8. What is the principle advantage of a synchronous counter over a ripple counter?

COUNTER ICS
AND APPLICATIONS

This chapter extends the basic principles of binary counter circuits to include commercially available counter IC packages. The applications cited here are only intended to illustrate the essential features of these counters, however; and the reader will find more practical applications in later chapters of this book.

9-1 Asynchronous Up Counters With Clear

The small family of asynchronous up counters listed in Fig. 9-1 represents the simplest counter ICs on the market today. Their simplicity does not restrict their usefulness to simple applications, however; it happens they are among the most popular counter packages in the digital business.

Aside from its inherent simplicity, this particular family is noted for two unique features: The output of FF-A is *not* internally connected to the CLK input of FF-B, and the family includes a special divide-by-12 counter.

The general block diagram for the 7490, '92 and '93 family in Fig. 9-1 shows that the output of FF-A is separated from the other three, thus giving the user the option of altering the modulus of the counter by clocking either the first or second flip-flop from an external source. The significance of this option is explained later in this section.

The 7492 counter package is unique in that it is internally wired for modulo-12 counting if clocked at FF-A and the *A* output is externally connected to CLK

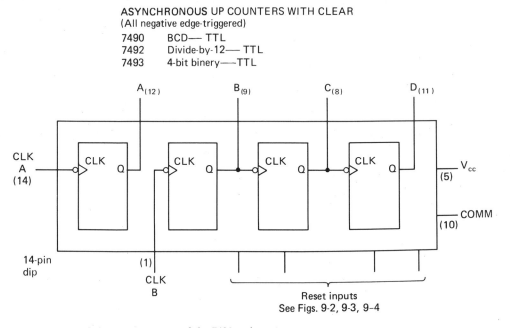

ASYNCHRONOUS UP COUNTERS WITH CLEAR
(All negative edge-triggered)

7490 BCD— TTL
7492 Divide-by-12— TTL
7493 4-bit binery——TTL

Figure 9-1 Basic internal structure of the 7490-series counters.

B (CLK input of FF-B). The same chip can be used for modulo-6 counting, how-ever, if the user passes over the first flip-flop and externally clocks FF-B. The purpose of this particular counter becomes clear when considering the fact that digital time clocks must reset to 0 at the end of 59 min (using a modulo-60 counter).

The outputs of the 7490 family are true. That is, they generate positive-logic outputs whereby a logic-1 output represents a binary 1 numeral.

9-1.1 The 7493 4-Bit Binary Counter

Figure 9-2 summarizes the pinout and some applications notes for the 7493 binary counter. Note in Fig. 9-2(a) that this chip has two RESET or CLEAR inputs designated R_{01} and R_{02}. These two inputs are internally NAND-ed together, creating a logic format where $\text{CLR} = R_{01} \cdot R_{02}$. In plain English, the counter runs normally as long as either or both of the reset inputs are at logic 0. Setting them both to logic 1 at the same time stops the count and resets the outputs to 0.

Figure 9-2(b) shows the external connections necessary for normal 4-bit binary (modulo-16) counting. Note that the external source of trigger pulses is fed to the CLK *A* input, while the output *A* is wired to CLK *B*. The counter can be cascaded with any other counter in the family by connecting the *D* output to the CLK input of the next counter down the line. In other words, the *D* output can serve as the chip's CARRY output, a feature made possible by the fact that the MSB of an up counter shows a negative-going edge only when it is being reset to 0.

It is possible to alter the modulus of the 7493 counter by bypassing the first flip-flop and applying the external source of clock pulses to CLK *B*. In this instance,

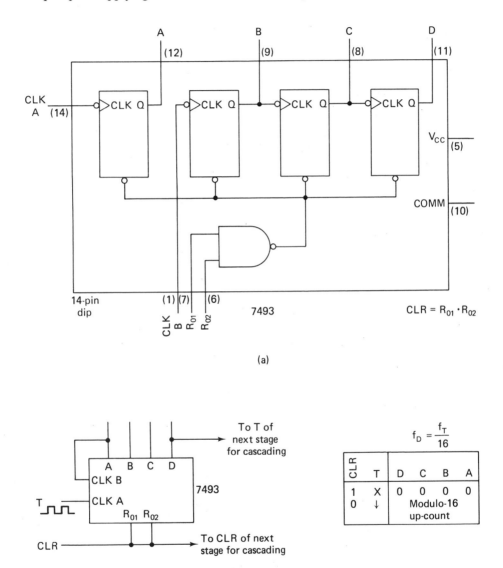

(a)

(b)

Figure 9-2 The 7493 4-bit binary counter IC. (a) Internal logic. (b) Interconnections for modulo-16 counting.

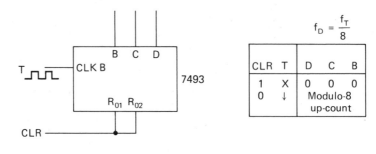

(c)

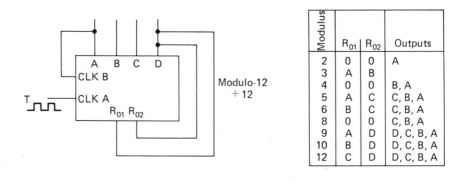

(d)

Figure 9-2 (*cont.*) (c) External connections for modulo-8 counting. (d) External connections for fixed-modulus counting.

the device has a modulus of 8 as seen from the *B*, *C* and *D* output connections. See Fig. 9-2(c).

The fact that the 7493 has a pair of NAND-ed reset inputs makes it rather easy to transform it into a fixed-modulus counter with asynchronous feedback. The circuit arrangement in Fig. 9-2(d), for instance, is wired for modulo-12 counting. As long as the outputs of this particular circuit are showing binary numbers less than binary 12, at least one of the reset inputs is at logic 0. When the count reaches binary 12, however, the reset inputs are both at logic 1, and the counter's output is immediately reset to 0. The table in Fig. 9-2(d) shows the connections between the reset inputs and the counter's outputs that yield ten different modulus formats. Resetting at numbers such as 7 and any number larger than 12 requires three connections to the outputs, and this is not possible with only two reset inputs available. Of course, an external NAND gate can be added to provide three reset inputs.

9-1.2 The 7490 BCD (Decade) Counter

Figure 9-3 shows the pinout and some basic applications procedures for the 7490 BCD counter chip. It is very similar to the 7493 4-bit binary counter in many respects, actually differing only in the basic counting format and the number of reset inputs.

Any BCD or decade counter must automatically reset to 0 after showing a binary count of 9. This is accomplished by internal asynchronous feedback circuitry associated with FF-B, FF-C and FF-D. In other words, FF-B, FF-C and FF-D are internally connected as a modulo-5 counter. So, if the output of FF-A (a modulo-2 counter in its own right) is connected to CLK B (the input of a modulo-5 counter), the overall result is a 2×5 or modulo-10 counter. See the basic modulo-10, BCD counting connections and waveforms in Fig. 9-3(b).

The R_{01} and R_{02} inputs to the 7490 perform exactly the same function as the corresponding reset inputs on the 7493; the counter's output is reset to 0 whenever R_{01} and R_{02} are set to logic 1 at the same time. The BCD counter, however, has an additional pair of reset inputs designated R_{91} and R_{92}. Setting both of these inputs to logic 1 at the same time resets the counter's output to binary 9. This feature is especially useful when working with 9's complement numerals in arithmetic logic circuits. For most purposes, however, the reset-to-9 inputs are simply tied to ground.

Connected as BCD or decade counters, 7490s can be cascaded to produce decade counter circuits having any desired number of decades. The circuit in Fig. 9-3(c) is an example of a four-decade counter that is capable of counting the binary equivalent of decimal numbers 0000 through 9999. Note that the D outputs of each decade serve as the clocking inputs for the stage that follows. Setting the common CLR bus to logic 1 stops the count and resets the outputs to 0.

The circuit and table in Fig. 9-3(d) show how the 7490 can be connected for modulo-5 counting. The circuit simply bypasses FF-A and applies the external source of clock pulses to CLK B.

Of course, it is possible to fix the modulus of the counter to numbers other than 5 or 10 by feeding back various combinations of outputs to the reset-to-0 inputs, but this is not usually done with the 7490 counter. Using the 7493 as shown in Fig. 9-2(d) is better for getting a fixed modulus other than the basic ones.

One of the unique features of the 7490 counter is that it can be externally wired as a *biquinary* counter. This special counting format is achieved by triggering CLK B from an external source and connecting the D output back to the CLK A input. See the diagram, table and waveforms in Fig. 9-3(e). The frequency of the A output in this circuit is equal to one-tenth the clock frequency at CLK B;

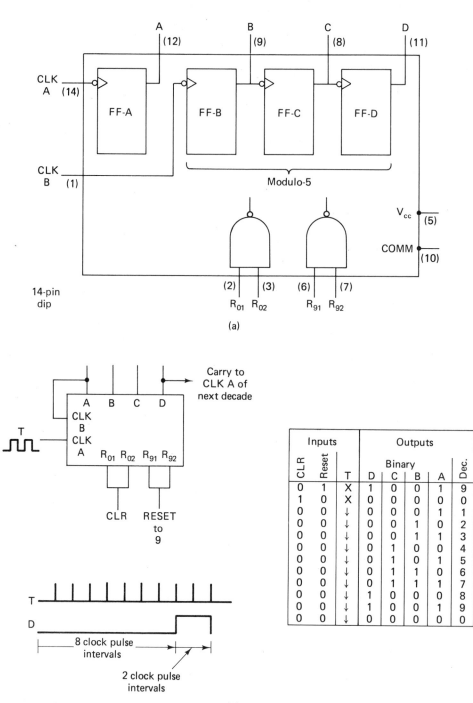

Figure 9-3 The 7490 BCD counter IC. (a) Simplified view of internal logic. (b) External connections for the basic BCD counting.

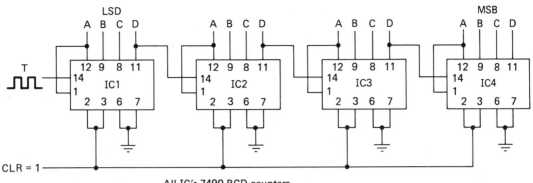

(c)

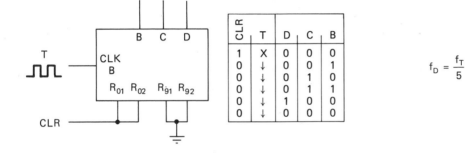

$$f_D = \frac{f_T}{5}$$

(d)

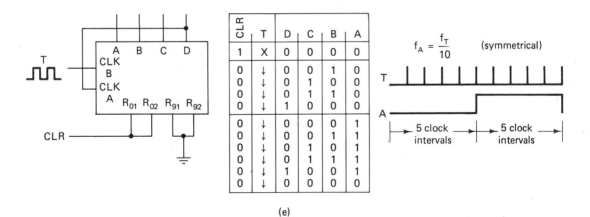

(e)

Figure 9-3 (*cont.*) (c) Cascading technique for a 4-digit BCD counter. (d) External connections and table for modulo-5 counting. (e) External connections and table for biquinary counting or symmetrical divide-by-10 frequency division.

205

but unlike the output of the divide-by-10 counter in Fig. 9-4(b), the biquinary output is perfectly symmetrical; it has a precise 50-percent duty cycle that can be very useful for many digital timing and control operations.

9-1.3 The 7492 Divide-by-12 Counter

Figure 9-4(a) shows the pinout and general internal structure of the 7492 divide-by-12 counter IC. This chip is similar to the 7490 BCD counter in the sense that the three most significant bit positions are internally connected for a fixed modulus, modulo-6 in this instance. Like the 7493, however, this counter has only a single pair of reset inputs, R_{01} and R_{02}.

Connected in its divide-by-12 mode, the 7492 generates an unusual binary output. As shown in the table accompanying the hookup diagram in Fig. 9-4(b), the counter's output skips decimal numerals 6 and 7. The D output does, indeed, generate a symmetrical waveform having a frequency equal to the CLK A frequency divided by 12; but under these circumstances, the circuit should not be used as a counter; it should be used only as a frequency divider.

Figure 9-4(c) shows the 7492 connected for modulo-6 counting. The external source of trigger pulses is fed directly to the CLK B input, bypassing the first

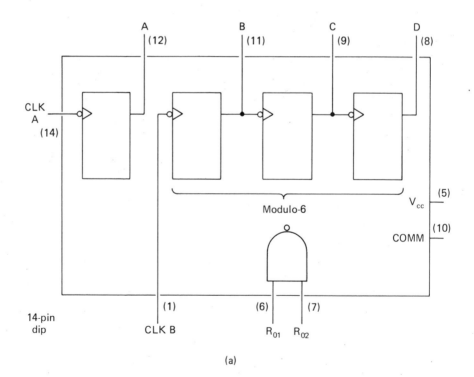

(a)

Figure 9-4 The 7492 divide-by-12 counter IC. (a) Simplified diagram of internal logic.

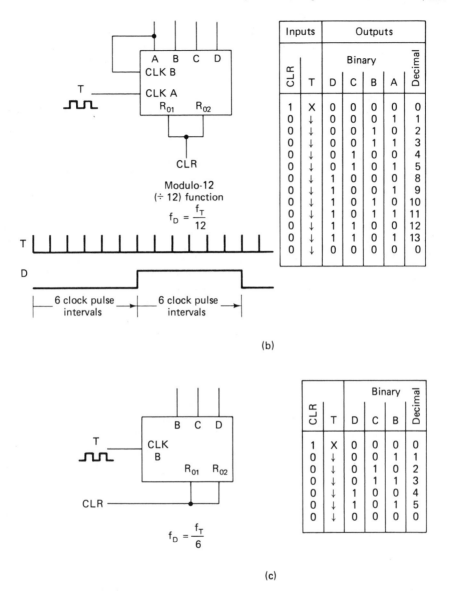

Inputs		Outputs				
		Binary				
CLR	T	D	C	B	A	Decimal
1	X	0	0	0	0	0
0	↓	0	0	0	1	1
0	↓	0	0	1	0	2
0	↓	0	0	1	1	3
0	↓	0	1	0	0	4
0	↓	0	1	0	1	5
0	↓	1	0	0	0	8
0	↓	1	0	0	1	9
0	↓	1	0	1	0	10
0	↓	1	0	1	1	11
0	↓	1	1	0	0	12
0	↓	1	1	0	1	13
0	↓	0	0	0	0	0

Modulo-12
(÷ 12) function

$$f_D = \frac{f_T}{12}$$

6 clock pulse intervals | 6 clock pulse intervals

(b)

		Binary			
CLR	T	D	C	B	Decimal
1	X	0	0	0	0
0	↓	0	0	1	1
0	↓	0	1	0	2
0	↓	0	1	1	3
0	↓	1	0	0	4
0	↓	1	0	1	5
0	↓	0	0	0	0

$$f_D = \frac{f_T}{6}$$

(c)

Figure 9-4 (*cont.*) (b) External connections and table for the basic modulo-12 counting or symmetrical divide-by-12 frequency division. (c) External connections and table for modulo-6 counting.

flip-flop in the package altogether. The table shows that the 3-bit output from B, C and D generates a valid 0-through-5 binary count.

In short, the 7492 can be used as a divide-by-12 frequency divider only [Fig. 9-4(b)] or a modulo-6 counter/frequency divider [Fig. 9-4(c)]. The primary application of this particular chip is demonstrated in the following section.

9-1.4 Using the 7490 Family in Digital Clocks

Before the advent of large-scale MOS clock chips, virtually all digital time clocks on the market were built around the 7490 BCD and 7492 divide-by-12 counter ICs. The simplified clock circuit in Fig. 9-5 illustrates the application of these popular ICs as both frequency dividers and counters.

The input to the circuit is 120 Hz trigger pulses that are derived from a full-wave rectified version of the standard 60-Hz utility line voltage. IC1 and IC2 form a divide-by-120 counter that yields a 1-Hz pulse to the CLK *A* input of IC3.

The BCD output of IC3 advances at a 1-Hz rate, providing a convenient source of numerals for a 1's seconds output. The *D* output of IC3 responds according to the divide-by-10 feature of the 7490, producing one pulse each 10 s to operate the CLK *B* input of IC4.

IC4 is a 7492 that is connected for modulo-6 counting and frequency division. Since its BCD outputs count 0 through 5, and then automatically reset to 0, these outputs represent a 10's seconds output.

Considered together, IC3 and IC4 make up a modulo-60 counter/frequency divider that is suitable for generating numerals representing 0 s through 59 s.

IC5 and IC6 work together as IC3 and IC4 do. The two BCD numerals from IC5 and IC6, however, advance at a rate of one increment per minute, thus making them capable of generating numerals representing 0 min through 59 min. The waveform appearing at the *D* output of IC6 is a pulse that drops from logic 1 to logic 0 once per hour.

The remainder of the circuit handles the "hours" portion of the digital clock system. IC7 is a BCD counter that generates numerals 0 through 9 while the 10's hours output is at 0. Whenever the hours outputs attempt to show binary 13, however, asynchronous reset circuitry, consisting of IC9-A and IC9-B, resets the hours counters back to 0. And each of these reset actions triggers the output of the AM/PM indicator, IC8-B, to its complemented state.

Digital clocks generally include a fast-forward and slow-forward feature that lets the user set the clock by running the counters at faster than normal rates. The setting circuitry isn't included in Fig. 9-5, but doing that particular job is a simple matter of passing the 120-Hz input to the 1's minutes CLK *A* input for fast set or to the CLK A input of the 1's seconds counter for slow set.

9-2 Asynchronous Up Counters With Preset and Clear

Figure 9-6 summarizes the essential characteristics of the 74176, '177, '196, '197 family of asynchronous presettable counters. Generally speaking, these counters are presettable versions of the 7490 and 7493 devices described in the previous section of this chapter. The presettable family does not include a divide-by-12 counter, however.

This family of presettable asynchronous counters is clearly divided into 35-MHz and 50-MHz versions as shown in Fig. 9-6(a), but they all have the same

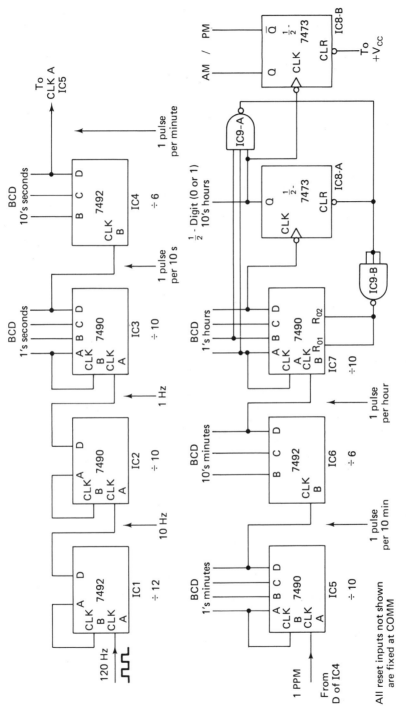

Figure 9-5 Applying the 7490 series ICs in a digital time clock.

ASYNCHRONOUS UP COUNTERS WITH PRESET AND CLEAR

(All negative edge-triggered)
74176   BCD, 35 MHz—TTL
74177   4-bit binary, 35 MHz—TTL
74196   BCD, 50 MHz—TTL
74197   4-bit binary, 50 MHz—TTL

(a)

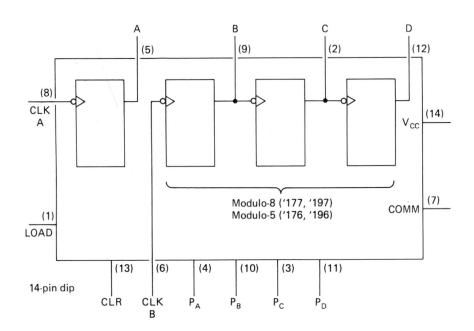

(b)

CLR	LOAD	CLK A or B	Mode
0	X	X	Clear to zero
1	0	X	Load P inputs
1	1	↓	Up count

(c)

Figure 9-6 Asynchronous up counters with preset and clear. (a) List of available IC packages. (b) Basic internal logic structure. (c) Basic function table.

basic internal structure and pinout as shown in Fig. 9-6(b). Of course, the '176 and '196 units have internal circuitry for generating a modulo-5 count at their B, C and D outputs, thus making them function with the first flip-flop as BCD counters.

The circuit and tables in Fig. 9-7 show how to connect these counters for their full 4-bit binary or BCD counting. Note in every instance that the external source of trigger pulses is applied to the CLK A input and that the A output is tied around to the CLK B input.

Setting the CLR input to logic 0 both stops the counting action and resets the outputs to 0. Setting the LOAD input to 0 (while CLR $-$ 1) both stops the count and loads any binary number at the preset inputs (PA through PD) into the counter's output. The CLR and LOAD actions are both asynchronous. Setting CLR and LOAD both to logic 1 permits normal up counting on the negative-going edge of each input T pulse.

Figure 9-8(a) shows how these counters can be connected for modulo-8 or modulo-5 counting. In such instances, the T input is applied directly to the CLK B input, bypassing the first flip-flop in the chip altogether. The 74176 and 74196 BCD counters can be wired for biquinary counting by externally clocking the CLK B input and returning the D output to CLK A. See Fig. 9-8(b).

9-2.1 The 74177, '197 as Modulo-n Counters

It is possible to transform the 74177, '197 4-bit binary counters into modulo-n counters by adding a 4-input NAND gate that automatically loads a preset binary number into the counter whenever the count reaches binary 15. See the circuit in Fig. 9-9(a).

Fixing the CLR input at logic 1 allows normal up counting on each negative-going edge of the T input waveform. The instant the counter reaches 1111 (binary 15) the output of the NAND gate drops to logic 0, satisfying the conditions for loading the binary number at the P inputs. The counter thus shows that binary number at its outputs for one clock-pulse interval, and then it allows the count to increment up through binary 14 again. As the counter attempts to show 1111 at the end of the next T pulse, the NAND gate resets the outputs to the preset number once again. The circuit thus continuously cycles between the preset number and binary 14.

The table in Fig. 9-9(b) summarizes all of the useful preset binary inputs and the resulting modulus and count cycles. Note that the counter's modulus is equal to $15 - P$, where P is the binary number set at the preset inputs. Rearranging this little mathematical relationship to solve for P, it turns out that $P = 15 - n$, where n is the counter's modulus.

The circuit in Fig. 9-9(a) can be used as a modulo-n counter or variable frequency divider. (Recall that a counter divides the input frequency by its modulus.) In this particular circuit, the pulse frequency at output D is equal to the frequency at the T input divided by the counter's modulus. A 24-kHz waveform at T, for

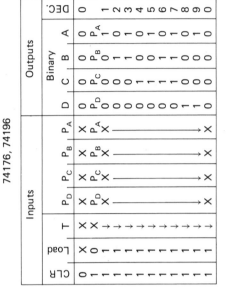

Figure 9-7 Basic external connections and operating tables for the 74170 and 74190 series counter ICs. (a) External connections. (b) 4-bit binary operating table. (c) BCD operating table.

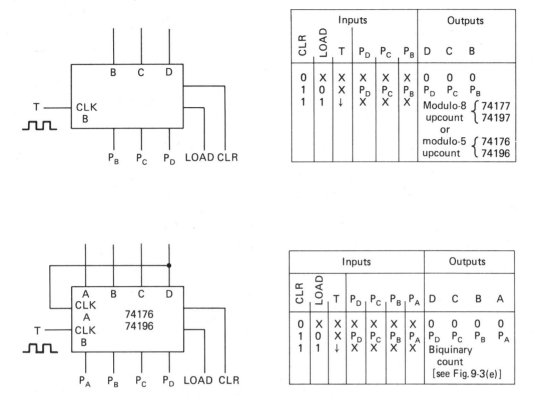

Inputs						Outputs		
CLR	LOAD	T	P_D	P_C	P_B	D	C	B
0	X	X	X	X	X	0	0	0
1	0	X	P_D	P_C	P_B	P_D	P_C	P_B
1	1	↓	X	X	X	Modulo-8 $\begin{cases} 74177 \\ 74197 \end{cases}$ upcount or modulo-5 $\begin{cases} 74176 \\ 74196 \end{cases}$ upcount		

Inputs							Outputs			
CLR	LOAD	T	P_D	P_C	P_B	P_A	D	C	B	A
0	X	X	X	X	X	X	0	0	0	0
1	0	X	P_D	P_C	P_B	P_A	P_D	P_C	P_B	P_A
1	1	↓	X	X	X	X	Biquinary count [see Fig.9-3(e)]			

Figure 9-8 Some special fixed-modulus applications of the 74170 and 74190 series counters. (a) Modulo-8 or modulo-5 counting. (b) Symmetrical divide-by-10 or biquinary counting.

instance, would appear as a 4-kHz waveform at output D if the preset inputs are set at binary 9. This feature has powerful implications as far as modern frequency synthesizers are concerned: This circuit is capable of transforming any one input frequency into 14 different output frequencies.

9-2.2 Fixed-Modulus Stop Counters

Digital systems often call for a counting operation that does not cycle continuously. Instead, the circuit begins counting from 0 and increments up to a certain number at which the counting action stops until the circuit is reset at some later time. Figure 9-10(a) shows how this can be done with a presettable counter such as the ones featured in this section.

The basic idea is to allow the counter to run normally up to a point at which it loads its own outputs into its preset inputs. Note in Fig. 9-10(a) that the A, B, C and D outputs are connected directly to their corresponding preset inputs: P_A, P_B, P_C and P_D.

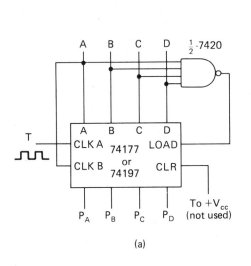

Preset inputs						
		Binary			Modulus (n)	Count cycle
Dec.	P_D	P_C	P_B	P_A		
0	0	0	0	0	15	0 → 14
1	0	0	0	1	14	1 → 14
2	0	0	1	0	13	2 → 14
3	0	0	1	1	12	3 → 14
4	0	1	0	0	11	4 → 14
5	0	1	0	1	10	5 → 14
6	0	1	1	0	9	6 → 14
7	0	1	1	1	8	7 → 14
8	1	0	0	0	7	8 → 14
9	1	0	0	1	6	9 → 14
10	1	0	1	0	5	10 → 14
11	1	0	1	1	4	11 → 14
12	1	1	0	0	3	12 → 14
13	1	1	0	1	2	13 → 14
14	1	1	1	0	Invalid	
15	1	1	1	1	Invalid	

(a)

(b)

Figure 9-9 Fixed-modulus applications. (a) External connections. (b) Operating truth table.

The NAND gate is used for detecting the binary count where the counter is supposed to stop. If the inputs to the NAND gate are connected to counter outputs *D* and *C*, for example, the counter will increment normally between 0 and binary 11. Upon reaching the count of 12, however, the output of the NAND gate drops to logic 0, loading the counter with its own output and effectively freezing the count until the CLR input is momentarily dropped to logic 0. Dropping the CLR input to logic 0 resets the counter's outputs to 0, allowing the output of the NAND gate to return to logic 1. When the CLR input is set to logic 1 again, counting begins from 0 and progresses up to 12 where it stops. See the waveforms in Fig. 9-10(b).

Of course, the stop-count code fed to the inputs of the NAND gate must be within the counting capacity of the counter circuit. A BCD counter, for instance, cannot possibly be used for stop-counting operations involving numbers greater than 9.

A stop counter like this can also be used for generating a waveform having a fixed number of pulses in it. Suppose the circuit is fixed for a stop count at binary 12 as in the previous example. Such a circuit would require exactly 12 input clock pulses at the *T* input to carry out its full cycle of operation, incrementing from 0 to the stop count of 12. In the process of running the count from 0 to 12, the *A* output would generate exactly 6 clock pulses: The *A* output would be low at the beginning of the count and it would be low at the end of the count, but 6 full pulses appear during the counting interval. The actual number of pulses appearing at the

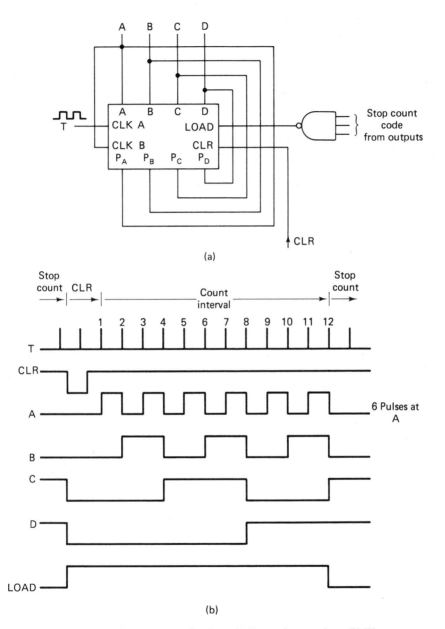

(a)

(b)

Figure 9-10   Stop-count applications. (a) External connections. (b) Waveforms for a fixed stop-count of 12.

A output between the time the circuit is cleared and it reaches its stop-count number depends on the stop code fed to the inputs of the NAND gate. Generally speaking, the number of pulses in the *A* output pulse waveform is equal to $k/2$, where k is the stop-count number.

9-3 Synchronous Up Counters With Preset and Clear

This section introduces synchronous counter IC packages. As described in Sec. 8-4, synchronous counters are noted for their higher operating frequencies relative to purely asynchronous, or ripple, versions. Synchronous counter ICs also include enable inputs and carry outputs that permit fully synchronous cascading of the counter packages.

Figure 9-11 summarizes the main features of the '160 series counters, one of the two major families of synchronous counter IC devices. This particular family can be subdivided in several different ways:

TTL or CMOS—The counters with a 74 prefix are built around a TTL technology; those with a 40 prefix are built around CMOS devices. The counters are otherwise identical, with the 74160, for instance, having the same general operating features and pinout as the 40160.

BCD or 4-bit binary—The '160 and '162 counters are internally wired for synchronous BCD counting; the '161 and '163 versions count out a full modulo-16, 4-bit binary code.

Asynchronous or synchronous CLEAR—All members of this counter family have synchronous PRESET inputs that allow any data at the preset inputs to be loaded into the counter only on the positive-going edge of the input clock pulse. Only the '162 and '163 ICs have a similar synchronous clearing feature. The '160 and '161 devices have an asynchronous CLEAR input that clears the outputs to 0, regardless of the status of the CLK input.

All members of this counter family have the pinout shown in Fig. 9-11(b). Outputs A, B, C and D are the counters' binary outputs; CLK is the clock-pulse input; inputs P_A, P_B, P_C and P_D are the preset data inputs; and L and CLR are the preset and clearing inputs, respectively.

These counters are unique because they feature two separate enable inputs, ENT and ENP. Setting either of these inputs to logic 0 stops the counting action asynchronously. The two count-enable inputs are different in that setting ENT to logic 0 not only stops the count, but it also inhibits the generation of a logic-1 level at the RC (RIPPLE CARRY) output whenever the counter reaches its highest count (binary 9 for BCD counters and binary 15 for 4-bit binary versions).

Figure 9-11(c) is a basic function table for this series of synchronous counter ICs. Note especially that the counters are all positive edge-triggered; the outputs change state and synchronous loading and clearing take place on the positive-going edge of the input CLK pulse. This is a marked difference from asynchronous counters that usually change state on the negative-going edge of their input CLK pulses.

9-3.1 Full-Modulus Up Counting

Any of the counters in this series can be programmed for full-modulus up counting by fixing L, ENT and ENP to logic 1 and applying an external source of trigger pulses to the CLK input. Since conditions at the P inputs aren't relevant

SYNCHRONOUS UP COUNTERS WITH PRESET AND CLEAR
(All positive edge-triggered)

74160	BCD with synchronous PRESET and asynchronous CLEAR—TTL
74161	4-bit binary with synchronous PRESET and asynchronous CLEAR—TTL
74162	BCD with synchronous PRESET and CLEAR—TTL
74163	4-bit binary with synchronous PRESET and CLEAR—TTL
40160	CMOS version of 74160
40161	CMOS version of 74161
40162	CMOS version of 74162
40163	CMOS version of 74163

(a)

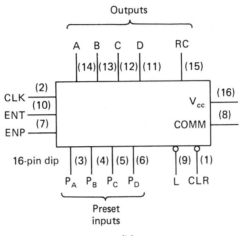

(b)

LOAD	ENP	ENT	CLR	CLK	Mode
0	X	X	1	↑	Preset
1	0	X	1	↑	Stop count
1	X	0	1	↑	Stop count, disable RC
X	X	X	0	See note 1	Reset to zero
1	1	1	1	↑	Up count (see note 2)

Note 1: CLK = X for '160, '161
CLK = ↑ for '162, '163

Note 2: Modulo-10 (BCD) for '160, '162
Modulo-15 (4-bit binary) for '161, '163

(c)

Figure 9-11 The 74160 and 40160 series counter ICs. (a) List of available counter devices. (b) Basic I/O format and pinout. (c) Basic operating truth table.

in this case, the counter's outputs (A, B, C and D) increment on every positive-going edge of the input CLK pulse.

Clearing the outputs to 0 is a matter of applying a logic-0 level at the CLR input. The counting stops and the outputs go to 0 immediately in the case of those counters having the asynchronous reset feature. Those having the synchronous reset feature will not clear to 0 until the next positive-going clock-pulse edge occurs. When attempting to clear the outputs of the counters having the synchronous clearing feature, the CLR input should be set to logic 0 while the CLK pulse is at logic 0. Otherwise, the count will increment one more number before the clearing action takes place.

Dropping either ENT or ENP to logic 0 stops the counting action immediately, leaving the last number showing at the A, B, C and D outputs.

9-3.2 Presetting the Up Count

Any binary number appearing at the P inputs can be loaded into the counter's output by setting the L input to logic 0. Since the loading feature is synchronous in all cases, the binary number at the P inputs will not appear at the outputs until the next positive-going edge of the input CLK pulse occurs.

A number loaded into the counter via the preset inputs remains fixed at the output until L is set to logic 1 and the next input CLK pulse occurs.

Returning the RC output through an inverter to the L input permits modulo-n up counting. The counter will increment up to its maximum count, at which time the RC output goes to logic 1. This output is inverted to a logic 0, then, preparing the circuit for preset loading on the next CLK pulse. At the moment the next positive-going CLK pulse edge occurs, the counter's outputs are loaded to the P number; and then remain at that number until the next clock pulse appears at the CLK input. When one of the 4-bit binary counters is used in this fashion, the modulus of the counter is equal to $16 - P$, where P is the binary number set at the P inputs.

When the modulo-n configuration is used as a frequency divider, it turns out that the frequency of the waveform at the RC output is equal to the input clock frequency divided by the modulus of the counter.

9-3.3 Fully Synchronous Cascading

All of the counters in this '160 series devices can be rather easily cascaded by disabling the ENT and ENP inputs (connecting them to logic 1) and connecting the RC output of one stage to the CLK input of the next. This is asynchronous cascading, however; and that means inserting at least one propagation delay between each counter stage. This is properly known as *semi-synchronous* counting: The counters, themselves, are synchronous, but the cascading circuitry is asynchronous.

The roles of the ENT, ENP and RC connections become apparent when connecting the counters for fully-synchronous cascading. See the 3-counter cascading circuit in Fig. 9-12.

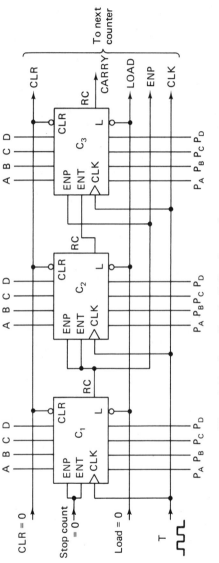

Figure 9-12 Cascading the '160 series synchronous counters.

Counter C1 is permitted to count continuously as long as its ENP and ENT inputs are fixed at logic 1. The *RC* output of C1 is normally at logic 0, however, effectively disabling counter package C2. Counter C2 is allowed to increment only when the *RC* output of C1 is at logic 1 (indicating a maximum count from C1) and the system sees a positive-going edge of the input clock pulse, *T*. The outputs of all the counters thus change state at precisely the same time; there is no asynchronous ripple effect to slow down the counting operation and generate counting spikes in the supply voltage.

9-4 Synchronous Up/Down Counters With Preset

Figure 9-13 summarizes the essential characteristics of a pair of "universal" counter ICs: the 74190 and the 74191. Both units are fully synchronous, they have preset inputs and they are capable of either up or down counting. The only real differece between these two counters is that the 74190 counts a modulo-10, BCD sequence and the 74191 counts a full modulo-16, 4-bit binary sequence.

The circuits are enabled—allowed to count—as long as the ENAB input is at logic 0 and *L* (load) input is at logic 1. According to the function table accompanying the pinout diagram in Fig. 9-13(a), the direction of count is determined by the logic level appearing at the U/D input terminal: Up counting occurs when U/D = 0 and down counting occurs when U/D is set to logic 1.

The count can be asynchronously stopped and held by setting the ENAB input to logic 1; and by the same token, any valid binary number appearing at the *P* inputs can be loaded into the counter by setting the *L* input to logic 0.

The RC (ripple clock) and MAX/MIN output combination is unique to these two counter ICs. The primary purpose of the MAX/MIN output is to detect the counters' maximum or minimum count. Whenever the counter is up counting, for instance, the MAX/MIN output is normally low, rising to logic 1 only during the interval the counting outputs show binary 15 (74191 4-bit counter) or binary 9 (74190 BCD counter). See the waveforms in Fig. 9-13(b). If, however, the circuit is set for down counting, the MAX/MIN output remains at logic 0 until the counter shows a 0 output.

Whether the MAX/MIN output goes high at the counters' largest output count or at 0 depends on which direction the counters are running. It serves the purpose of an overflow detector while up counting and an underflow detector while down counting.

The RC output works in much the same way as the MAX/MIN output, but it is the result of NAND-ing MAX/MIN with CLK. The RC output is thus normally high, dropping to logic 0 only when the counter reaches a MAX/MIN point and the CLK input is low. See the RC waveforms in Fig. 9-13(b).

The circuit in Fig. 9-14(a) is a basic demonstrator circuit for the 74190/'191 IC counters. The circuit is clocked from an external source of trigger pulses such as those generated by the 555-type oscillator circuit in Fig. 8-2. The U/D and ENAB

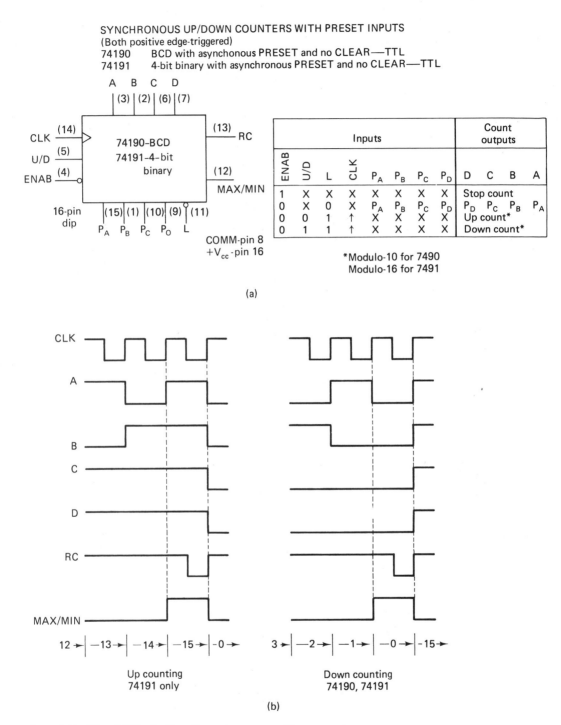

SYNCHRONOUS UP/DOWN COUNTERS WITH PRESET INPUTS
(Both positive edge-triggered)
74190   BCD with asynchonous PRESET and no CLEAR—TTL
74191   4-bit binary with asynchronous PRESET and no CLEAR—TTL

	Inputs								Count outputs			
ENAB	U/D	L	CLK	P_A	P_B	P_C	P_D	D	C	B	A
1	X	X	X	X	X	X	X	Stop count			
0	X	0	X	P_A	P_B	P_C	P_D	P_D	P_C	P_B	P_A
0	0	1	↑	X	X	X	X	Up count*			
0	1	1	↑	X	X	X	X	Down count*			

*Modulo-10 for 7490
Modulo-16 for 7491

(a)

Up counting
74191 only

Down counting
74190, 74191

(b)

Figure 9-13 The 74190 family of synchronous up/down counters.
(a) Basic I/O format and pinout. (b) MAX/MIN and RIPPLE CARRY
output waveforms.

221

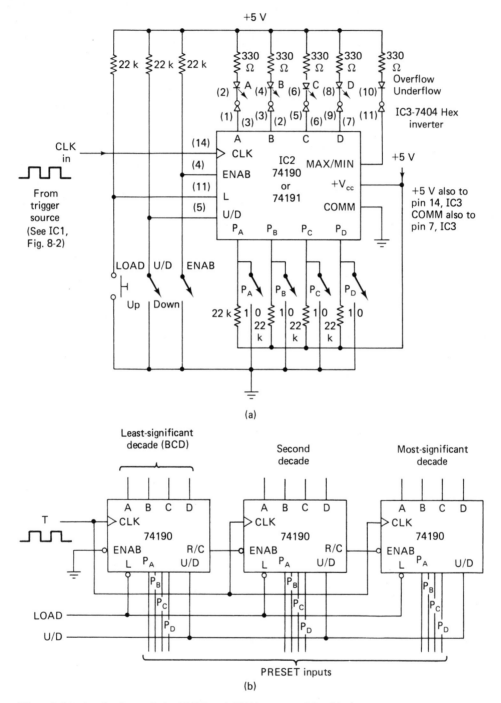

(a)

(b)

Figure 9-14 Applications of the 74190 and 74191 counter ICs. (a) A practical demonstration circuit. (b) Cascading the 74190 BCD counters.

switches let the user select the direction of counting and start and stop the count at will. The counter can be preset to any valid number by first setting the desired binary number on the P switches and then depressing the LOAD push button.

Since the outputs of these counter circuits are active-high (true), they must be inverted before applying them to the cathodes of LED indicator lamps. This inversion operation is carried out by means of the logic inverters in a 7404 hex inverter IC package.

The LED indicators in Fig. 9-14(a) thus indicate the binary output count. The OVERFLOW/UNDERFLOW indicator lamp responds to the counter's MAX/MIN output, either lighting to indicate an overflow while up counting or lighting to indicate an underflow while down counting.

The circuit in Fig. 9-14(b) shows how three 74190 BCD counters can be cascaded for fully synchronous decade counting. Note that the CLK inputs are all tied together at a common CLK bus. The RC output of each counter goes directly to the ENAB input of the following stage, effectively disabling the counter until the preceding one is showing its maximum (or minimum) output count.

The counters can be reset to 0 only by loading 0000 at the P inputs.

9-5 Synchronous Up/Down Counters With Preset and Clear Inputs

The 74192, 74193 family of counter ICs is summarized in Fig. 9-15. They are very similar to the 74190 and 74191 counters in many respects, but there are some important differences that call for special explanation.

First note that these BCD and 4-bit binary counters are available in both TTL and CMOS packages. The 40192, for instance, is a CMOS version of the 74192 TTL BCD counter.

According to the function table in Fig. 9-15, these counters count up or down, depending on which CLK input is being driven by an external source of clocking pulses. The unused CLK input must be pulled up to logic 1 before proper up or down counting can take place.

The counters can be asynchronously cleared by pulling the CLR input up to logic 1; and they can be likewise loaded with any valid binary number appearing at the P inputs by setting the L (load) input terminal to logic 0.

The CARRY and BORROW outputs work much the same way as the RC (ripple carry) output featured on the 74190 and 74191 counter ICs described in Sec. 9-4. These two outputs are normally at logic 1; but the CARRY output drops to logic 0 only when (1) the counter is operating from the CLK UP input, (2) the counter is showing its maximum-count numeral and (3) the CLK UP input is at logic 0. The BORROW output remains at logic 1 as long as the circuit is operating from the CLK UP input.

Running the counter in its down-count mode (by clocking the CLK DN input) enables the BORROW function and disables the CARRY function just described.

SYNCHRONOUS UP/DOWN COUNTERS WITH PRESET AND CLEAR
(All positive edge-triggered)

74192	BCD with asynchronous PRESET and CLEAR—TTL
74193	4-bit binary with asynchronous PRESET and CLEAR—TTL
40192	CMOS version of 74192
40193	CMOS version of 74193

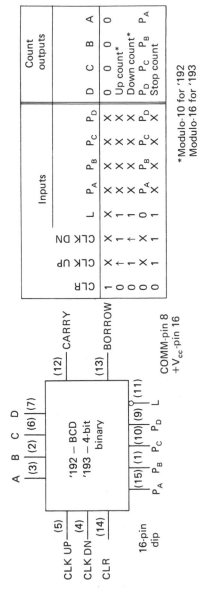

Inputs								Count outputs			
CLR	CLK UP	CLK DN	L	P_A	P_B	P_C	P_D	D	C	B	A
1	X	X	X	X	X	X	X	0	0	0	0
0	↑	1	1	X	X	X	X	Up count*			
0	1	↑	1	X	X	X	X	Down count*			
0	X	X	0	P_A	P_B	P_C	P_D	P_D	P_C	P_B	P_A
0	1	1	1	X	X	X	X	Stop count			

*Modulo-10 for '192
Modulo-16 for '193

CLK UP (5)
CLK DN (4)
CLR (14)

A B C D
(3) (2) (6) (7)

'192 — BCD
'193 — 4-bit binary

(12) CARRY
(13) BORROW

(15) (1) (10) (9) (11)
P_A P_B P_C P_D L

16-pin dip

COMM-pin 8
+V_{cc}-pin 16

Figure 9-15 The 74192 and 74193 counters.

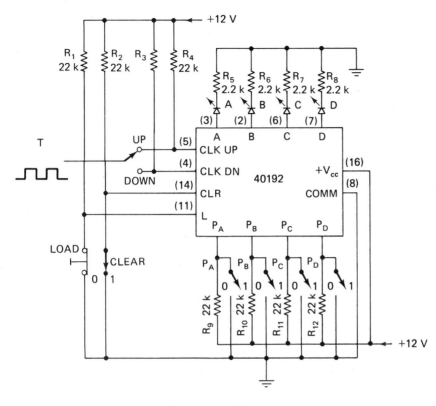

(a)

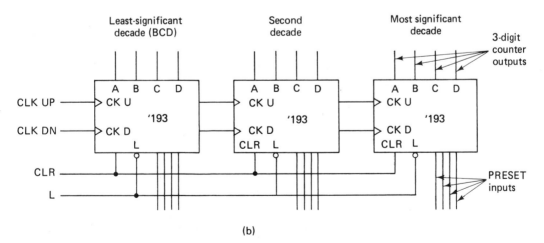

(b)

Figure 9-16 Applications of the 74192 and 74193 counter ICs. (a) A practical demonstration circuit. (b) Cascading the 74193 BCD counters.

In this case, the BORROW output remains at logic 1 until the count reaches 0 and the CLK DN input falls to logic 0.

Figure 9-16(a) shows a circuit for demonstrating the essential features of these counter circuits. The counter in this instance is the 40193 CMOS 4-bit binary version. Of course, the power supply voltage, output logic levels and the values of the LED dropping resistors can be modified to suit the 74194 TTL version.

The U/D switch determines whether the circuit counts up or down. Note that the unused CLK input is always pulled up to $+V_{cc}$ potential. The counter can be cleared to 0 at any time by setting the CLR switch to its logic-1 position; and any valid binary number entered on the P switches can be loaded into the counter by depressing the LOAD push button.

The circuit can be made into a fixed-modulus counter by applying the appropriate outputs, A, B, C and D, to the inputs of a NAND gate and connecting the output of that gate to the L input. Modulo-12 up counting is possible by connecting outputs A, B and D to the NAND gate and setting 0000 at the P inputs. The count will then increment from 0 through binary 11 and then automatically reset (load) to 0 to begin the cycle all over again.

Suppose, however, the user wants a fixed modulo-12 down-counting action. In this instance, outputs A, B, C and D go to the inputs of the NAND gate, the output of the NAND gate goes to the L input, and the P inputs are programmed with binary 11. Clocking the CLK DN input, the count begins at binary 11 and decrements through 0. When the outputs attempt to show binary 15 on the next positive-going edge of the CLK DN pulse, it automatically resets (loads) binary 11 to begin the cycle again.

Figure 9-16(b) shows how to interconnect these counters for fully synchronous, cascaded counting. Note that the BORROW outputs are connected to the CLK DN inputs of the following counter stages. Similarly, the CARRY outputs are connected to the following CLK UP input. Although this arrangement might appear to be semi-synchronous, it is not. The CARRY and BORROW outputs are normally at logic 1; and as shown in the last line of the function table in Fig. 9-15, feeding these 1's to the CLK inputs of the next counter effectively disables its count. The second and third counters in this circuit are enabled only during the one clock-pulse interval that the CARRY or BORROW outputs are low.

Exercises

1. What is the simplest procedure for externally connecting the 7493 4-bit binary up counter for modulo-5 counting?

2. How is it possible to wire the 7493 4-bit binary counter for symmetrical divide-by-10 counting?

3. How is it possible to add some external circuitry to make the 7493 4-bit binary counter into a binary down counter?

4. What is the primary application of the 7492 divide-by-12 counter IC?

5. What is unusual about the basic counting format for the 7492 divide-by-12 counter?

6. What are the similarities and differences between the CLEAR and LOAD functions of a digital counter?

7. What is the main distinguishing feature of a cascaded, fully synchronous counter chain?

8. Why are the logic inverters necessary with the circuit in Fig. 9-14(a) but not necessary with the circuit in Fig. 9-16(a)?

CODE CONVERTERS
AND DISPLAY DRIVERS

Digital systems, no matter how simple or complex they might be, are virtually useless without some interfacing mechanisms between them and the real world. Human operators must be able to communicate with the circuits at the input end, and the system must be able to present its output information in a fashion that is intelligible to the human operator.

One of the most important jobs in this matter of I/O (input/output) interfacing involves translating one kind of digital code into another. At the input end, for instance, the user might want to enter information or instructions into the circuit by depressing one of ten switches labeled 0 through 9. It would be far simpler for the circuit, however, to deal with that information in terms of a 4-bit BCD code, as opposed to a more complex 10-line decimal code. Hence the need for a decimal-to-BCD code converter at the circuit's input interface.

Looking at the output end of the process, suppose that same circuit generates its output information in a BCD format. Experienced digital engineers and technicians have little trouble interpreting BCD-coded outputs, but most users would rather see the output information displayed as 7-segment decimal numerals; hence the need for a BCD-to-7 segment code converter.

These are only a couple of examples of I/O code conversion operations, but they are among the most common ones in use today. The converters described in this chapter are decimal-to-BCD, BCD-to-binary, binary-to-BCD, BCD-to-decimal and BCD-to-7 segment converters.

There are, of course, more elaborate code converter systems, such as input converters that translate the depression of one key of an alphanumeric keyboard into a standard 7-bit ASCII code, and there are output converters that translate 7-bit ASCII into a format suitable for displaying the characters on a TV screen. Such converters are rather complex, however, calling for MOS LSI circuits that are beyond the scope of this book.

And as a final introductory note, it must be pointed out that there is no real reason why the code converters described in this chapter must be used only for I/O code-translation operations. As shown in Fig. 10-10, some of these code converters find their way into data selector and control circuits deep within a digital system. This chapter emphasizes I/O code conversion operations, however.

10-1 Decimal-to-BCD Code Conversion

One of the most common and useful input mechanisms for digital circuitry is a set of switches that generates 1 or 0 logic levels in response to turning them "on" or "off." Very often the situation involves energizing one of ten different switches that represents a decimal numeral between 0 and 9; but for the sake of convenience and economy, the ten separate data lines required for a full unit of decimal counting is converted to a simpler 4-line BCD format.

The table in Fig. 10-1(a) shows how ten active-high inputs representing decimal numerals 0 through 9 can be converted into active-high or active-low BCD outputs. Energizing the "3" decimal input, for example, ought to yield a BCD output of 0011 (active-high binary 3) or 1100 (active-low binary 3).

The logic equations accompanying the table in Fig. 10-1(a) show that the process is basically an OR-ing operation. For active-high BCD outputs, for instance, output A (the least-significant bit) goes to logic 1 whenever inputs 1, 3, 5, 7 or 9 are energized. Similarly, output B should go to logic 1 whenever inputs 2, 3, 6 or 7 are energized.

The circuit in Fig. 10-1(b) is the classic diode matrix circuit for decimal-to-BCD conversion. The diodes, arranged in this particular matrix pattern, perform the OR-ing operations prescribed by the logic equations in Fig. 10-1(a). Both the inputs and outputs are active-high. This circuit has been around for many, many years, and it can still be found in a lot of operational digital equipment.

In the context of this age of digital integrated circuits, it would seem more appropriate to build a decimal-to-BCD converter from OR logic gates or their NAND equivalents. Figure 10-1(c) shows how NAND gates can be applied to this particular conversion operation. Using NAND gates for performing OR-ing functions calls for inverted data inputs; therefore, this particular circuit has active-low inputs and active-high outputs.

The most convenient way to perform the decimal-to-BCD conversion operation today involves the application of a 74174 TTL priority encoder IC. As its name implies, this device was originally designed for ordering the priority of nine different inputs. The truth table accompanying the pinout diagram in Fig. 10-1(d)

Decimal	Active-high inputs										Active-high output				Active-low output			
	0	1	2	3	4	5	6	7	8	9	D	C	B	A	D	C	B	A
0	1	0	0	0	0	0	0	0	0	0	0	0	0	0	1	1	1	1
1	0	1	0	0	0	0	0	0	0	0	0	0	0	1	1	1	1	0
2	0	0	1	0	0	0	0	0	0	0	0	0	1	0	1	1	0	1
3	0	0	0	1	0	0	0	0	0	0	0	0	1	1	1	1	0	0
4	0	0	0	0	1	0	0	0	0	0	0	1	0	0	1	0	1	1
5	0	0	0	0	0	1	0	0	0	0	0	1	0	1	1	0	1	0
6	0	0	0	0	0	0	1	0	0	0	0	1	1	0	1	0	0	1
7	0	0	0	0	0	0	0	1	0	0	0	1	1	1	1	0	0	0
8	0	0	0	0	0	0	0	0	1	0	1	0	0	0	0	1	1	1
9	0	0	0	0	0	0	0	0	0	1	1	0	0	1	0	1	1	0

$$A = 1 + 3 + 5 + 7 + 9$$
$$B = 2 + 3 + 6 + 7$$
$$C = 4 + 5 + 6 + 7$$
$$D = 8 + 9$$

Logic equations

(a)

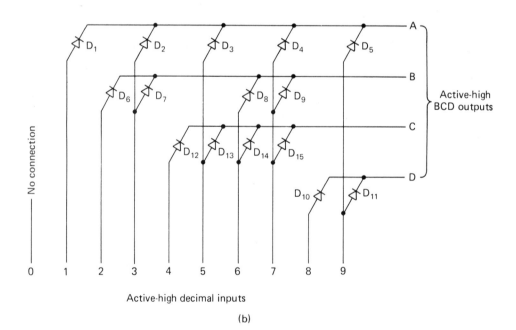

(b)

Figure 10-1 Decimal-to-BCD conversion. (a) Basic truth table and logic equations. (b) Conventional diode matrix circuit.

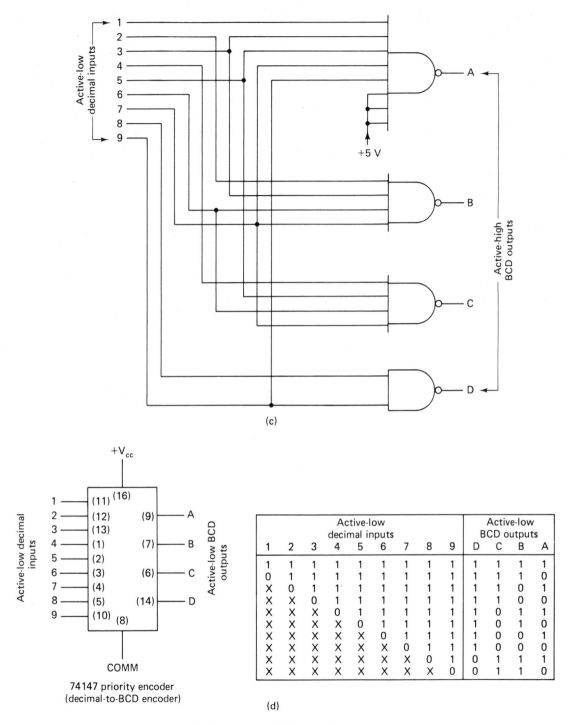

(c)

74147 priority encoder
(decimal-to-BCD encoder)

	Active-low decimal inputs									Active-low BCD outputs			
1	2	3	4	5	6	7	8	9	D	C	B	A
1	1	1	1	1	1	1	1	1	1	1	1	1
0	1	1	1	1	1	1	1	1	1	1	1	0
X	0	1	1	1	1	1	1	1	1	1	0	1
X	X	0	1	1	1	1	1	1	1	1	0	0
X	X	X	0	1	1	1	1	1	1	0	1	1
X	X	X	X	0	1	1	1	1	1	0	1	0
X	X	X	X	X	0	1	1	1	1	0	0	1
X	X	X	X	X	X	0	1	1	1	0	0	0
X	X	X	X	X	X	X	0	1	0	1	1	1
X	X	X	X	X	X	X	X	0	0	1	1	0

(d)

Figure 10-1 (*cont.*) (c) NAND gate circuit. (d) 74147 priority encoder as a decimal-to-BCD encoder circuit.

shows that energizing any one of the inputs (active-low input format) overrides the effect of energizing any other inputs having a numeric designation. Setting input 5 to logic 0, for instance, produces a binary 5 output, even if any of the lower-numbered inputs (4, 3, 2 and 1) are energized at the same time.

Although priority encoding is a highly specialized operation that is normally reserved for internal digital control and logic functions, the fact that this IC also generates valid, active-low BCD outputs makes it quite suitable for input decimal-to-BCD code conversion.

The demonstration circuit in Fig. 10-2 takes advantage of the code-conversion characteristics of the 74147 priority encoder IC. Note the use of pull-up resistors at the switch inputs. These resistors ensure a good logic-1 input until a switch is closed. Closing one of the input switches pulls the corresponding input down to logic 0, the active condition for this active-low input format.

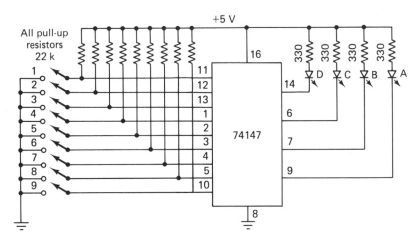

Figure 10-2 A practical decimal-to-BCD encoder using the 74147 priority encoder.

The active-low nature of the circuit's output makes it rather easy to interface it with LED indicator lamps. Whenever any one of the outputs goes to logic 0, the corresponding LED lights up; so as far as the user is concerned, depressing one of the decimal input switches causes the LEDs to indicate the corresponding BCD version of that numeral.

10-2 BCD-to-Decimal Conversion

Digital engineers and technicians soon learn to read raw BCD codes rather well, but a vast majority of the people in the world do not care to read the outputs of digital equipment in that particular format. Just as most people prefer to enter numerals into a digital system via a decimal format, they also prefer to see the outputs in a decimal format. One approach to this situation is an output BCD-to-decimal conversion scheme.

The truth table in Fig. 10-3 summarizes the logic relationships between any active-high BCD input and active-high or active-low outputs.

The basic scheme involved in the BCD-to-decimal conversion process is a set of logic AND-ing functions. Active-high decimal 6, for instance, is equal to $\bar{D}CB\bar{A}$, while a decimal 7 active-high output is equal to $\bar{D}CBA$. The job can be done in a rather straightforward manner with ten separate 4-input NAND gates. Such a circuit would directly yield active-low outputs, but it is far simpler and less expensive to use one of the IC devices especially designed for BCD-to-decimal conversion.

Table 10-1 lists the more popular BCD-to-decimal converter ICs on the market today. Note that most of the TTL converter packages are listed as decoder/drivers, implying that they are capable of both decoding the BCD inputs and driving output loads at higher than normal power levels. These decoder/driver packages feature open-collector outputs that can sink as much as 80 mA in two instances and operate loads having supply voltages as high as $+60$ V in another instance. In spite of their unusually high output voltage and current ratings, the internal decoding logic for these TTL circuits requires the normal $+5$-V power supply.

The 60-V output rating of the 74141 BCD-to-decimal decoder/driver makes it especially useful for driving neon-filled Nixie™ tubes. Such tubes have cathodes bent into the shape of decimal numerals 0 through 9 and a single anode. See the circuit in Fig. 10-4(a). The IC decodes the active-high BCD input, pulling down one of the active-low outputs near ground potential. This action creates a 50-V potential difference between the energized cathode and the anode, so that particular cathode lights up to display its decimal numeral. Note that the 74141 still requires a $+5$-V power connection at V_{cc}.

Table 10-1 AVAILABLE BCD-TO-DECIMAL CONVERTER/DRIVER ICs

7442	BCD-to-decimal decoder; totem-pole output: TTL
7445	BCD-to-decimal decoder/driver; open-collector output rated at 30 V, 80-mA sink current: TTL
74141	BCD-to-decimal decoder/driver; open-collector output rated at 60 V, 7-mA sink current: TTL
74145	BCD-to-decimal decoder/driver; open-collector output rated at 15 V, 80-mA sink current: TTL
4028	BCD-to-decimal decoder; active-high outputs rated at V_{DD}, 8-mA sink or source current: CMOS

Notes: 1. All TTL decoders have active-high inputs and active-low outputs.
2. See Appendix A for pinouts and output responses to invalid (binary 10 through 15) inputs.

The demonstration circuit in Fig. 10-4(b) uses the 7442 BCD-to-decimal decoder package, the only one in the entire TTL family that has a normal totem-pole output. The decimal outputs from this IC are active-low, making them suitable for driving the cathodes of a common-anode LED display. The BCD inputs are

Decimal	Active-high BCD inputs				Active-high decimal outputs										Active-low decimal outputs									
	D	C	B	A	0	1	2	3	4	5	6	7	8	9	0	1	2	3	4	5	6	7	8	9
0	0	0	0	0	1	0	0	0	0	0	0	0	0	0	0	1	1	1	1	1	1	1	1	1
1	0	0	0	1	0	1	0	0	0	0	0	0	0	0	1	0	1	1	1	1	1	1	1	1
2	0	0	1	0	0	0	1	0	0	0	0	0	0	0	1	1	0	1	1	1	1	1	1	1
3	0	0	1	1	0	0	0	1	0	0	0	0	0	0	1	1	1	0	1	1	1	1	1	1
4	0	1	0	0	0	0	0	0	1	0	0	0	0	0	1	1	1	1	0	1	1	1	1	1
5	0	1	0	1	0	0	0	0	0	1	0	0	0	0	1	1	1	1	1	0	1	1	1	1
6	0	1	1	0	0	0	0	0	0	0	1	0	0	0	1	1	1	1	1	1	0	1	1	1
7	0	1	1	1	0	0	0	0	0	0	0	1	0	0	1	1	1	1	1	1	1	0	1	1
8	1	0	0	0	0	0	0	0	0	0	0	0	1	0	1	1	1	1	1	1	1	1	0	1
9	1	0	0	1	0	0	0	0	0	0	0	0	0	1	1	1	1	1	1	1	1	1	1	0

BCD-to-decimal   Logic table

Logic equations
(active-high output)

$0 = \bar{D}\,\bar{C}\,\bar{B}\,\bar{A}$ $5 = \bar{D}\,C\,\bar{B}\,A$

$1 = \bar{D}\,\bar{C}\,\bar{B}\,A$ $6 = \bar{D}\,C\,B\,\bar{A}$

$2 = \bar{D}\,\bar{C}\,B\,\bar{A}$ $7 = \bar{D}\,C\,B\,A$

$3 = \bar{D}\,\bar{C}\,B\,A$ $8 = D\,\bar{C}\,\bar{B}\,\bar{A}$

$4 = \bar{D}\,C\,\bar{B}\,\bar{A}$ $9 = D\,\bar{C}\,\bar{B}\,A$

Figure 10-3 BCD-to-decimal conversion: truth table and logic equations.

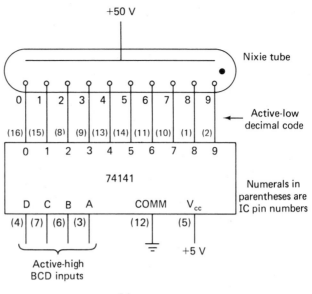

(a)

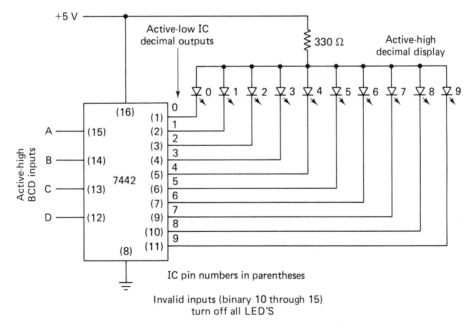

(b)

Figure 10-4 Applications of BCD-to-decimal converter/driver ICs. (a) 74141 as a Nixie driver. (b) 7442 as an LED decoder/driver.

active-high, however; and they can be energized from a set of four switches or the outputs of any of the TTL BCD counters described in Chap. 9.

The 4028 BCD-to-decimal decoder is a CMOS version of the TTL 7442. The CMOS decoder has active-high outputs, though; and it can drive the anodes of a common-cathode LED decimal display assembly. The current at the output of the IC must be limited to 8 mA by means of a ballast resistor, one serving the same function as the 330-Ω resistor in Fig. 10-4(b).

These decoder/drivers aren't limited to optical display assemblies. The 80-mA drive capabilities of the 7445 and 74145 ICs make this very suitable for driving other kinds of low-power devices such as relay coils and the base connections of common-emitter transistor amplifiers.

A final question regarding the operating features of these BCD-to-decimal decoders concerns the way the outputs respond to 4-bit binary inputs that represent numerals larger than 9 (invalid or overrange BCD inputs between binary 10 and 15). The 74141 Nixie decoder/driver, for instance, responds to input numerals larger than binary 9 by restarting the decimal count all over again: A binary 10 input energizes the decimal 0 output, a binary 11 input energizes the decimal 1 output and so on. The 7442 decoder, however, responds to overrange inputs by blanking the outputs altogether.

10-3 BCD-to-7-Segment Converters

Before digital integrated circuits came into being there was a time in the history of electronics technology when virtually all decimal displays called for either ten separate readout lamps, each labeled with a numeral between 0 and 9, or else a Nixie-type display. In either case, the large amount of vacuum-tube or transistor circuitry required for the BCD-to-decimal decoding kept the price and complexity of the readout system rather high.

The idea of using 7-segment displays is as old as the conventional decimal displays, but the greater complexity of vacuum-type or transistor circuitry required for BCD-to-7 segment decoding forced such a scheme out of the question. It was only after modern MSI (medium-scale integrated) circuits became a part of every-day technology that 7-segment displays and BCD-to-7 segment decoders became practical.

In the context of today's IC-oriented electronics, it turns out that 7-segment display assemblies are actually simpler and less costly than their older, discrete-lamp or Nixie counterparts. Almost all digital equipment calling for in-line numeric readout uses 7-segment displays and the appropriate decoding circuitry.

10-3.1 Principles of 7-Segment Displays

Figure 10-5(a) shows the standard arrangement of segments for a 7-segment display assembly. The segments themselves can be individual LED devices, fluour-escent-gas elements, liquid-crystal elements, neon bulbs or even groups of incan-

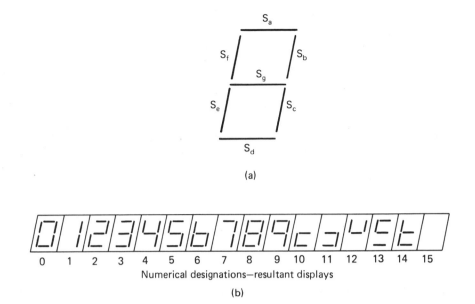

(a)

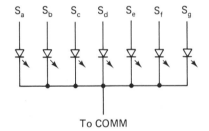

| 0 | 1 | 2 | 3 | 4 | 5 | 6 | 7 | 8 | 9 | 10 | 11 | 12 | 13 | 14 | 15 |

Numerical designations—resultant displays

(b)

Common-cathode
active-high segment inputs

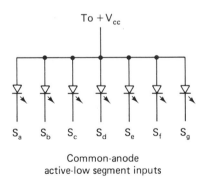

To COMM

To +V_{cc}

Common-anode
active-low segment inputs

(c)

Figure 10-5 Seven-segment LED displays. (a) Standard segment designation. (b) Standard character format. (c) Internal 7-segment connections for common-cathode and common-anode arrays.

237

descent lamps. In any case, the segments are designated S_a through S_g, running in a clockwise direction as viewed from the front of the display assembly.

Figure 10-5(b) shows how the separate elements are energized to produce the basic decimal numeric characters between 0 and 9. The characters that result from overrunning the BCD input to the code converter ICs are described in the next section of this chapter.

The two LED circuits in Fig. 10-5(c) show two techniques for interconnecting the seven LED elements within 7-segment display packages. The common-cathode version has all the cathodes connected together at a common grounding point The individual LEDs are then energized by applying active-high logic levels to their anodes. Making the common-cathode display show numeral 2, for instance, is a matter of grounding the COMM connection and applying forward-biasing positive voltages to the anodes of S_a, S_b, S_d, S_e and S_g.

The common-anode configuration works on the same basic principle, but the anodes are connected together to a positive voltage source and the individual cathode connections are energized by 0-going or active-low logic levels. Making the common-anode display show numeral 2 is a matter of connecting S_a, S_b, S_d, S_e and S_g to common, while the anodes are fixed at a suitable positive voltage level.

10-3.2 BCD-to-7-Segment Decoder ICs

The truth table in Fig. 10-6 shows how the segment outputs of a BCD-to-7-segment decoder should respond to binary input numerals 0 through 15. This table generates the characters shown in Fig. 10-5(b); and the active-high outputs section of the table is the one most appropriate for common-cathode LED displays; the active-low section applies to common-anode displays [compare the circuits in Fig. 10-5(c)].

In principle at least, it is possible to build up a BCD-to-7-segment decoder

	Active-high BCD inputs				Active-high 7-seg. outputs							Active-low 7-seg. outputs							
	D	C	B	A	a	b	c	d	e	f	g	a	b	c	d	e	f	g	
0	0	0	0	0	1	1	1	1	1	1	0	0	0	0	0	0	0	1	
1	0	0	0	1	0	1	1	0	0	0	0	1	0	0	1	1	1	1	
2	0	0	1	0	1	1	0	1	1	0	1	0	0	1	0	0	1	0	Standard numerals
3	0	0	1	1	1	1	1	1	0	0	1	0	0	0	0	1	1	0	
4	0	1	0	0	0	1	1	0	0	1	1	1	0	0	1	1	0	0	
5	0	1	0	1	1	0	1	1	0	1	1	0	1	0	0	1	0	0	
6	0	1	1	0	0	0	1	1	1	1	1	1	1	0	0	0	0	0	
7	0	1	1	1	1	1	1	0	0	0	0	0	0	0	1	1	1	1	
8	1	0	0	0	1	1	1	1	1	1	1	0	0	0	0	0	0	0	
9	1	0	0	1	1	1	1	0	0	1	1	0	0	0	1	1	0	0	
10	1	0	1	0	0	0	0	1	1	0	1	1	1	1	0	0	1	0	Overrun characters
11	1	0	1	1	0	0	1	1	0	0	1	1	1	0	0	1	1	0	
12	1	1	0	0	0	1	0	0	0	1	1	1	0	1	1	1	0	0	
13	1	1	0	1	1	0	0	1	0	1	1	0	1	1	0	1	0	0	
14	1	1	1	0	0	0	0	1	1	1	1	1	1	1	0	0	0	0	
15	1	1	1	1	0	0	0	0	0	0	0	1	1	1	1	1	1	1	Off code

Figure 10-6 BCD-to-7-segment truth table.

circuit from AND/OR combinations of standard logic gates. Although this might be a good exercise in logic circuit design, the final system turns out to be a very cumbersome and relatively expensive one. Without question, the best approach to converting a BCD code to the standard 7-segment format is by using one of the ICs designed specifically for that purpose.

Figure 10-7(a) lists the three most common TTL BCD-to-7-segment decoder ICs. The diagram in Fig. 10-7(b) shows the pinout arrangement that is identical for all three decoder packages.

7446A	BCD-to-7 segment decoder/driver; active-low, open-collector output rated at 30 V, 40-mA sink current—TTL
7447A	BCD-to-7 segment decoder/driver; active-low, open-collector output rated at 15 V, 40-mA sink current—TTL
7448	BCD-to-7 segment decoder; active-high, with internal 2-k pull-up resistor; output rated at 5.5 V, 6-mA sink current

Notes:
1. All of these decoders have active-high BCD inputs.
2. All have the same pinout. See Fig. 10-7.
3. All require external current-limiting resistors when driving LED segments.

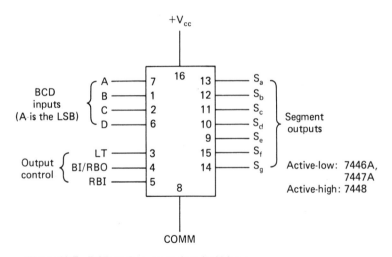

Figure 10-7 BCD-to-7-segment decoder/drivers.

The 7446A and '47A decoders generate the active-low segment data as shown in the table in Fig. 10-6, and they have 80-mA open-collector outputs that are capable of driving 30-V and 15-V loads, respectively. The 7448 decoder, however, generates active-high data and an internal pull-up resistor that limits the output sink current to about 8 mA.

The basic decoding applications of these ICs are rather straightforward, standing between the circuit that generates a BCD code and the 7-segment display assembly they drive. The BCD code is applied directly to terminals A, B, C and

D (input *A* being the least-significant bit), and the seven segment outputs are wired through current-limiting resistors to the corresponding terminals on a 7-segment display assembly. The relatively low output drive capacity of the 7448 often makes it necessary to apply its segment outputs through a special segment-driver circuit, however.

The output control terminals on these three decoder ICs require some special explanation. These terminals, labeled LT, BI/RBO and RBI, are used for programming the operating modes of the decoder system. Generally speaking, the LT (LAMP TEST) input is used only for lighting up all the segments in the display for segment-testing purposes. The RBI (RIPPLE BLANKING INPUT) and BI/RBO (BLANKING INPUT/RIPPLE BLANKING OUTPUT) terminals are used together for blanking the display under certain special operating conditions.

The circuit and table in Fig. 10-8 illustrate the basic operating modes available with these three BCD-to-7-segment decoders. Especially note that the BI/RBO terminal can be used as either a blanking input (BI) or a ripple blanking output (RBO); this input/output feature is unique in the TTL family of digital circuits.

According to the table in Fig. 10-8(b), the decoder is fixed for normal decoding whenever LT = RBI = 1 and the BI/RBO terminal is used as an output terminal. Any BCD number applied to the BCD inputs, including 0, appears properly decoded at the circuit's segment outputs. This is the normal operating mode of the circuit as it is shown in Fig. 10-8(a).

Depressing the LAMP TEST push button puts the system into its LAMP TEST mode by pulling the LT input down to logic 0. Under this particular set of circumstances, all of the segment outputs go to their active state, lighting all the segments in the 7-segment display. The BI/RBO lamp should not light up, indicating a logic-1 output from that terminal.

Depressing the DISPLAY BLANK push button pulls the BI/RBO terminal down to logic 0, making that terminal serve as an input point. The display should then switch off altogether, but the BI/RBO lamp should light up.

Depressing the ZERO BLANKING switch pulls the RBI input down to logic 0. Under this condition, the segment outputs will generate the data for normal 7-segment decoding for all BCD inputs *except 0*. Whenever the BCD inputs are indicating binary 0 (0000), the 7-segment display should switch off and the BI/RBO lamp should go on. Normal decoding resumes the instant the BCD inputs are changed from 0.

The two sections that follow illustrate some useful applications of the blanking modes available with these 7-segment decoder ICs.

10-3.3 Multi-Digit Displays with Leading-Zero Blanking

Suppose a piece of digital counting equipment has a 4-digit decimal display that is capable of showing any decimal numeral between 0 and 9999. It is very often desirable to connect the 7-segment decoders so that any numeral less than

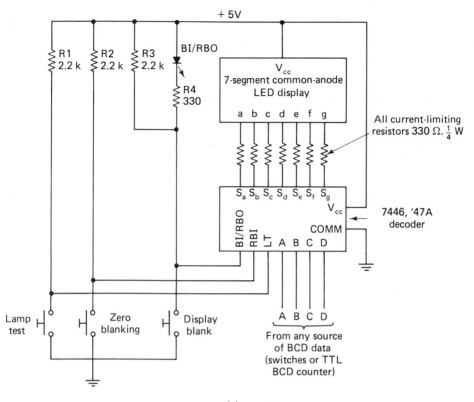

(a)

LT	RBI	BI/RBO	BCD Inputs	Segment Output Display Mode
0	X	1 out	Don't care	LAMP TEST (all outputs active)
X	X	0 input	Don't care	DISPLAY BLANK (all outputs inactive)
1	1	1 out	Any number	Normal decoding
1	0	Normally at logic-1 out; goes to logic 0 during zero-blanking interval	Any number	Normal decoding with zero blanking

(b)

Figure 10-8 A BCD-to-7-segment decoder/display demonstration circuit.
(a) Schematic diagram. (b) Function table.

1000 will not show leading zeros. A count of 40, for instance, should appear as 40, not as 0040 (the two leading zeros in this case should be blanked off altogether). And whenever the counter is cleared, the display should show 0 instead of 0000. The diagram in Fig. 10-9 shows the decoder blanking connections necessary for suppressing or blanking any leading zeros in the 7-segment display assembly.

The four ICs in Fig. 10-9 are 7446A or '47A BCD-to-7-segment decoders. Their inputs are BCD information from some sort of BCD generator, such as a 4-decade BCD counter circuit. The outputs from the decoders all go to common-anode 7-segment LED displays.

Recall that setting the RBI input to logic 0 places the decoder into its zero-blanking mode so that it decodes all input BCD numerals *except* 0; and whenever the BCD inputs are at 0 and the circuit is in the zero-blanking mode, the RBO terminal drops to logic 0. Setting the RBI input to logic 1, however, lets the circuit decode all input BCD numerals, including 0. And while in this nonblanking mode, the RBO output is at logic 1 at all times.

IC1 in Fig. 10-9, for instance, is permanently fixed for zero-blanking decoding; its RBI input is tied to COMM. This particular decoder thus shows all numerals except 0; and whenever the zero input occurs, its RBO terminal (pin 4 of IC1) drops to logic 0.

By contrast, IC2 can operate in either its zero-blanking or normal decoding mode. Since its RBI terminal is normally at logic 1, this IC is normally in the normal decoding mode, showing all BCD input numerals, including 0. But when IC1 is blanking a 0 appearing at the MSD BCD input, the RBI terminal of IC2 is set to logic 0, and IC2 operates in its zero-blanking mode, decoding all numerals except 0. If both the BCD inputs to IC1 and IC2 are 0, for example, both circuits blank off their display outputs.

IC3 has a similar relationship with IC2. As long as IC2 is in its normal decoding mode, or as long as IC2 is in its zero-blanking mode and showing a numeral other than 0, IC3 is in its normal decoding mode. But when IC1 and IC2 are both blanking zeros, IC3 is switched over to its zero-blanking mode because its RBI input (pin 4 of IC3) is at logic 0; and it, too, blanks a zero BCD input.

IC4, however, does not play any part in the zero-blanking operations because its RBI terminal is permanently fixed at logic 1. Any set of BCD inputs having three leading zeros will result in a blanked display for those digit positions. IC4 decodes its BCD inputs under any condition, including a 0 at the LSD BCD input.

The table accompanying the diagram in Fig. 10-9 summarizes the inputs, outputs and operating modes of the four decoders when six different kinds of BCD numerals are entered into the decoding system. The first line of the table shows what happens when all four input decades are at 0. The zero-blanking effect "ripples" down the line from IC1 to IC2 and IC3. All three of these decoders blank their respective displays, leaving a single 0 showing in the LSD position.

Entering a 4-decade numeral having no leading zeros, such as the numeral 4000 in the second line of the table, lets IC1 decode the numeral 4; and since it does

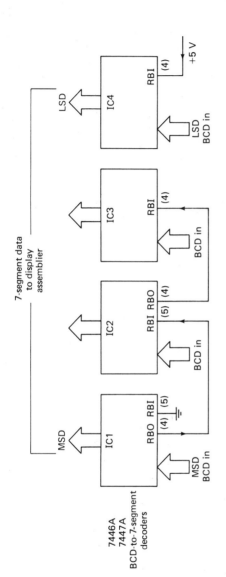

7-segment data to display assembler

MSD · IC1 · IC2 · IC3 · IC4 · LSD

7446A 7447A BCD-to-7-segment decoders

IC1 RBO (4) RBI (5) ⏚ MSD BCD in
IC2 RBI RBO (5) (4) BCD in
IC3 RBI (4) BCD in
IC4 RBI (4) +5 V · LSD · LSD BCD in

Overall BCD in. (decimal equiv.)	IC1 RBI	IC1 D	IC1 C	IC1 B	IC1 A	IC1 RBO	IC1 Mode	IC2 RBI	IC2 D	IC2 C	IC2 B	IC2 A	IC2 RBO	IC2 Mode	IC3 RBI	IC3 D	IC3 C	IC3 B	IC3 A	IC3 RBO	IC3 Mode	IC4 RBI	IC4 D	IC4 C	IC4 B	IC4 A	IC4 RBO	IC4 Mode	Display
0000	0	0	0	0	0	0	Z Blank	0	0	0	0	0	0	Z Blank	0	0	0	0	0	0	Z Blank	1	0	0	0	0	NC	Norm	0
4000	0	0	1	0	0	1	Norm	1	0	0	0	0	1	Norm	1	0	0	0	0	1	Norm	1	0	0	0	0	NC	Norm	4000
0400	0	0	0	0	0	0	Z Blank	0	0	1	0	0	1	Norm	1	0	0	0	0	1	Norm	1	0	0	0	0	NC	Norm	400
0040	0	0	0	0	0	0	Z Blank	0	0	0	0	0	0	Z Blank	0	0	1	0	0	1	Norm	1	0	0	0	0	NC	Norm	40
0004	0	0	0	0	0	0	Z Blank	0	0	0	0	0	0	Z Blank	0	0	0	0	0	0	Z Blank	1	0	1	0	0	NC	Norm	4
4444	0	0	1	0	0	1	Norm	1	0	1	0	0	1	Norm	1	0	1	0	0	1	Norm	1	0	1	0	0	NC	Norm	4444

NORM = normal decoding
Z Blank = decode all numbers except zero

Figure 10-9 Block diagram of a 4-digit BCD-to-7-segment display assembly with leading-zero blanking.

not zero blank IC2 under this particular set of circumstances, the other three decoders are in their normal decoding modes. The display thus shows 4000.

If a numeral such as 0400 is entered, IC1 blanks its 0 output and sets up IC2 for its zero-blanking mode. IC2 is seeing a BCD 4 input, however; therefore, it properly decodes that numeral and allows IC2 to operate in its normal decoding mode. The overall result is a display assembly showing 400.

IC1 also blanks its 0 output when a numeral with two leading zeros appears at the inputs. Similarly, IC2 blanks its 0 output and sets up IC3 for zero blanking. But the BCD input to IC3 is the numeral 4; thus it properly decodes that figure, and the display shows 40.

Whenever a 4-decade numeral such as 0004 is entered, IC1, IC2 and IC3 all blank their outputs, leaving a single numeral 4 showing in the LSD position. The display, in other words, simply shows the single numeral, 4; all three leading zeros are blanked off.

10-3.4 Conserving Power with Multiplexed Displays

Every 7-segment LED display assembly contains at least seven LEDs that can drain as much as 20 mA apiece at full brightness. When displaying the numeral 8, then, the drain on the power supply can go as high as 160 mA; and if there are four decades of display, the total current drain can rise as high as 640 mA. LED displays aren't normally operated at maximum current; but even so, a nonmultiplexed LED display assembly consumes a major portion of the power from the circuit's power supply. And if the power supply happens to be batteries, the LED display alone drastically reduces the available operating life of the supply.

A popular technique for reducing the overall current drain of the display assembly is to multiplex the display—to energize the output digits one at a time and in rapid succession. The result is that only one display is energized at any given instant, and the overall current drain is equal to that of a single 7-segment display.

Figure 10-10 shows the circuits and waveforms for a multiplexed 4-decade display system. The BCD data is fed to the decoders in the usual fashion, but the decoders (IC1 through IC4) are normally blanked off, switching on for only one-quarter of each multiplexing cycle.

IC7 is a 555-type oscillator that operates anywhere in the range of 10 kH to 30 kH. In this particular circuit, the frequency is set at about 20 kHz. The oscillator's output clocks a 2-bit up counter that counts binary 0, 1, 2, 3. The counter in this particular circuit is either a 7490 BCD or 7493 4-bit binary counter; it makes no difference which one is used because the circuit requires only the two lower-order bit outputs. In fact, IC6 can be any sort of 2-bit binary counter circuit, including a pair of cascaded J-K flip-flops.

At any rate, the 2-bit output from this continuously running counter goes to the *A* and *B* inputs of a 7442 BCD-to-decimal decoder, IC5. While the counter is cycling between binary 0 and 3, the outputs of the decoder drop to 0 for one

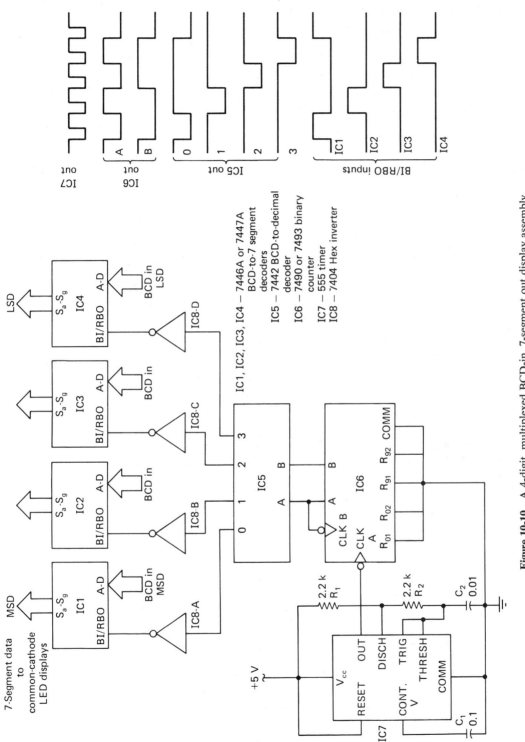

Figure 10-10 A 4-digit, multiplexed BCD-in, 7-segment out display assembly.

245

clock-pulse interval in succession. See the IC5 outputs in the waveforms in Fig. 10-10.

The four lower-order outputs of the decoder go to inverters and then to the BI/RBO inputs of the four BCD-to-7-segment decoders. The display decoders are thus energized one at a time, and in a sequence running from IC1 through IC4. Each display decoder is thus energized for one clock-pulse interval (about 50 μs in this case) and at a repetition rate of about 5 kHz.

Each display is operating at about a 25 percent duty cycle and will appear somewhat dimmer than it would in a nonmultiplexed display. It is possible to brighten the display, however, by reducing the value of the current-limiting resistors [see Fig. 10-8(a)] to 100 Ω.

It is generally impractical to multiplex LED displays that have only one or two decades, but the tradeoffs between power drain and the cost of the extra circuitry become significant for displays having four or more decades.

Expanding the multiplexing system to accommodate more than four decades of display is a simple matter of expanding the counting modulus of the counter circuit (IC6) and the BCD-to-decimal decoder (IC5). Using a 3-bit output from IC6, for example, lets the same circuit in Fig. 10-10 multiplex up to eight decades of display.

There are other good reasons for multiplexing the output of a digital system, but for the time being it is sufficient to say that display multiplexing is a technique for reducing the power consumption of the display assembly to that of a single-digit display. Other reasons for multiplexing outputs will be described in connection with multiplexer ICs.

10-3.5 An LED Display Brightness Control

As indicated in the previous section of this chapter, the brightness of an LED display is proportional to the unblanking duty cycle: The higher the operating duty cycle of any one display, the brighter the display will appear to the human eye. That particular feature was considered something of a disadvantage in a multiplexed display system, but it is possible to take advantage of it whenever it is necessary to adjust the brightness of the display under varying ambient light conditions.

The problem of ambient light and display brightness is especially critical in digital readout equipment intended for use in aircraft cockpits. Bright sunlight falling on the display tends to wash out the numerals, making it necessary to adjust the circuit for maximum display brightness. At night, however, very bright LED or fluourescent displays can annoy the pilot.

Of course, it is possible to make the brightness of an LED adjustable by inserting a potentiometer in series with the current-limiting resistors between the 7-segment decoders and the display assembly; but that approach is rather impractical, considering the numer of current-limiting resistors required for multi-digit displays. And then, too, a manual brightness control gives the pilot one more control to adjust—there are often too many controls already.

The answer is a rather simple control circuit that automatically adjusts the brightness of any LED or fluourescent readout. See the circuit and waveforms in Fig. 10-11.

The circuit in Fig. 10-11 controls the brightness of a display assembly by varying the duty cycle of positive pulses applied to the BI/RBO inputs of the display decoder ICs. The display decoders aren't included in this particular drawing, but they can be the same four decoders shown in Fig. 10-10 (without the inverters and multiplexing control circuitry). IC1-A in Fig. 10-11 is one section of a 556 dual timer IC that is connected as a free-running multivibrator. The operating frequency is determined by the values of C_2, R_3, R_2, R_1 and the photoresistor, PR.

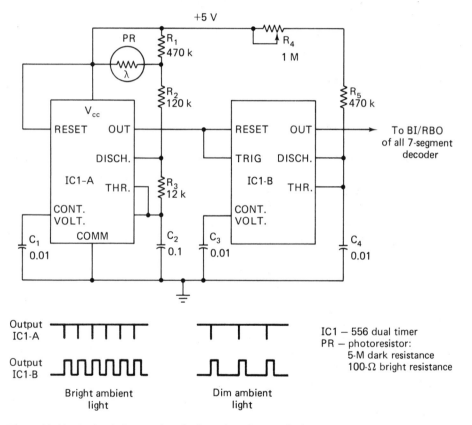

Figure 10-11 A circuit for varying the intensity of LED displays under varying ambient light conditions.

The photoresistor is situated near the display assembly. It does not respond to the light from the display itself; it responds to any ambient light falling on it. The brighter the ambient light level, the lower the resistance of the photoresistor, and thus the higher the oscillator's operating frequency. Compare the waveforms for the output of IC1-A as shown for bright and dim ambient light conditions.

IC1-B is connected as a monostable or one-shot multivibrator that is triggered to its *on* state by negative-going pulses from the oscillator. The timing of the monostable is determined by the values of C_4, R_5 and R_4. R_4 is a service adjustment that fixes the circuit's brightness response; and once it is set by the maintenance technician, it should not have to be changed again.

Once the circuit is set up and operating, it responds only to ambient light conditions. Increasing the amount of light falling on the photoresistor increases the operating frequency of IC1-A. The duration of the pulses from the monostable is fixed, however; so the overall effect is a larger duty cycle and a brighter display. Decreasing the amount of ambient light falling on the photoresistor decreases the oscillator's operating frequency and ultimately decreases the duty cycle to dim the display.

The frequency range for IC1-A in this particular circuit is between 20 Hz in absolute darkness and about 100 Hz in full sunlight. The monostable can be set for an output duration of about 10 ms; and if this is the case, the output duty cycle varies between 20 percent and 100 percent.

10-4 BCD-to-Binary Converters

A BCD format is an especially convenient and useful format for input and output operations. Many digital operations, especially those calling for memory addressing and arithmetic operations, however, are more convenient and simpler if they operate from a binary format. Thus the need for circuits that translate multi-decade BCD inputs into straight binary. Most output display decoders work from a BCD input, though; therefore, there is also a need for circuits that convert binary into multi-decade BCD.

This section deals with BCD-to-binary converters. The subject of binary-to-BCD converters is taken up in the following section.

Figure 10-12 shows the pinout and pin designations for the 74184 TTL BCD-to-binary converter IC. The BCD inputs are applied at terminals A through E; and if the strobe input G is held at logic 0, the binary outputs appear at terminals Y_1 through Y_5 (output terminals Y_6, Y_7 and Y_8 aren't used for BCD-to-binary conversion). Incidentally, this particular pinout is the same one used for the 74185A binary-to-BCD converter described in the following section.

The basic truth table for the 74184 BCD-to-binary converter IC appears in Fig. 10-13(a). This table doesn't really make much sense, however, until it is used for analyzing the operation of the simple BCD-to-binary converter circuit in Fig. 10-13(b). The table actually takes into account the fact that the least-significant bit of the least-significant BCD digit bypasses the converter altogether and appears as the least-significant bit in the binary output.

The circuit in Fig. 10-13(b) accepts two BCD numerals: a full decade at circuit inputs A_1, B_1, C_1 and D_1 and two bits of a second BCD input at A_2 and B_2. This particular converter is thus capable of converting BCD inputs 00 through 49. The

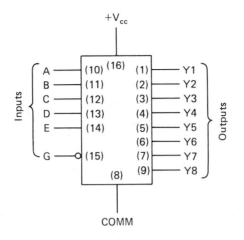

74184 TTL BCD-to-binary converter
74185A TTL binary-to-BCD converter

Figure 10-12 Pinout for the 74184 and '185A BCD-to-binary and binary-to-BCD converter ICs.

Inputs						Outputs				
E	D	C	B	A	G	Y_5	Y_4	Y_3	Y_2	Y_1
0	0	0	0	0	0	0	0	0	0	0
0	0	0	0	1	0	0	0	0	0	1
0	0	0	1	0	0	0	0	0	1	0
0	0	0	1	1	0	0	0	0	1	1
0	0	1	0	0	0	0	0	1	0	0
0	1	0	0	0	0	0	0	1	0	1
0	1	0	0	1	0	0	0	1	1	0
0	1	0	1	0	0	0	0	1	1	1
0	1	0	1	1	0	0	1	0	0	0
0	1	1	0	0	0	0	1	0	0	1
1	0	0	0	0	0	0	1	0	1	0
1	0	0	0	1	0	0	1	0	1	1
1	0	0	1	0	0	0	1	1	0	0
1	0	0	1	1	0	0	1	1	0	1
1	0	1	0	0	0	0	1	1	1	0
1	1	0	0	0	0	0	1	1	1	1
1	1	0	0	1	0	1	0	0	0	0
1	1	0	1	0	0	1	0	0	0	1
1	1	0	1	1	0	1	0	0	1	0
1	1	1	0	0	0	1	0	0	1	1
X	X	X	X	X	1	1	1	1	1	1

(a)

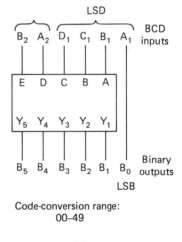

Code-conversion range:
00–49

(b)

Figure 10-13 Applying the 74184 BCD-to-binary converter IC. (a) 74184 truth table. (b) A $1\frac{1}{2}$-decade BCD-to-binary converter circuit.

249

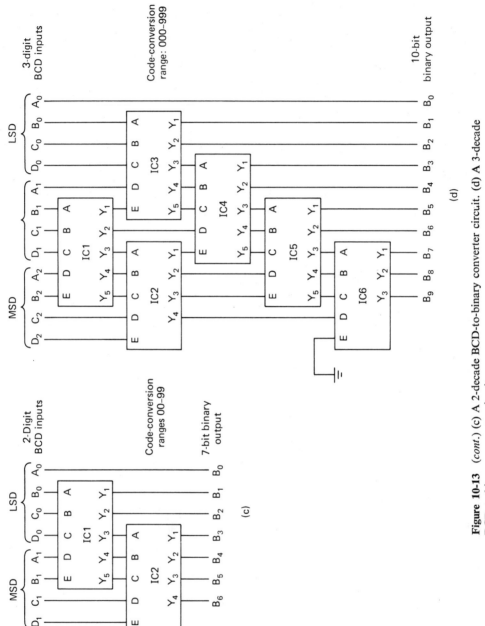

Figure 10-13 (*cont.*) (c) A 2-decade BCD-to-binary converter circuit. (d) A 3-decade BCD-to-binary converter circuit.

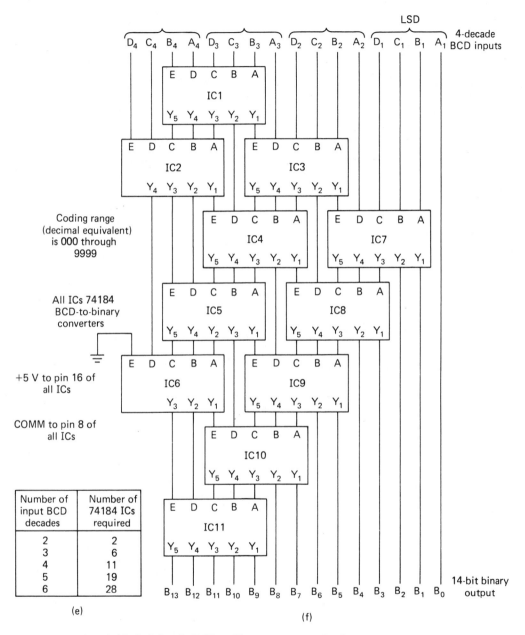

Figure 10-13 (*cont.*) (e) A 4-decade BCD-to-binary converter circuit. (f) Number of converter packages as a function of the number of BCD decade inputs.

conversion range cannot exceed 49 for the simple reason that there are only two inputs available for the most-significant BCD input.

The circuit can be expanded to accommodate two full decades of BCD input as shown in Fig. 10-13(c). Adding a second converter IC expands the code-conversion range to 99.

Figures 10-13(d) and 10-13(e) show the package interconnections required for 3-decade and 4-decade conversion, respectively. In all instances, the G input is tied to COMM to enable the converter IC operation.

The table in Fig. 10-13(f) shows the number of 74184 BCD-to-binary converter ICs required for a given number of input BCD decades. Analyzing and designing these circuits are tremendous exercises in interpreting truth tables; and of course the engineer or technician must have a complete understanding of BCD and binary coding.

10-5 Binary-to-BCD Converters

Binary-to-BCD conversion is necessary wherever a digital system works primarily on a straight binary format and the output must be translated into a format that is compatible with BCD-oriented output schemes (BCD-to-7 segment decoders for numeric displays, for instance).

The 74185A IC has been designed specifically for binary-to-BCD conversion. It has the same pin designations as the BCD-to-decimal counterpart illustrated in Fig. 10-12. For the binary-to-BCD converter, however, the binary inputs are connected to terminals A through E, and the BCD outputs are taken from terminals Y_1 through Y_6. (Output terminals Y_7, Y_8 and Y_9 are always at logic 1 and are not used for the binary-to-BCD conversion operation.)

The table in Fig. 10-14(a) is the truth table for the 74185A binary-to-BCD converter package. The table itself does not indicate the fact that the LSB of the least-significant BCD output word is the same as the LSB of the binary input word. Using this table to analyze the basic converter circuit in Fig. 10-14(b) shows how the basic converter scheme works.

With six binary inputs to the circuit in Fig. 10-14(b), its code-conversion range, expressed in decimal equivalents, is 00 through 63. Counting 0 through 63 in the output BCD format calls for one full LSD decade and the three lower-order bits of a second decade. The same circuit can be used with a 4-bit binary output and thereby generate a 4-bit, single-digit BCD output by grounding inputs E and D on the IC and taking the BCD output from the LSD connections shown in the diagram.

Figures 10-14(c) and 10-14(d) show binary-to-BCD converters expanded for 8- and 16-bit binary inputs, respectively. In both instances the unused E inputs should be tied to COMM and all the ICs must be enabled by connecting their G inputs to COMM.

The table in Fig. 10-14(e) summarizes the number of 74185 packages required for any number of binary inputs between 4 and 16.

Binary inputs						BCD outputs					
E	D	C	B	A	G	Y_6	Y_5	Y_4	Y_3	Y_2	Y_1
0	0	0	0	0	0	0	0	0	0	0	0
0	0	0	0	1	0	0	0	0	0	0	1
0	0	0	1	0	0	0	0	0	0	1	0
0	0	0	1	1	0	0	0	0	0	1	1
0	0	1	0	0	0	0	0	0	1	0	0
0	0	1	0	1	0	0	0	1	0	0	0
0	0	1	1	0	0	0	0	1	0	0	1
0	0	1	1	1	0	0	0	1	0	1	0
0	1	0	0	0	0	0	0	1	0	1	1
0	1	0	0	1	0	0	0	1	1	0	0
0	1	0	1	0	0	0	1	0	0	0	0
0	1	0	1	1	0	0	1	0	0	0	1
0	1	1	0	0	0	0	1	0	0	1	0
0	1	1	0	1	0	0	1	0	0	1	1
0	1	1	1	0	0	0	1	0	1	0	0
0	1	1	1	1	0	0	1	1	0	0	0
1	0	0	0	0	0	0	1	1	0	0	1
1	0	0	0	1	0	0	1	1	0	1	0
1	0	0	1	0	0	0	1	1	0	1	1
1	0	0	1	1	0	0	1	1	1	0	0
1	0	1	0	0	0	1	0	0	0	0	0
1	0	1	0	1	0	1	0	0	0	0	1
1	0	1	1	0	0	1	0	0	0	1	0
1	0	1	1	1	0	1	0	0	0	1	1
1	1	0	0	0	0	1	0	0	1	0	0
1	1	0	0	1	0	1	0	1	0	0	0
1	1	0	1	0	0	1	0	1	0	0	1
1	1	0	1	1	0	1	0	1	0	1	0
1	1	1	0	0	0	1	0	1	0	1	1
1	1	1	0	1	0	1	0	1	1	0	0
1	1	1	1	0	0	1	1	0	0	0	0
1	1	1	1	1	0	1	1	0	0	0	1
X	X	X	X	X	1	1	1	1	1	1	1

(a)

Figure 10-14 Applying the 74185A binary-to-BCD converter IC. (a) 74185A truth table.

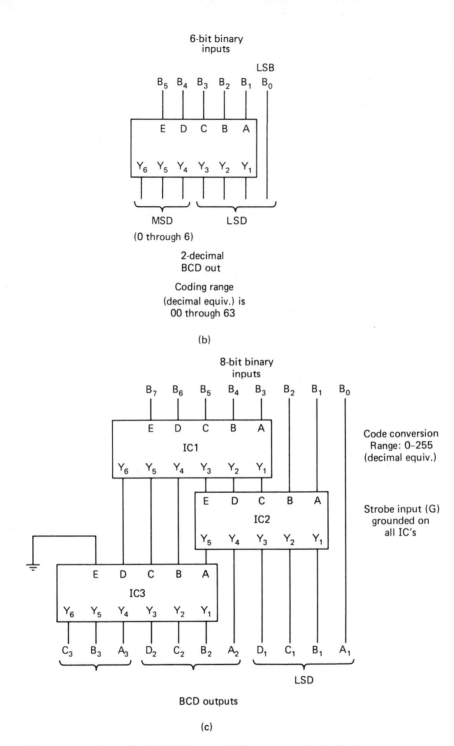

Figure 10-14 (*cont.*) (b) A 6-bit binary-to-BCD converter circuit. (c) An 8-bit binary-to-BCD converter circuit.

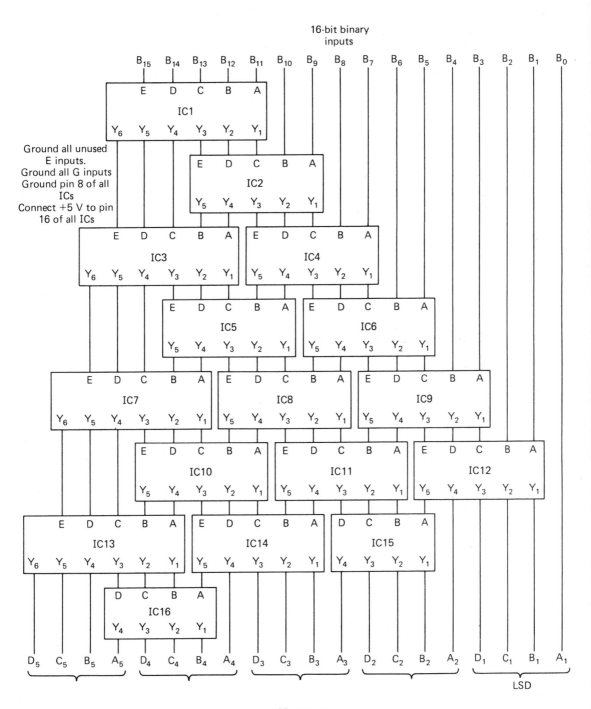

Figure 10-14 *(cont.)* (d) A 16-bit binary-to-BCD converter circuit.

Binary bits in	Number of 74185A packages
4–6	1
7, 8	3
9	4
10	6
11	7
12	8
13	10
14	12
15	14
16	16

(e)

Figure 10-14 (*cont.*) (e) Number of converter packages as a function of the number of binary bits input.

10-6   Octal and Hexadecimal Converters

The tables in Fig. 10-15 show the basic logic relationships between octal and hexadecimal numbers and their binary equivalents. The octal code [Fig. 10-15(a)] consists of eight decimal numbers between 0 and 7. These numbers can be translated into active-high or active-low BCD equivalents.

The hexadecimal code has 16 different characters, ranging from decimal 0 through 9, and then extending alphabetically from *A* through *F*. These characters can be translated into a 4-bit binary word as shown in Fig. 10-15(b).

Octal and hexadecimal counting formats are often used at the inputs of digital equipment that calls for manually entering long binary words. A single octal number, for example, represents three binary bits. So an octal number such as 503 can represent the 9-bit binary word 101   000   011. Using an octal representation certainly makes the number easier to use and remember.

The idea behind the hexadecimal format is the same as the octal version, but with one exception: The hexadecimal character represents four binary bits instead of just three. A hexadecimal sequence such as 5F2, for instance, represents the 12-bit binary word 0101   1111   0010.

The 74148 priority encoder IC package illustrated in Fig. 10-16 is a handy device for making octal or hexadecimal conversions to a binary format. This particular IC is an octal counterpart of the 74147 priority encoder that was described as a decimal-to-BCD converter in Sec. 10-1.

A single 74148 priority encoder very naturally generates an active-low binary version of any active-low octal number entered at inputs 0 through 7. Enabling the chip by pulling *EI* down to logic 0 and entering an active-low octal input yields an active-low version of the octal number as well as some carry-out data that is useful for cascading the chips.

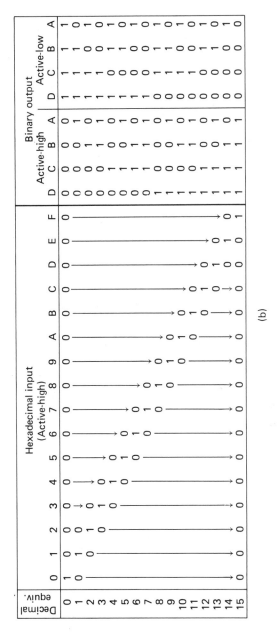

Figure 10-15 Standard octal and hexadecimal truth tables. (a) Active-high octal to active-high and active-low binary. (b) Active-high hexadecimal to active-high and active-low binary.

257

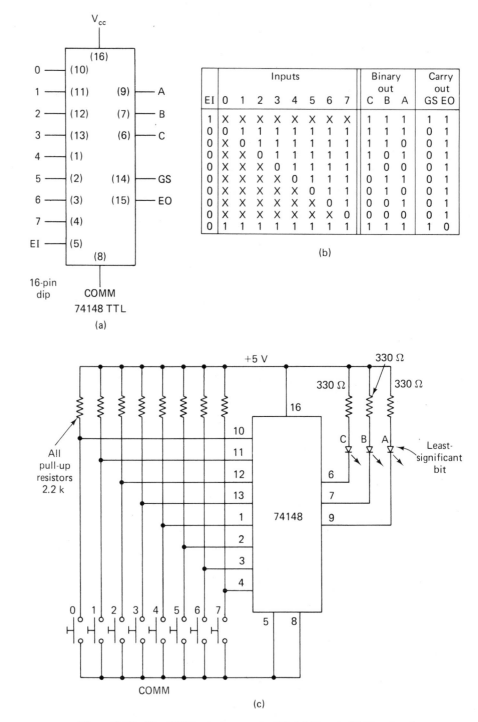

Figure 10-16 The 74184 priority encoder IC. (a) Pinout. (b) Functional truth table. (c) An octal-to-binary demonstration circuit.

According to the truth table in Fig. 10-16, entering a number such as octal 2 (often written 2_8), yields the output 101, an active-low or inverted version of binary 2 (010).

The circuit in Fig. 10-16(c) is a practical octal-to-binary converter circuit. The seven switches at the inputs provide active-low octal inputs, while the LEDs connected at the outputs show the corresponding binary equivalent. This is a straightforward application of a single 74148 priority encoder IC.

Figure 10-17 shows how it is possible to cascade a pair of octal-to-binary

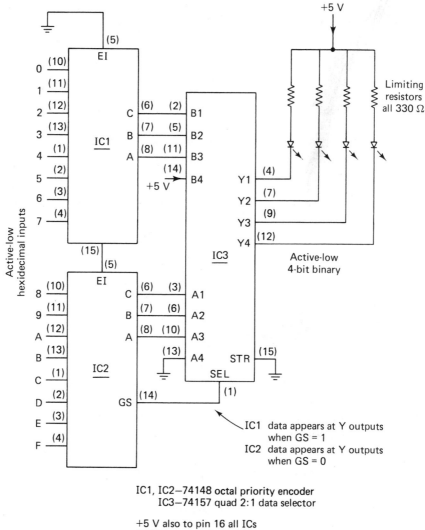

Figure 10-17   A hexadecimal-to-binary converter demonstration circuit.

converters to expand the number of available inputs to 16, thereby making it possible to achieve hexadecimal-to-binary conversion.

IC1 in Fig. 10-17 handles hexadecimal inputs 0 through 7, while IC2 handles inputs 8 through *F*. Bear in mind that these inputs must be active-low; and according to the basic 74148 truth table in Fig. 10-16(b), pulling any one of the inputs at IC1 to logic 0 causes an active-low binary number to appear at outputs *A*, *B* and *C*, *and* that action disables IC2 by setting its *EI* connection to logic 1. Pulling any one of the inputs at IC2 to logic 0 likewise causes an active-low binary number to appear at outputs *A*, *B* and *C* of that chip, provided, of course, none of the inputs to IC1 are energized at the same time.

Binary numbers thus appear at the outputs of IC1 and IC2, depending on which one has energized inputs. The problem is then to select the set of outputs that are, indeed, generating a valid binary number and direct that number to the indicator LEDs. IC3, a data selector chip, does this particular job.

Note that IC3 has two sets of 4-bit inputs but only one 4-bit output. One of the two sets of inputs will be seen at the output, depending on the logic level appearing at the SEL (SELECT) terminal of IC3. Whenever the SEL input is at logic 1, IC3 delivers data from its *B* inputs to its *Y* outputs. Setting the SEL input to logic 0, however, makes the *A* inputs of IC3 appear at its *Y* outputs.

Now when the use is entering a logic-0 level at one of the inputs of IC1, IC2 is disabled; and according to the truth table in Fig. 10-16(b), the *GS* connection of IC2 will output a logic 1 level to SEL of IC3. IC3 is thus set to deliver the outputs from IC1 to the LEDs. Using one of the inputs to IC2, however, enables that chip and sets the SEL input of IC3 to logic 0, a condition that makes the outputs of IC2 appear at the LEDs.

To put it simply, energizing any one of the 16 hexadecimal inputs to IC1 and IC2 makes the 4 LEDs light up in a pattern representing the 4-bit binary equivalent.

Exercises

1. Write the BCD equivalent of decimal numbers 0 through 9.

2. Devise a 1-decade BCD-to-decimal converter circuit using ten 4-input NAND gates and four logic inverters. The inputs should be active-high and the outputs active-low.

3. Why is it necessary to use current-limiting resistors between the outputs of a 7-segment decoder/driver and a 7-segment LED display assembly? See Fig. 10-8a for instance.

4. Convert the following decimal numbers to their binary equivalents: (a) 12; (b) 120; (c) 56; (d) 1234.

5. Convert the following binary numbers first to their decimal and then to their BCD equivalents: (a) 001001; (b) 11010; (c) 101010; (d) 11101.

6. Convert the following BCD numbers to their binary equivalents: (a) 01 1001; (b) 0010 1000; (c) 0101 1000 0010.

7. Express the following binary numbers in an octal format: (a) 001001; (b) 110100; (c) 111001010010.

8. Convert the following hexadecimal numbers first to their binary and then to their decimal equivalents: (a) 0A; (b) 12; (c) 2FA; (d) DDD.

DIGITAL MEASURING SYSTEMS

This chapter represents the culmination of all the basic principles of digital counting and display circuits presented in Chaps. 8, 9 and 10. These counting and display principles, combined with the basics of flip-flops, timers and combinatorial logic, are sufficient for building up some of the most useful digital measurement and control instruments in the digital business today.

The first sections of this chapter deal with digital measuring instruments, including event counters and a host of instruments that convert pulse frequencies and voltage levels into a digital readout format. The last section of the chapter takes up the subject of pulse-duration measurements that are appropriate for understanding instruments such as digital thermometers and light meters.

11-1   Power Supply and Digital Display

All of the digital systems described in this chapter use the same power supply and digital display assembly. See Figs. 11-1 and 11-2. The power supply in this case is capable of providing a regulated $+5$ V V_{cc} at current levels up to 1 A. Of course, many of the circuits can be converted for battery operation; TTL and CMOS circuits both work rather well from a standard $+6$-V battery supply, for instance.

The display assembly cited throughout this chapter has nonblanked, 4-decade, 7-segment LED readouts. The display can be easily modified for any number of decades, however.

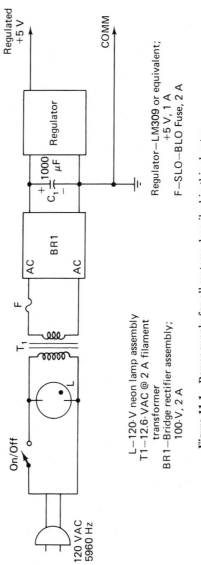

On/Off

120 VAC
5960 Hz

T_1

F

AC

BR1

AC

C_1 + 1000 μF —

Regulator

Regulated +5 V

COMM

L—120-V neon lamp assembly
T1—12.6-VAC @ 2 A filament
    transformer
BR1—Bridge rectifier assembly;
    100-V, 2 A

Regulator—LM309 or equivalent;
    +5 V, 1 A
F—SLO—BLO Fuse, 2 A

Figure 11-1 Power supply for all systems described in this chapter.

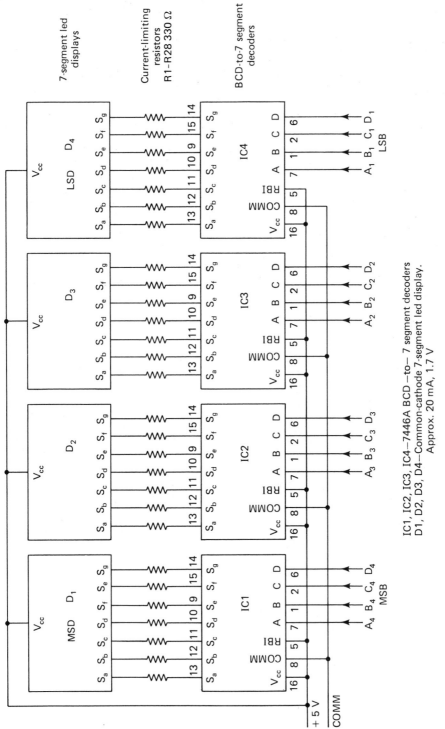

Figure 11-2 A 4-decade BCD decoder and LED display assembly that can be used with all systems described in this chapter.

IC1, IC2, IC3, IC4—7446A BCD —to— 7 segment decoders
D1, D2, D3, D4—Common-cathode 7-segment led display.
Approx. 20 mA, 1.7 V

The power supply illustrated in Fig. 11-1 uses a filament transformer to step down ordinary 120-VAC, 50/60-Hz utility power to about 12.6 VAC. The bridge rectifier assembly and filter capacitor provide a fairly well-filtered +18 VDC to the regulator assembly. The regulator package in this particular example is an LM309 5-V, 1-A, 3-terminal regulator. The power supply can be modified for higher current operation by using a transformer, fuse, rectifier assembly and regulator having higher current ratings.

The display assembly shown in Fig. 11-2 is the basic model for all the circuits described in this chapter. The display units in this case are standard 7-segment LED displays having a rating of about 1.7 V at 20 mA for each segment. The displays must be of the common-anode variety to make them compatible with the active-low outputs from the BCD-to-7 segment decoder/drivers, IC1 through IC4. The current-limiting resistors in the segment lines are all 330-Ω, $\frac{1}{4}$-W resistors.

11-2   Event Counters

The simplest kind of digital counting device is one that simply counts the number of pulses applied to it. Any event in the real world that can be translated into an electrical pulse can be counted with such a system. Such systems can be used for counting the number of objects moving along an assembly line, accumulating the number of atomic particles that energize a geiger tube and keeping track of the number of white blood cells in a blood sample placed into a special electropical microscope.

The heart of an event counter is a set of cascaded BCD counters. A suitable counter arrangement, shown in Fig. 11-3(a), uses four 74190 TTL asynchronous BCD up counters. IC5 is the least-significant-digit counter, and it is thus triggered by the incoming event pulse. The counters can be reset to 0 by pulling the CLR bus up to logic 1. Capacitors C_2 and C_3 are despiking capacitors that filter out high-frequency power-supply ripple that is generated by any asynchronous counter assembly.

The counter assembly in Fig. 11-3(a) is fully compatible with the display assembly and power supply in Figs. 11-2 and 11-1, respectively.

IC9 and IC10-A in Fig. 11-3(b) make up an overflow detector circuit. The output of IC10-A is normally at logic 0, rising up to logic 1 only when the counter reaches the BCD equivalent of decimal 9999, the largest possible count for a 4-decade counter system.

The output of IC10-A is applied to one input of a NOR-gate R-S flip-flop. The other input to this flip-flop scheme is the counter's CLR bus logic level. An analysis of this flip-flop scheme will show that the output of IC10-C is normally at logic 1, thus keeping the OVERFLOW lamp turned off. Whenever an overflow condition occurs (the counter assembly reaches or exceeds 9999), the output of IC10-C is set to logic 0, and the OVERFLOW lamp switches on. The OVERFLOW lamp then remains on, in spite of continued counting, until the CLR bus is pulled up to logic 1 to reset the counter output to 0000.

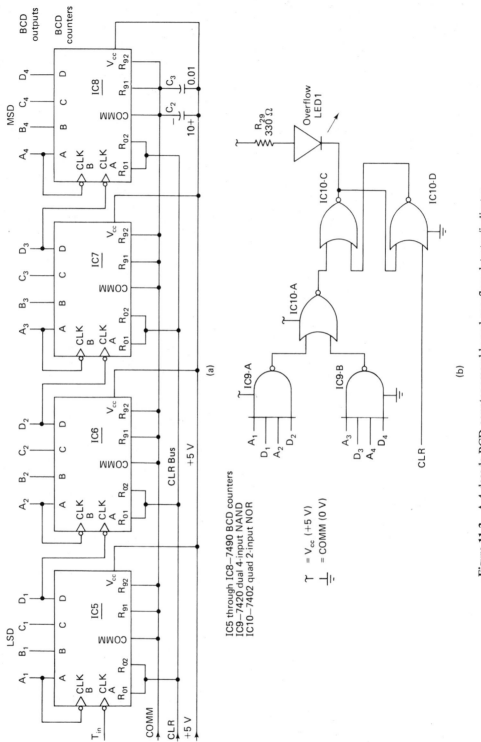

Figure 11-3 A 4-decade BCD counter assembly and overflow detector/indicator.

In short, the OVERFLOW indicator lights up whenever an overflow count occurs, and it remains lighted until the user resets the counter to 0. Any count appearing at the outputs must be considered invalid as long as the OVERFLOW lamp is on.

The block diagram in Fig. 11-4 shows how the counter assembly can be combined with the 4-decade display and power supply to make up a complete event counter system. The interface block merely represents any sort of interfacing required for matching the characteristics of the pulse to be counted with the TTL characteristics of the counter assembly. See Chap. 5 for the appropriate interfacing techniques.

11-3   A Simple Frequency Counter

An event counter can be transformed into a simple frequency counter by providing control circuitry that first clears the counter to 0 and then allows the event pulses to increment the counter for a precise period of time. Suppose, for example, the events are occurring at a 10-Hz rate. If the counter is first reset to 0 and then allowed to count for exactly 1 s, the counter outputs should show the numeral 10 at the end of that interval; in other words, the event frequency is 10 per second or 10 Hz.

As another example, suppose the events to be counted are occurring at a 100-Hz rate. Clearing and then enabling the counter for exactly 1 s will let the counter increment to 100, a clear indication the events are occurring at a 100-Hz rate. If the counter is allowed to run for 0.1 s, incidentally, it follows that it will show a count of 10 whenever the events are occurring at a 100-Hz rate.

Figure 11-5 shows the circuitry that can be added to the event counter described in the previous section of this chapter. Adding this circuitry to the counter block diagram in Fig. 11-4 effectively transforms it into a frequency counter.

In this particular instance, the user has access to a RESTART push button. Assuming event pulses are arriving continuously at the input interface circuit, the user starts the frequency-counting cycle by momentarily depressing the RESTART button. As shown in the top waveform in Fig. 11-5(b), depressing the RESTART button causes the output of IC10-B to go to logic 1. This positive-going edge has no effect on the 555-type monostable timer, IC12. The logic-0 output from the timer thus gates off IC13-A, preventing event pulses from reaching the event counter assembly. Depressing the RESTART button does, however, immediately reset the event counters to 0.

As long as the RESTART push button is depressed, then, the event counter assembly is cleared to 0 and no event pulses are arriving at the trigger input of the counter assembly.

Releasing the push button creates a negative-going pulse at the trigger input of the monostable circuit, IC11; and that device responds by generating a logic-1 output that lasts for a period of time equal to 1.1(R32)(C6). As a result, event pulses pass through IC13-A to the counter assembly.

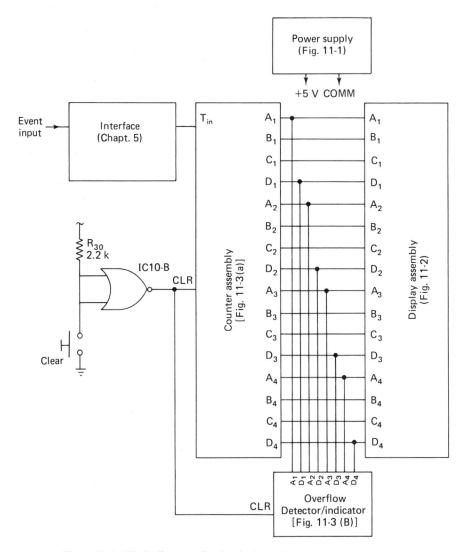

Figure 11-4 Block diagram of a simple 4-decade event counter.

Releasing the push button also removes the clearing voltage level from the counters; and in short, the event counter responds to incoming event pulses as long as IC11 is timing.

The instant the monostable timing interval is over, the output of IC11 returns to logic 0, gating off IC13-A so that event pulses no longer increment the counter assembly. The counter thus stops counting, indicating the number of events that occurred during the timing interval. That number remains on the display assembly until the user depresses the RESTART push button to clear the counter and initiate another counting cycle.

Suppose, for instance, the monostable circuit is fixed for a timing interval of 1 s and event pulses are occurring at the rate of 1 kHz. The user depresses the RESTART push button and notes that the display is cleared to 0. The instant the user releases the button, the display responds by counting to 1000 within a 1-s interval. The user sees a mad jumble of numbers during that timing interval; but as soon as the timing interval is over, the counter will show that 1000 pulses have occurred within that time. The frequency of the event pulses, in other words, was 1000 per second or 1 kHz.

The user can update the reading any time he or she chooses by momentarily depressing the RESTART push button once again.

This particular counter is capable of measuring frequencies up to 10 MHz with no trouble at all. The only limiting factor is the switching speed of the event-counter ICs. If the user is attempting to measure a 1-MHz source of pulses while the timer is set for 1 s, the user is going to encounter an overflow condition. A count of 1 million will certainly overflow a four-decade counter.

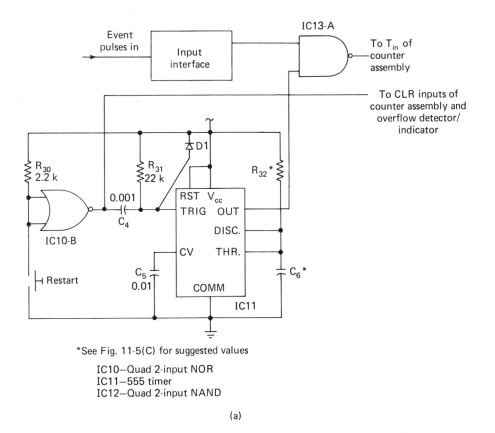

*See Fig. 11-5(C) for suggested values

IC10—Quad 2-input NOR
IC11—555 timer
IC12—Quad 2-input NAND

(a)

Figure 11-5   A simple frequency counter with manual updating and five different operating ranges.

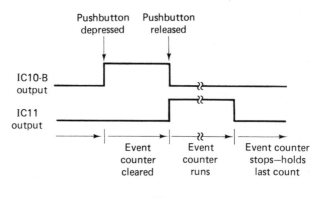

(b)

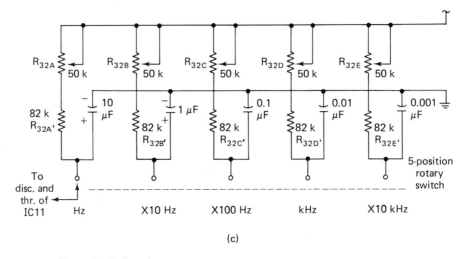

(c)

Figure 11-5 (*cont.*)

Measuring frequencies in excess of 9999 Hz without overflowing the counter assembly is a simple matter of shortening the timing interval of IC11. The user can switch in a smaller value for C_6, for instance.

Perhaps the events are occurring at a rate of 10 MHz. In this case, reducing the timing interval of IC11 to 100 μs lets the counter assembly increment to 1000 during that time. The counter does not overflow; and the user can read the actual event frequency by multiplying the figures on the display by 10^4.

Figure 11-5(b) shows how it is possible to replace R_{32} and C_6 in the timer circuit with a 5-position rotary switch that lets the user select an appropriate multiplier scale. Note that changing the setting of the rotary switch changes the values of the timing capacitor. The variable resistors in each section of the multiplier selector circuit are used for calibrating the timing to make up for inaccuracies caused by the tolerance of the capacitor values.

So if a particular counting cycle results in an overflow condition, the user merely sets the multiplier selector switch to a higher multiplier scale. Eventually the user should find a point that gives a good reading on the display assembly without causing the overflow lamp to go on. Determining the frequency is then a matter of adjusting the number on the display by the units listed on the multiplier selector switch.

11-4 Frequency Counters with Automatic Updating

The frequency counter system described in the previous section has some useful applications in situations calling for only an occasional updating of the frequency reading. Updating the reading is a simple matter of depressing the RESTART push button. The frequency reading then appears on the display assembly after the system completes one timing interval, and the reading remains fixed on the display until the user inititates another timing cycle to update the display.

Manually updating the reading can be a troublesome affair under many other kinds of measuring conditions, however. Suppose, for example, the frequency to be read changes continuously, making it necessary to update the reading every second or so. Obviously, it would be far better to include some provisions for automatically updating the display at regular intervals.

Figure 11-6(a) shows a count control circuit that is very similar to the one in Fig. 11-5(a). In this case, however, the RESET push button scheme is replaced with a free-running multivibrator, IC14. This multivibrator generates rather brief negative-going pulses having a rather short duration (about 15 μs) and a repetition rate that determines the system's updating frequency. See the top waveform in Fig. 11-6(b).

The pulse waveform from IC14 is inverted by IC13-B and applied to the CLR bus of the event-counter assembly and the overflow detector/indicator. See Fig. 11-3. The display assembly is thus cleared to 0 each time IC14 generates its negative-going pulse.

IC11 is a basic 555-type monostable multivibrator that is connected in such a way that its timing interval begins on the positive-going edge of the waveform from IC14. This positive edge-triggering is accomplished by connecting the RESET input to TRIGGER as described in Sec. 7-2.1. Compare the middle and bottom waveforms in Fig. 11-6(b).

IC11 thus generates a positive-going timing pulse that gates on IC13-A, thereby allowing event pulses to increment the counter assembly for a fixed period of time. The theory of operation at this point is identical to that of the manually updated frequency counter described in the previous section of this chapter.

At the end of the timed counting interval, the display shows the event frequency until the next negative-going clock pulse from IC14 occurs.

The duration of the clearing pulse from IC14 is determined by the values of R_{32} and C_7; the update repetition rate is fixed by the values of R_{33}, R_{34}, R_{35} and C_7. R_{32} and C_6 determine the count timing interval as described in connection with the circuit in Fig. 11-5.

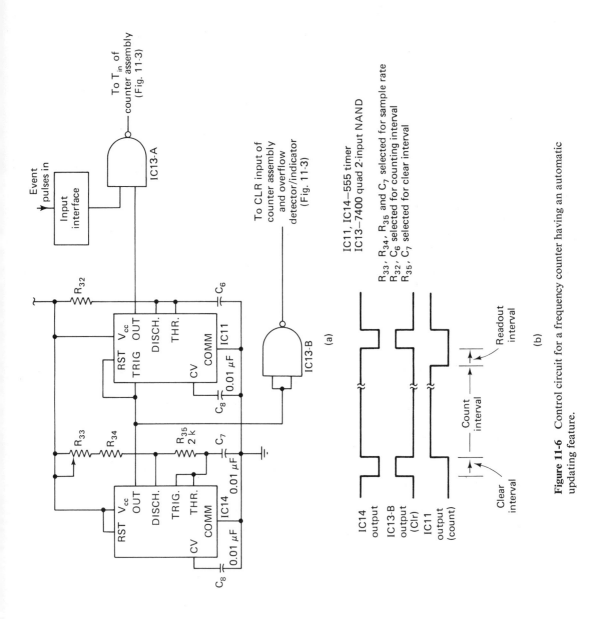

Figure 11-6 Control circuit for a frequency counter having an automatic updating feature.

This circuit suffers from one serious shortcoming. Indeed, it is much more convenient to use than a frequency counter that calls for manual updating; but in this case the automatic updating can cause a bigger problem than it cures.

Note from the waveforms in Fig. 11-6(b) that the display assembly shows the actual event frequency only between the time one counting interval ends and the next clearing pulse occurs. This doesn't pose any serious problem as long as the counting interval is rather short relative to the period between successive clearing pulses. But when the counting interval is almost as long as the time between successive clearing pulses, the user spends more time looking at a mad jumble of numbers than a fixed and readable display indicating the actual event frequency.

This particular readout problem arises because the display assembly is allowed to show the event-counting process in action. The problem could thus be cured if there were some way to store or latch the results of one counting interval while the next counting interval is taking place. The user would thus see only the result of each event-counting interval; the user would not see the event-counting process itself, which is meaningless anyway. The display would not flicker at all during the event-counting interval. It might change slightly each time the display is updated, but even then the event frequency would have to be changing at the time.

There is a way to latch the results of each counting interval. The basic idea is to insert a set of 4-bit data latches between the counter and display assemblies. By means of appropriate control circuitry, the latches can be loaded at the end of each event-counting interval and then put into their memory mode while the counter assembly is cleared and allowed to count the next segment of event pulses. The system can use the power supply, display assembly and counter assembly already described and illustrated in Figs. 11-1, 11-2 and 11-3, respectively. The revised scheme, however, calls for inserting data latches between the counter and display assemblies and it requires a slightly different timing control circuit.

Figure 11-7 shows a four-decade, 4-bit data latch assembly. The ICs are 4-bit latches of the level-D variety (Sec. 6-4) that load their data inputs (*A, B, C* and *D*) directly to their respective *Q* outputs as long as the LOAD input is at logic 1. Returning the LOAD inputs to logic 0 effectively latches the data into place at the *Q* outputs, and any change occurring at the inputs has no influence on the outputs.

The significance of these latching operations becomes clear after considering two facts: (1) The latch assembly shown in Fig. 11-7 is connected between the outputs of the event counter assembly and the display decoders. (2) The latch assembly is loaded only for a very brief time at the *end of each event-counting sequence*. The latches thus latch the largest-counted number, the number representing the frequency of the incoming event pulses. And, furthermore, the latches hold that number fixed at the display outputs until the counters complete another counting sequence, thus updating the display.

The control circuit and waveforms for the latched frequency counter system are illustrated in Fig. 11-8. IC14 is connected as an astable multivibrator, and its output frequency determines the rate at which the display is updated. The updating

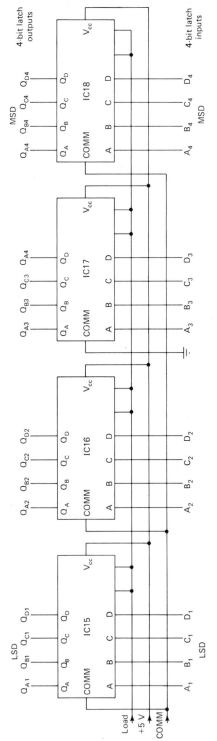

Figure 11-7 A 4-decade BCD latch assembly.

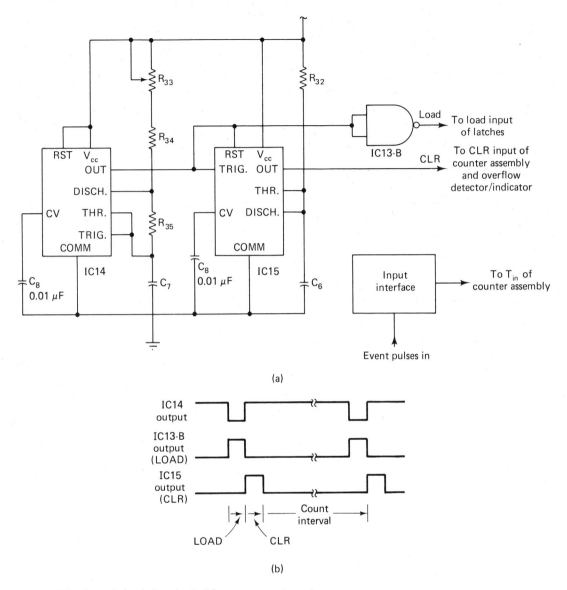

(a)

(b)

Figure 11-8 Control circuit for a latched frequency-counter system.

period can be as long as once every two or three seconds or as short as 0.1 s. Regardless of the frequency of the updating operation, short negative-going pulses from IC14 are inverted and sent directly to the LOAD bus of the data latch assembly. This pulse is responsible for loading whatever numbers happen to be present at the outputs of the counter assembly at the time.

The positive-going edge of the pulse waveform from IC14 initiates a short clearing pulse from another 555-type circuit connected as a monostable multivibrator. The duration of the output from this circuit determines the clearing interval of the counters.

Each time IC14 generates a negative-going pulse, then, two things happen. First, an inverted version of that pulse loads the latches (and readout assembly) with whatever number happens to be present at the counter's output. At the end of that same pulse, IC15 generates a positive-going pulse that clears the counters to 0. The display, however, does not clear to 0 because it has been previously loaded with the count that existed just prior to the clearing operation. See the waveforms in Fig. 11-8(b).

Carefully note that the timers in this system merely control the load/clear sequence. The timers have nothing at all to do with gating the event pulses delivered to the counter assembly. Event pulses are delivered continuously to the counter assembly, and except during the brief clearing interval, the counters run at all times.

The operating cycle follows this pattern: clear, count and latch; clear, count and latch; and so on. The clearing operation stops the counters and clears them to 0. The instant the clear pulse is removed, the counter begins incrementing in response to incoming event pulses. The counters continue incrementing through the latching interval, but since the latching interval is limited to about 10 μs or so, the user cannot possibly notice any flickering effect during that load interval. The latch circuitry thus picks up the number representing the frequency of the incoming event pulses and holds it in the display through the entire clearing operation and most of the counting operation.

The accuracy and precision of the system are dictated by the accuracy and precision of the overall counting interval, the interval between the end of the clearing pulse and the end of the following load pulse. The waveforms in Fig. 11-8(b) show that this critical counting interval is equal to the period of the waveform from IC14 less the load interval. Although the period of the waveform from IC14 might be changed to prevent counter overflow when measuring relatively high frequencies, it is usually impractical to adjust the load-pulse interval a corresponding amount. The load-pulse interval could thus represent a source of error in systems featuring a range selector [a circuit similar to that shown in Fig. 11-5(c)]. Any load-interval error, however, can be compensated out of the timing by means of the timing calibration potentiometer, R_{33}.

Figure 11-9 shows a complete block diagram of the improved frequency counter. This scheme represents the most popular frequency-counting technique in the digital business today. The next section of this chapter illustrates some of the great range of applications for frequency counters of this type.

11-5 Extending the Usefulness of Frequency Counters

The most straightforward application of a frequency counter with automatic updating is in laboratory frequency-measuring equipment. A range multiplier switch can be used for decreasing the timing interval from the count-control assembly by a

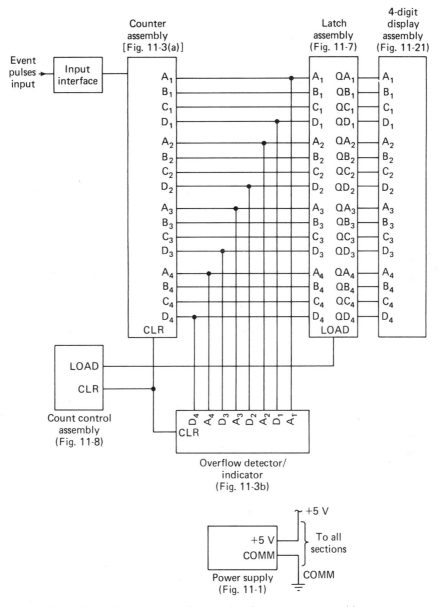

Figure 11-9   Block diagram of a complete frequency counter with automatic updating and display latching features.

factor of ten each time the switch is incremented from one range scale to another that is ten times larger. The input interface can include a voltage-divider network and a high-impedance buffer that matches the TTL counter assembly with any sort of input voltage and impedance level.

And without further modification, the circuit outlined in Fig. 11-9 could be used as a tachometer for auto, marine and motorcycle engines. Calibrating the tach would be a matter of running the engine at a known revolutions per minute and then adjusting the timing interval of the count-control assembly so that the known revolutions per minute figure appears on the digital readout. Once calibrated at one known revolutions per minute, the system should be accurate for all others.

Virtually any physical parameter that can be converted to a frequency can be digitized by a circuit comparable to the one in Fig. 11-9. If a circuit is devised for generating a pulse each time the drive shaft in an auto makes one revolution, the pulses can be used for operating the frequency counter. The result is a digital speedometer. In a similar fashion, picking up a pulse each time the front wheel of a bicycle turns provides the type of input necessary for making a digital bicycle speedometer. Of course, there should be no need for a counting having more than two readout digits in this case, unless the auto or bicycle happens to run at speeds in excess of 99 mph.

Any physical parameter that can be first changed into a rotary motion and then into electrical pulses having a frequency proportional to the rate of rotation can be measured in digital terms. Consider digital devices for measuring the revolutions per minute of a motor, wind speed, rate of flow of fluids and distance (a digital "tape measure").

It is possible to carry the usefulness of a frequency-counter system even further by considering that any voltage can be converted into a proportional frequency by means of a voltage-controlled oscillator (VCO). A digital voltmeter, for example, can include a VCO circuit that generates a frequency proportional to the dc voltage to be measured. That frequency can then be applied to a frequency counter to yield a digital readout that indicates the input voltage level.

As diagrammed in Fig. 11-10, any physical parameter that can be converted into a proportional voltage level can be measured digitally by means of a frequency-counter system. Sensing devices such as solar cells, piezoelectric materials and thermocouples generate voltages proportional to light intensity, pressure or strain and temperature, respectively. Applying buffered and amplified versions of these

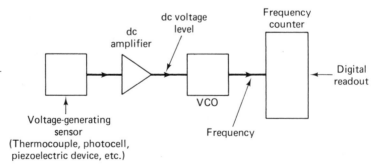

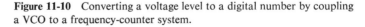

Figure 11-10 Converting a voltage level to a digital number by coupling a VCO to a frequency-counter system.

voltage levels to a VCO, followed by an appropriately scaled frequency counter, results in instruments such as digital light meters, strain gauges and thermometers.

Although a good many sensing devices generate a voltage, the more common types respond with a change in resistance. Photoresistors and thermistors, for example, show an inverse change in resistance to the medium that energizes them, namely, light or heat.

This change in resistance can be translated into a change of frequency by making the sensing device itself part of the RC timing network in an astable multivibrator. A photoresistor, for instance, can replace one of the two timing resistors in a 555-type astable multivibrator. [See Fig. 11-11(a)]. The result is an output frequency that would increase with any decrease in the resistance of the sensor; or to carry the analysis a step further, the output frequency would increase with any increase in the amount of light falling onto the photoresistor.

Most resistance-changing sensing devices show a reasonably linear response to

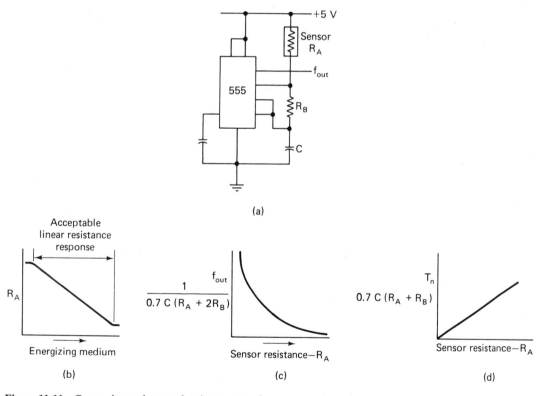

(a)

(b)

(c)

(d)

Figure 11-11   Converting a change of resistance to a frequency and proportional pulse period. (a) A 555-type RC oscillator using a resistance-changing sensor as one of the timing elements. (b) Linear response of the sensor to changes in the amount of energy that influences it. (c) Nonlinear frequency response of the circuit. (d) Linear output period response of the oscillator circuit.

any change in the intensity of the medium that energizes them. See Fig. 11-11(b). But while the resistance might change in a linear fashion, the oscillator's output frequency does not. Note that the frequency from this particular oscillator circuit is given by the nonlinear equation:

$$f_{out} = \frac{1}{0.7C(R_A + 2R_B)}$$

where R_A is the resistance of the sensing device. Figure 11-11(c) shows the circuit's output frequency as a function of the values of timing resistance and capacitance.

Whenever it is necessary to digitize parameters that are sensed by resistance-changing devices, the idea of applying the output frequency to a frequency counter is wholly inappropriate; the digital readout would be a nonlinear and useless function of the amount of energy applied to the sensing device.

The period of the waveform from the oscillator in Fig. 11-11(a), however, is a linear function of the value of R_A. See Fig. 11-11(d). Digitizing the response of the sensor is thus a matter of measuring the period, rather than the frequency, of the output waveform.

The following section of this chapter describes how the basic circuitry of a frequency counter can be rearranged to produce an interval-measuring instrument. Such an instrument can measure the period of an oscillator's output waveform and thereby properly digitize the response of resistance-changing sensors.

11-6 Pulse-Duration Counters

A counter circuit is at the heart of any of the frequency-counting schemes described thus far in this chapter. An internally generated timing pulse determines when the counter is cleared to 0 and how long it is allowed to be incremented in response to the frequency being monitored. If the internal control pulse happens to last 1 ms and the input frequency happens to be 10 kHz, for example, the counter will increment to 10, indicating that the input frequency is 10 kHz.

The roles of the input frequency and internally generated timing pulse are reversed in the case of pulse-duration counters. In this instance, the waveform being monitored is used as the control pulse, a waveform that determines when and how long the counter runs. The counter itself is incremented from an internally generated clock frequency. The result is a digital readout that actually shows the number of clock pulses that occur during the input pulse period.

Suppose, for example, the internal clock frequency is fixed at exactly 1 MHz; and further suppose the pulse to be monitored happens to last 100 μs. That means the counter will increment 100 steps while the input pulse is being applied to the circuit. If the counter has been cleared to 0 before this input pulse occurs, it follows that the counter's display assembly will show a figure of 100 at the end of the input pulse interval. That particular readout indicates that a hundred 1-μs clock pulses occurred during the input pulse interval; this is an exact measure of the duration of the input pulse scaled in microseconds.

If, as another example, the readout happens to show 1234 (and the internal clock interval is still set at 1 μs), the input pulse must have a duration of 1234 μs.

Exercises

1. What is the essential difference between the tasks performed by an event counter and a frequency counter?

2. Why is the 2-phase latch-and-clear operation so important to the operation of a frequency counter?

3. Does an electronic digital clock work more like a frequency counter or an event counter?

4. Does a digital tachometer work more like a frequency counter or an event counter?

5. How could a frequency counter be used as part of an A/D (analog-to-digital) converter?

6. An engineer might have a choice of using a thermocouple or a thermistor as the sensing element in a digital thermometer. Which one of these sensors would be used with a VCO/frequency-counter combination? An RC oscillator/pulse-duration counter combination?

12
DATA SELECTORS:
AOIs, MULTIPLEXERS
AND DEMULTIPLEXERS

The preceding chapters dealt with basic logic circuits, flip-flops, astable and mono-stable multivibrators, counter circuits and code converters. Although these circuits in themselves are adequate for performing a great variety of important digital operations, they are often used as subsystems within larger and more sophisticated kinds of digital equipment.

The IC devices described in this chapter are most often used for coordinating the operations of two or more subsystems. Although each of these IC devices does, indeed, have some simpler applications in its own right, its true power can be fully realized only when viewing it as a data controller or a data selector.

As its name implies, a *data selector* is a circuit that is capable of looking at two or more different sources of digital data and selecting only one of those sources that is to be sent to another part of the digital system. In a very real sense, a data selector circuit works as a multi-position rotary or selector switch. Of course, the selection process is a purely electrical one that replaces the mechanical switching mechanisms with solid-state logic circuits.

This chapter introduces the most commonly used data-selector ICs, showing them first as data selectors in larger digital systems and then as handy devices for performing some simpler logic operations that would otherwise call for using a number of different logic-gate packages.

12-1 AND-OR-INVERT (AOI) Functions

Figure 12-1(a) shows a very popular combination of AND and NOR gates. A logic-equation analysis of this circuit shows that output $E = \overline{AB + CD}$. The circuit first ANDs together two pairs of inputs, ORs the results and finally inverts the data at the output. This popular AND-OR-INVERT sequence is commonly abbreviated as AOI.

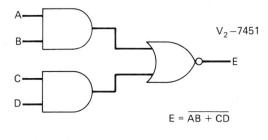

$$E = \overline{AB + CD}$$

(a)

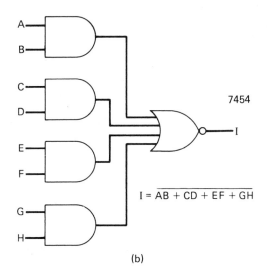

$$I = \overline{AB + CD + EF + GH}$$

(b)

Figure 12-1 Basic AOI functions. (a) 2-wide, 2-input. (b) 4-wide, 2-input.

The three basic logic gates in an AOI combination are included in a single IC package. The user has access to all four inputs and the one output, but the connections between the AND gates and NOR gate are inaccessible.

The AOI combination in Fig. 12-1(a) is available in a 14-pin TTL DIP package. There are actually two identical AOI combinations in this 7451 package; therefore, it is properly called a dual AOI. And since each of the AOI combinations has two

2-input connections, it is further called a 2-wide, 2-input AOI. The 7451 is thus called a dual 2-wide, 2-input AOI.

Figure 12-1(b) shows the internal elements of a TTL 7454 package known as a 4-wide, 2-input AOI. The logic equation for this particular AOI function is $I = \overline{AB + CD + EF + GH}$.

The TTL family of AOIs includes a pair of expandable AOI packages and an AOI expander. An expandable AOI is one that follows the general logic equation $E = AB + CD + X$, where X is a function from another expandable AOI or an expander gate. In essence, an expandable AOI is one that can be expanded to include an additional term, X in this case. See the 2-wide, 2-input expandable AOI circuit in Fig. 12-2(a).

The EXPANDER (X) inputs can come from any TTL-compatible source of logic levels, but they are more often derived from an expander circuit. One such expander gate, shown in Fig. 12-2(b), is simply a 4-input AND gate with a pair of complemented outputs, X and $\bar{X}$. The outputs from this gate are $X = RSTU$ and $\bar{X} = \overline{RSTU}$.

Figure 12-2(c) shows how an expandable AOI and an expander gate are interconnected to yield the function $E = \overline{AB + CD + RSTU}$. Interconnecting expandable AOIs and expander gates is thus a simple matter of connecting together their respective X and $\bar{X}$ terminals. The IC packages used in this particular instance are the expandable section of a 7450 dual AOI and one-half of a dual 4-input expander, 7460.

The EXPANDER connections on an expandable AOI are rather unusual in that they can serve as either inputs or outputs. It is possible, then, to build a complicated 8-wide, 2-input AOI function from a pair of expandable 4-wide, 2-input AOI packages. See Fig. 12-3. Building up such functions is a simple matter of connecting together the corresponding X and $\bar{X}$ terminals on each package.

Whenever the EXPANDER connections aren't to be used, incidentally, they should be left completely unconnected. With the EXPANDER connections thus uncommitted, an expandable AOI functions as its non-expandable counterpart.

Table 12-1 summarizes the most popular TTL-family AOI and AOI expander packages.

Table 12-1 Available TTL AOI Devices

7451	Dual 2-wide, 2-input AOI
7454	4-wide, 2-input AOI
7450	Dual 2-wide, 2-input AOI with one expandable section
7453	Expandable 4-wide, 2-input AOI

	AOI Expanders

7460	Dual 4-input gate expander

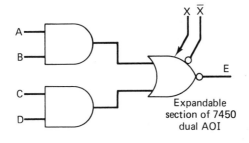

$E = \overline{AB + CD}$
when X, $\overline{X}$ inputs are uncommitted

$E = \overline{AB + CD + X}$
when X, $\overline{X}$ inputs come from gate expander

Expandable
section of 7450
dual AOI

(a)

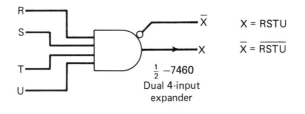

$X = RSTU$

$\overline{X} = \overline{RSTU}$

$\frac{1}{2}$ −7460
Dual 4-input
expander

(b)

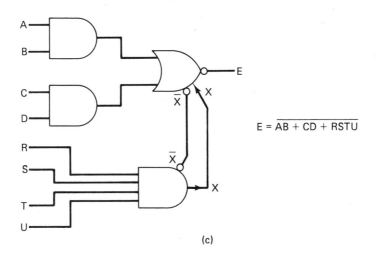

$E = \overline{AB + CD + RSTU}$

(c)

Figure 12-2 AOI gate expansion. (a) An expnadable 2-wide, 2-input AOI. (b) A 4-input gate-expander package. (c) A pair of expandable 2-wide, 2-input AOIs interconnected to perform the expanded function of a single 4-wide, 2-input AOI.

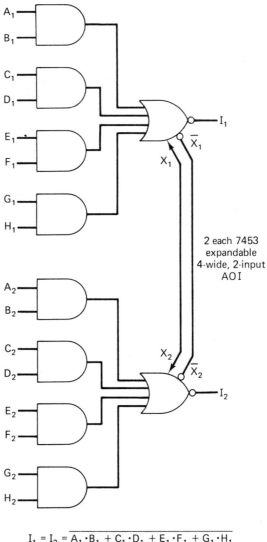

$$I_1 = I_2 = \overline{\begin{array}{l} A_1 \cdot B_1 + C_1 \cdot D_1 + E_1 \cdot F_1 + G_1 \cdot H_1 \\ + A_2 \cdot B_2 + C_2 \cdot D_2 + E_2 \cdot F_2 + G_2 \cdot H_2 \end{array}}$$

Figure 12-3 A pair of expandable 4-wide, 2-input AOIs expanded to perform the function of an 8-wide, 2-input AOI.

12-1.1 AOIs as Data Selectors

AOI packages, both expandable and non-expandable, can serve as data selectors. Their application as data selectors ought to be limited to simpler data-selection circuits, however, because digital multiplexers (described in Sec. 12-2) are far better suited for handling large amounts of data.

Figure 12-4(a) shows a 2-wide, 2-input AOI used in conjunction with a logic inverter to produce a 2:1 data selector. Just as a SPDT mechanical switch is capable of selecting one of two input signals, so is this little circuit. Whenever the S (select) input is at logic 0, AND gate A_2 is effectively gated off. At the same time, however, AND gate A_1 is gated on because the logic-0 level at the S input is inverted to a logic 1 at the lower input of A_1. Thus the output of A_2 is logic 0 and the output of A_1 is D_0. NOR-ing these two logic levels yields $\overline{D_0}$ at output M. In short, setting S to logic 0 makes an inverted version of D_0 appear at output M; the D_1 input is not relevant at all.

Changing the S input to logic 1 gates on AND gate A_2, but it gates off A_1. As a result, the output at terminal M is an inverted version of input D_1.

The circuit in Fig. 12-4(a) is thus capable of selecting data from either input D_0 or D_1, depending on the logic level at S. The function is exactly the same as that of a SPDT switch that is followed by a logic inverter. Selecting the data in the case of the AOI circuit is a matter of changing logic levels at S as opposed to changing the position of a mechanical switch assembly. In a sense, the AOI 2:1 data selector is a solid-state SPDT switch.

The 4:1 data selector circuit in Fig. 12-4(b) uses a 4-wide, 2-input AOI and a binary data converter circuit to mimic the action of a 4-position rotary switch. This circuit delivers an inverted version of any one of the data inputs (D_0, D_1, D_2 or D_3) to output M, depending on the 2-bit binary code at address inputs A_0 and A_1.

The principle of operation is identical to the simpler 2:1 data selector in Fig. 12-4(a), but it is complicated by the need for selecting one of four data inputs. The selection circuitry is made up of the four NAND gates and six inverter circuits.

The selection circuit is actually a modified binary-to-decimal decoder. The inputs, however, are limited to two binary bits, and the outputs are restricted to four decimal units (S_0 through S_3). Whenever the address inputs at A_0 and A_1 are binary 0, select line S_0 is the only one at logic 1; the other three select lines are at logic 0. The overall effect is that an inverted version of the D_0 input appears at output M. See the first two lines in the accompanying truth table.

Setting the address inputs to binary one causes select line S_1 to rise to logic 1 and sets S_0, S_2 and S_3 to logic 0. An inverted version of any data at input D_1 thus appears at M.

In a similar fashion, setting the address inputs to binary 2 and 3 causes inverted versions of D_2 and D_3, respectively, to appear at output M.

Now imagine what would happen if the address inputs, A_0 and A_1, are connected to a 2-bit counter that is running at a 10-kHz rate. The four channels of data at inputs D_0 through D_3 would appear at output M in rapid sequence at a cycle rate of 2.5 kHz, one-fourth the counter's clock frequency. Output M would, for instance, show an inverted version of input D_0 for 100 μs, followed by a look at D_1 for 100 μs. Input D_2 would then appear at M for 100 μs, followed by a 100-μs look at D_3. The whole cycle would then repeat. It would be like connecting

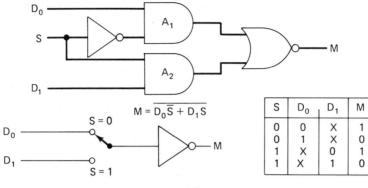

$$M = \overline{D_0\overline{S} + D_1 S}$$

S	D_0	D_1	M
0	0	X	1
0	1	X	0
1	X	0	1
1	X	1	0

(a)

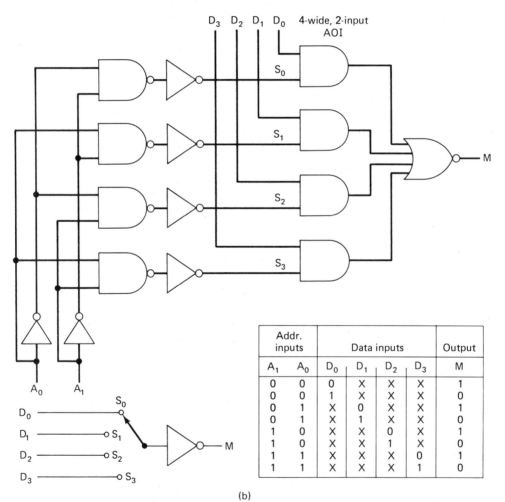

Addr. inputs		Data inputs				Output
A_1	A_0	D_0	D_1	D_2	D_3	M
0	0	0	X	X	X	1
0	0	1	X	X	X	0
0	1	X	0	X	X	1
0	1	X	1	X	X	0
1	0	X	X	0	X	1
1	0	X	X	1	X	0
1	1	X	X	X	0	1
1	1	X	X	X	1	0

(b)

Figure 12-4 AOIs as data selectors. (a) A 2:1 data selector. (b) A 4:1 data selector circuit and function table.

the shaft of an incredible 1,500,000-rpm motor to the selector knob of a 4-pole rotary switch.

12-1.2 AOIs as Logic Gates

Any 2-wide, 2-input AOI gate can be used for working out logic expressions of the general form $E = \overline{AB + CD}$. That happens to be the logic definition of an AOI function. The same system can be used for solving other logic expressions, however. Figure 12-5 shows AOIs used for performing four general classes of logic functions.

The circuit in Fig. 12-5(a) shows how the two pairs of inputs can be connected together to make the AOI behave as an ordinary 2-input NOR gate. This is not to suggest that one should always substitute an AOI gate for a common 2-input NOR gate under every design situation; but the fact that it is possible to perform such a function with an AOI package can be helpful in those instances we which a spare AOI function is available and a 2-input NOR gate is not.

The function in Fig. 12-5(b) can be more useful under a wider variety of design situations. Here the result is a negated AND/OR function of three variables. It is still a rather straightforward application of the general AOI function.

The functions expressed in Figs. 12-5(c) and 12-5(d), however, are less obvious AOI operations. In both instances the inputs are negated or expressed in an active-low form, and the results are active-high AND/OR functions that occur rather frequently in digital circuits. Both circuits rely on DeMorgan's theorems for their derivations. The proofs of these two particular functions are left as exercises at the end of this chapter.

The list of possible functions that can be performed with an expanded AOI or a 4-wide, 2-input AOI package is very extensive. Again, some of the functions that are possible with these more complex AOI forms are left as exercises at the end of the chapter.

Although AOI packages can, indeed, perform a wide variety of combinatorial logic functions, a circuit designer should not exert too much effort trying to force-fit a complex function to an AOI format. Whenever it is possible to do a logic operation with common NAND and NOR gates in a straightforward manner, the designer ought to use that approach.

12-2 Multiplexers

A multiplexer circuit is a type of data selector. The term *multiplexer*, however, is normally reserved for data selectors that are capable of handling at least four inputs at the same time.

The simplest commercially available multiplexer package, for example, works like a 4-pole, double-throw toggle switch. See Fig. 12-6(a). Whenever the ganged switches are in one position, data inputs A_0, A_1, A_2 and A_3 appear at outputs M_0, M_1, M_2 and M_3, respectively. Setting the switch to its opposite position, however, connects the M outputs to the corresponding B inputs.

$\overline{F + G}$ Functions of the form
                            $\overline{F + G}$

(a)

$\overline{FG + H}$ Functions of the form
                            $\overline{FG + H}$

(b)

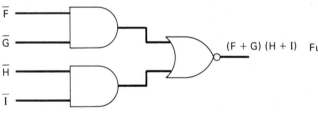

$(F + G)(H + I)$ Functions of the form
                   $(F + G)(H + I)$

(c)

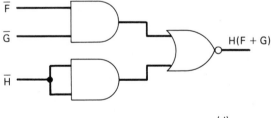

$H(F + G)$ Functions of the form
                    $H(F + G)$

(d)

Figure 12-5   Four classes of logic operations possible with a 2-wide, 2-input AOI.

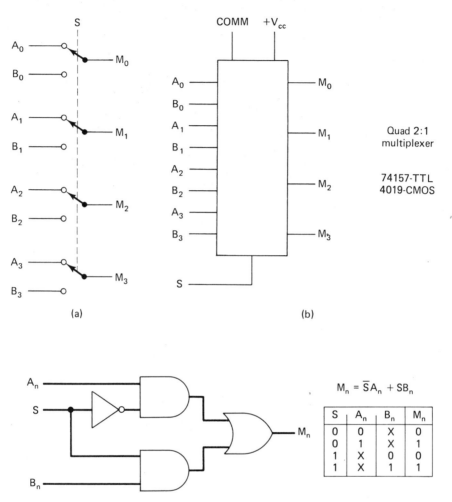

Figure 12-6 A quad 2:1 multiplexer function. (a) Switch-equivalent circuit. (b) Input, output and strobe connections for the 74157 (TTL) and 4019 (CMOS) quad 2:1 multiplexer/data selector. (c) Equivalent logic circuit and truth table for each of the four sections.

Since this particular circuit can select data from one of two sources, it is called a 2:1 multiplexer; in a sense, it reduces two lines of incoming data to a single output line. And since there are four of these 2:1 multiplexers in a single package, it is properly known as a quad 2:1 multiplexer.

Figure 12-6(b) shows an IC layout for a quad 2:1 multiplexer such as the 74157 TTL or 4019 CMOS packages. In either case, setting the S input to logic 0 makes the A inputs appear at the M outputs. Setting the S input to logic 1 directs the B inputs to the M outputs. The circuit is thus capable of directing one of two

4-bit binary words to the outputs, depending on the logic level applied at the S input.

Figure 12-6(c) shows an equivalent logic diagram of one of the four sections in the quad 2:1 multiplexer IC. It is basically a 2-wide, 2-input AND-OR function that directs the A_n input to M_n whenever $S = 0$ and the B_n input to M_n whenever $S = 1$. See the rather simple logic equation and truth table accompanying the circuit in Fig. 12-6(c).

The 74153 TTL IC device is known as a dual 4:1 multiplexer. As the name implies, it contains two identical multiplexer sections, each one capable of selecting one of four different data inputs. The circuit in this instance works like a 2-deck, 4-position rotary switch. See Fig. 12-7(a). The switch is "rotated" by applying dif-

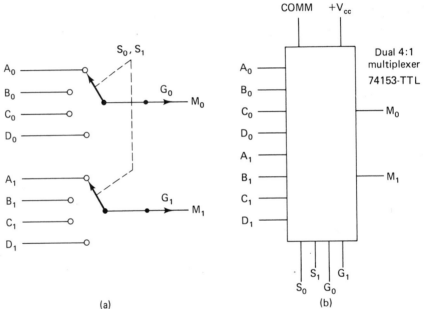

(a)

(b)

	Select inputs		Output
G_n	S_1	S_0	M_n
1	X	X	0
0	0	0	A_n
0	0	1	B_n
0	1	0	C_n
0	1	1	D_n

(c)

Figure 12-7 A dual 4:1 multiplexer function. (a) Switch-equivalent circuit. (b) Input, output and strobe connections for the 74153 TTL dual 4:1 multiplexer. (c) Function table for each of the two sections.

ferent 2-bit binary words to the S_0 and S_2 inputs. Whenever $S_0 = S_1 = 0$, for instance, the switch is in a position to deliver inputs A_0 and A_1 to M_0 and M_1, respectively. Changing the S inputs to $S_0 = 1$, $S_1 = 0$, however, changes the switch position to the B inputs, and so on.

Figure 12-7(b) shows the basic IC layout for the 74153 4:1 data multiplexer IC. Note the presence of a G (STROBE) input. According to the function table in Fig. 12-7(c), the G input is capable of turning off the multiplexer action altogether. Setting $G = 1$ sets the M outputs to logic 0, regardless of the status of the data inputs and the selector terminals. This STROBE feature is especially useful for expanding multiplexer functions as described later in this section.

Table 12-2 summarizes the multiplexer ICs now available in the TTL and CMOS families. The quad 2:1 and dual 4:1 packages have been described here in some detail. The 8:1 and 16:1 multiplexers work in a similar fashion. The 8:1 multiplexer package requires three separate select lines, however, in order to select one of eight different 1-bit inputs. The 16:1 unit selects one of 16 different 1-bit inputs by means of four different select lines. All multiplexers, with the notable exception of the 2:1 versions, have a G (STROBE) input that disables the entire operation whenever $G = 1$.

Table 12-2 AVAILABLE TTL/CMOS MULTIPLEXER DEVICES

74157	Quad 2:1 multiplexer——TTL
74153	Dual 4:1 multiplexer——TTL
74151A	8:1 multiplexer——TTL
74150	16:1 multiplexer——TTL
4019	Quad 2:1 multiplexer (AND-OR select)——CMOS

Figure 12-8 is a novel circuit that clearly demonstrates how a multiplexer circuit can be used for scanning (time multiplexing) a number of different input signals. The input signals in this particular instance are four different frequencies in the audio range. The select inputs on the multiplexer are connected to a modulo-4 counter that generates a continuous string of 2-bit binary numbers cycling between decimal 0 and 3.

The circuit requires two 555-type astable multivibrators, one operating at about 4 kHz and the other running at about 1 Hz. (See Sec. 7-2.2.1 for information on the design of these rather simple oscillator circuits.) The 4-kHz oscillator provides the main input frequency that is subsequently divided by 2, 4, 8 and 16 by means of an ordinary 4-bit binary counter, IC2. These four audio frequencies make up the data inputs to the multiplexer circuit.

The multiplexer in this case is one-half of a dual 4:1 multiplexer package. The frequency reaching the M_0 output at pin 7 of IC3 is determined by the binary number appearing at the S_0 and S_1 select inputs. Whenever $S_0 = S_1 = 0$, for instance, the output of the multiplexer is connected to the A_0 input, the 2-kHz frequency source. And when the select inputs change to $S_0 = 1$, $S_1 = 0$, the multiplexer outputs the signal appearing at B_0, the 1-kHz frequency source. Five-hun-

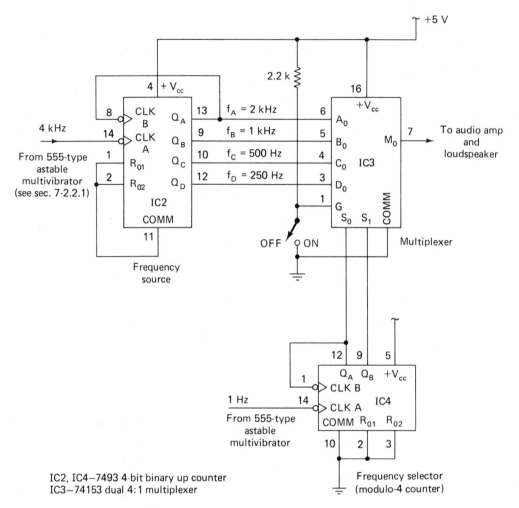

Figure 12-8   A simple circuit for demonstrating the data-selecting features of a 4:1 multiplexer.

dred hertz and 250-Hz signals appear at M_0 as the select inputs change to $S_0 = 1$, $S_1 = 0$ and then $S_0 = S_1 = 1$, respectively.

The select inputs come from a simple modulo-4 (2-bit binary) counter which is, in turn, clocked by the second 555-type astable multivibrator. This multivibrator runs at a 1-Hz rate, making the modulo-4 counter dwell at each count for about 1 s.

In short, IC2 provides four different audio frequencies at the inputs of the 4:1 multiplexer, IC3. These inputs are scanned at a 1-Hz rate by the outputs of IC4. The overall effect is that the M_0 output of IC3 is a sequence of audio frequencies changing at a 1-Hz rate from 2 kHz, to 1 kHz, 500 Hz, 250 Hz and then back to 2 kHz.

The output can be switched off and on by means of the switch connected to the G input of IC3. Setting the switch to the position that grounds G allows the multiplexer to run normally. Setting that switch to its OFF position lets the 2.2-k resistor pull the G input up to logic 1, thereby fixing the M_0 output at logic 0 and stopping the scanning action. (Actually, the scanning action will continue, but no signal will appear at M_0.)

The circuit is a simple sort of music maker, a circuit capable of playing a sequence of four audio tones. Although it can be fun to play with this circuit, bear in mind that it is also a good example of how a multiplexer can be used as a data scanner.

Such systems are commonly used for time multiplexing any number of data channels onto a single output channel. The whole point of the process is to make data transmission more efficient by replacing a large number of data channels with a single time-multiplexed channel. The discussion of demultiplexer circuits in the next section of this chapter shows how the time-multiplexed data can be "unscrambled" at the receiver end of the data link.

Multiplexers can thus be used as data selectors or data scanners. The multiplexer can be considered a data-selector circuit if it is possible to select any one of the outputs in any order and at any desired time. Replacing IC4 in Fig. 12-8 with two switches connected to the S_0 and S_1 inputs, for instance, would give the user the option of manually selecting any one of the four frequencies in any order and for any desired length of time. The user could then select the signal to appear at the multiplexer output.

Using a free-running counter circuit as shown in Fig. 12-8 ties the circuit to one particular multiplexing sequence and switching rate. Aside from being able to turn the circuit off by means of the switch at the G input of the multiplexer, a user has no control over the scanning operation.

12-2.1 Expanding Multiplexer Functions

It is possible to reduce the multiplexing range of any standard multiplexer IC by simply restricting the range of the input selector scheme. An 8 : 1 multiplexer, for example, can be reduced to the function of a 4 : 1 multiplexer by permanently fixing the most-significant bit of the select inputs to either logic 0 or logic 1. This would allow only two active select lines capable of addressing only four of the data inputs. If the most-significant select bit is connected to logic 0, the two remaining select inputs would select any one of the first four data input positions. If, however, the most-significant select bit is connected to logic 1, the two remaining select lines would address any one of the four higher-order data inputs.

As a specific example of restricting the data-selecting capability of a multiplexer, consider the effect of connecting S_1 of IC3 in Fig. 12-8 to COMM. The only multiplexing that could take place would then be between the A_0 and B_0 inputs— 2 kHz and 1 kHz. If S_1 is fixed at +5 V (logic 1), the counter input at S_0 could select only one of the two inputs at C_0 and D_0.

Although it is sometimes useful to restrict the data-selection or data-scanning capability of a standard multiplexer circuit, the need is more often to expand the selection range of available multiplexer ICs.

There are two principal kinds of multiplexer-expansion situations. The first is to extend the number of available data inputs. In this case, it might be necessary to use a pair of 8 : 1 multiplexers to build up the function of a single 16 : 1 multiplexer. The other kind of multiplexer-expansion situation arises whenever there is a need to multiplex digital words having more than one bit. It is, for instance, often necessary to use four 8 : 1 multiplexer ICs to select one of eight different 4-bit words, in essence, performing the function of a quad 8 : 1 multiplexer.

All of the examples cited here use the 74151A 8 : 1 multiplexer IC as a model for expanding multiplexer functions. The principles, however, apply equally well to any of the standard multiplexer devices.

Figure 12-9 shows the technique for building up a 16 : 1 multiplexer function from a pair of 8 : 1 multiplexer ICs. The first 8 inputs go directly to the 8 data inputs of IC1 and the second 8 inputs go to the corresponding inputs of IC2. Thus there are 16 different data inputs connected to the two 8 : 1 multiplexer ICs.

A single 8 : 1 multiplexer device has only three select inputs, however; and four select inputs are required for operating a 16 : 1 multiplexer function. Note in Fig. 12-9 that the three select inputs (S_0, S_1 and S_2) are connected together on the two ICs and go directly to the corresponding S_0, S_1 and S_2 data-selection inputs of the overall system.

Now notice that the fourth select input to the system, S_3, goes directly to the *ST* (STROBE) input of IC1. That same select input, however, goes to the *ST* input of IC2 only after passing through a NAND-gate version of a logic inverter. The STROBE inputs of the two multiplexer ICs are thus always complements of one another.

Recall that a multiplexer of this type is disabled whenever its STROBE input is set to logic 1 and enabled only when that same input is set to logic 0. IC1 is thus enabled only when $S_3 = 0$; and by the same token, IC2 is enabled only when $S_3 = 1$. Only one of the two 8 : 1 multiplexers is enabled at any given time, depending on the logic level at select input S_3.

The function table accompanying the drawing in Fig. 12-9 shows all 16 possible select-input combinations. While $S_3 = 0$, inputs S_0, S_1 and S_2 actively select one of the 8 inputs to IC1, data inputs *A* through *H*. IC2 is being addressed by S_0, S_1 and S_2 at the same time, but it is disabled by virtue of the fact that its STROBE input is seeing a logic-1 level from S_3.

When select-input S_3 goes to logic 1, inputs S_0, S_1 and S_2 actively select one of the eight inputs to IC2, data inputs *I* through *P*. IC1 is completely disabled under this set of conditions because its STROBE input is set to logic 1 by S_3.

The system's output is thus any one of the 16 different inputs; and the data is coming from IC1 as long as $S_3 = 0$ and from IC2 as long as $S_3 = 1$. The only problem that remains is to OR together the outputs from the two multiplexers.

It so happens the 74151A 8 : 1 multiplexer device has an inverted output avail-

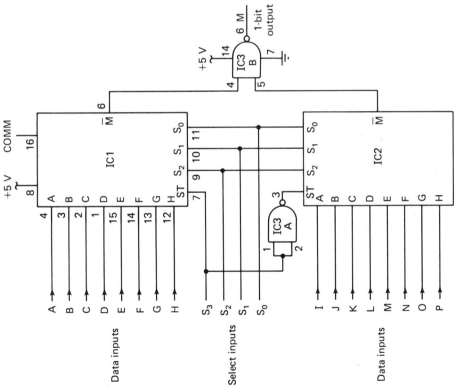

Select inputs				Output M
S_3	S_2	S_1	S_0	
0	0	0	0	A
0	0	0	1	B
0	0	1	0	C
0	0	1	1	D
0	1	0	0	E
0	1	0	1	F
0	1	1	0	G
0	1	1	1	H
1	0	0	0	I
1	0	0	1	J
1	0	1	0	K
1	0	1	1	L
1	1	0	0	M
1	1	0	1	N
1	1	1	0	O
1	1	1	1	P

IC1 enabled (ST = 0) / IC2 disabled (ST = 1)
IC1 disabled (ST = 1) / IC2 enabled (ST = 0)

Notes: (1) IC1, IC2—74151A 8:1 multiplexer
IC3—7400 quad 2-input NAND
(2) Pin numbers for IC2 identical to IC1
(3) ST = strobe input

Figure 12-9 Expanding two 8:1 multiplexer ICs to do the job of a 16:1 multiplexer.

297

able. Whenever IC1 is enabled, for instance, an inverted version of one of its eight inputs appears at the output, pin 6. Similarly, an inverted version of one of the data inputs I through P appears at the $\overline{M}$ output of IC2 whenever that IC is enabled by setting S_3 to logic 1.

These inverted data outputs, coming from either IC1 or IC2, can be easily OR-ed together by means of a 2-input NAND gate. (Recall the DeMorgan OR function that is possible with inverted or active-low data applied to the inputs of a NAND gate.)

The output of the system, terminal M at pin 6 of IC3-B, is thus equal to one of the 16 different inputs as specified in the function table. This circuit, in all respects, performs the function of a single 16:1 multiplexer.

Especially note the technique of expanding the number of select or address lines by means of the IC's STROBE inputs. This same procedure will be used for expanding memory functions described in a later chapter.

Now consider an example of an entirely different sort of multiplexer expansion. In this case, there is a need for interconnecting four 8:1 multiplexers to perform the function of a quad 8:1 multiplexer system. This particular situation arises whenever it is necessary to select or scan a set of eight different 4-bit binary words. All of the examples cited thus far select or scan 1-bit words.

The circuit in Fig. 12-10 might appear extremely complicated, but there is a definite pattern to the connections; and once that pattern is recognized, the circuit doesn't appear so imposing.

Note that there are eight different 4-bit inputs. The first 4-bit input is designated A_0, A_1, A_2 and A_3, with A_0 being the least-significant bit. The second 4-bit input is represented by B terms, B_0, B_1, B_2 and B_3. And in a similar fashion, the remaining six 4-bit inputs carry prefixes C, D, E, F, G and H; each one consists of bits 0, 1, 2 and 3.

So the system has eight different 4-bit inputs. The problem is to select one of these eight words and make it appear at the 4-bit outputs, M_0, M_1, M_2 and M_3. Whenever the system is selecting the A inputs, for instance, $M_0 = A_0$, $M_1 = A_1$, $M_2 = A_2$ and $M_3 = A_3$. And, similarly, selecting the H inputs should make $M_0 = H_0$, $M_1 = H_1$, $M_2 = H_2$ and $M_3 = H_3$. The M outputs, in other words, should be equal to one of the eight different 4-bit inputs at any given moment.

Much of what seems so complicated about the circuit in Fig. 12-10 is simply a lot of lines that are necessary for getting each of the four bits at the inputs to the corresponding multiplexer IC. Note that all of the 0 bits at the inputs go to the M_0 multiplexer, IC1. Similarly, all the 1 bits go to the M_1 multiplexer, IC2; all of the 2 bits go to the M_2 multiplexer, IC3; and all of the 3 bits go to the M_3 multiplexer, IC4.

Now see that like select inputs of the four multiplexer ICs are connected together: All of the S_0 inputs are connected together to the system's S_0 select input; all S_1 inputs are connected to the system's S_1 input; and all of the S_2 inputs are connected to the main S_2 input. In the same manner, the ST (STROBE) inputs are connected together. In short, all four multiplexer ICs are addressed

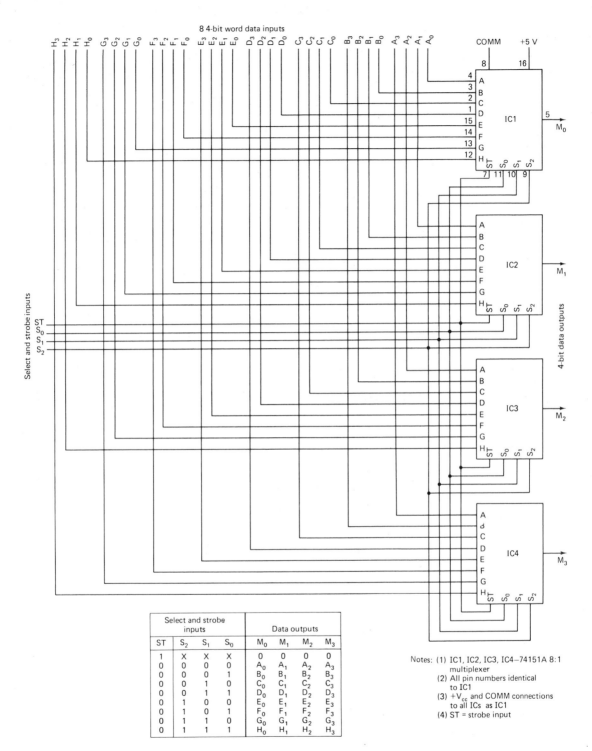

Figure 12-10 Expanding four 8:1 multiplexers to perform the function of a quad 8:1 multiplexer system.

299

in parallel; whatever input one of the multiplexers is selecting, all the other multiplexers are also selecting that input.

The function table accompanying the circuit in Fig. 12-10 summarizes the action of this quad 8:1 multiplexer function. Whenever ST is set to logic 1, for instance, all four multiplexers are disabled and their M outputs are at logic 0, regardless of the status of S_0, S_1 and S_2.

Setting ST to logic 0, however, enables all four ICs at the same time, allowing the select inputs to select one of the inputs at each of the multiplexer devices. If $S_0 = S_1 = S_2 = 0$, for example, all four multiplexers output whatever logic level appears at their A inputs; and since their A inputs are connected to the four components of the A input word, it follows that the circuit's outputs will show $M_0 = A_0$, $M_1 = A_1$, $M_2 = A_2$ and $M_3 = A_3$.

The four multiplexer ICs are addressed or selected in parallel, thus delivering the same input position to their M outputs at any given time.

These two examples of expanding multiplexer functions are only examples; they merely indicate the techniques for expanding multiplexer functions indefinitely. By properly selecting the STROBE inputs, a number of separate multiplexer packages can be cascaded to make up an n: 1 multiplexer function, where n is any desired number. And as illustrated in Fig. 12-10, any number of multiplexer packages can be paralleled to make up systems capable of multiplexing words having two or more bits each. Then, too, there is the possibility of applying both techniques at the same time, both expanding the number of inputs and paralleling the functions to handle words of any size.

12-2.2 Logic Functions from Multiplexers

A multiplexer is a data-selection device that has a number of data inputs that can be selected or addressed one at a time in any sequence. This feature makes it possible to use data multiplexers for performing some relatively complicated logic functions.

Consider, for example, the logic equation and truth table in Fig. 12-11(a). This is a typical digital design problem in which it is necessary to perform the function $E = A\bar{B}\bar{C} + \bar{A}\bar{B}C + ABC$. Figure 12-11(b) shows how the function can be executed by means of four 3-input NAND gates. If it is assumed that both inverted and true versions of the A, B and C inputs are already available, this particular NAND-gate approach requires two IC packages: all the NAND gates in one triple 3-input NAND package and one gate in a second package of the same type.

The very same function, however, can be worked out with a single 8:1 multiplexer as shown in Fig. 12-11(c). Note that the multiplexer's select inputs are connected to logic sources A, B and C. Whenever inputs A, B and C are all at logic 0, then, the multiplexer is addressing the D_0 input; and if that input is connected to logic 0, the first line of the truth table in Fig. 12-11(a) is satisfied.

In a similar fashion, setting $A = C = 0$ and $B = 1$ addresses input D_2; and

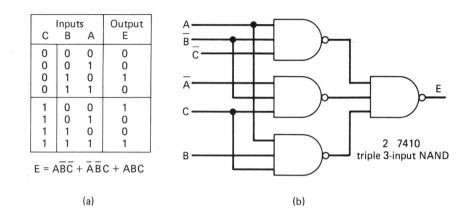

| Inputs | | | Output |
C	B	A	E
0	0	0	0
0	0	1	0
0	1	0	1
0	1	1	0
1	0	0	1
1	0	1	0
1	1	0	0
1	1	1	1

$E = \overline{A}B\overline{C} + A\overline{B}\overline{C} + ABC$

(a)

(b)

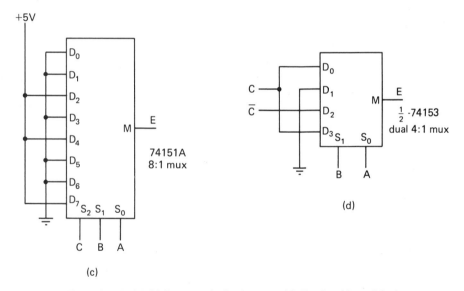

(c)

(d)

Figure 12-11   Multiplexers as logic elements. (a) Truth table and logic equation for a particular logic problem. (b) A NAND gate solution to the problem. (c) An 8:1 multiplexer solution to the logic function. (d) An efficient 4:1 multiplexer solution to the same logic function.

since the D_2 input of the multiplexer is connected to logic 1, the situation satisfies the third line of the truth table. Select any of the eight possible combinations of logic inputs A, B and C and you will see that the corresponding multiplexer D inputs are connected to logic 0 or logic 1 as specified by the E column in the truth table.

The solution to any logic equation, no matter how simple or complex it might be, is either a logic 0 or a logic 1; and by using the select inputs of a multiplexer device as the logic inputs the multiplexer's data inputs can be programmed to provide the appropriate 0 or 1 output. There is often quite a savings in the number of IC inputs, and there is always a considerable amount of time saved in designing and setting up the multiplexer-based logic circuit.

Another important advantage of multiplexer-based logic circuits is that it is so easy to modify the logic function. Suppose, for example, the circuit designer builds the NAND-gate solution to the truth-table situation in Fig. 12-11(a). After connecting up the circuit, the designer suddenly realizes that the first line of the truth table should read $E = 1$—$E = 1$ when $A = B = C = 0$. Modifying the existing NAND-gate version would call for a considerable amount of reworking and would result in a lot of inconvenience and a sloppy circuit.

Reworking the multiplexer version of the circuit, however, would call for simply changing the D_0 connection in Fig. 12-11(c) from the common bus to the +5-V bus.

The simplicity and low cost of multiplexer-solved logic equations make sophisticated digital mapping techniques all but obsolete. Whereas digital engineers once had to spend countless hours attempting to simplify and optimize the efficiency of logic designs by using conventional combinatorial logic functions, they can now use this multiplexer scheme to get the circuit operational in less time than it takes to set up the mappings.

Although the multiplexer-based logic system in Fig. 12-11(c) is, indeed, simple and quick to implement, it is possible to simplify the procedure even more. Actually, the 3-input logic function specified in Fig. 12-11(a) can be performed with a 4:1 multiplexer device as shown in Fig. 12-11(d). In this case, inputs A and B are connected to the multiplexer's two select inputs, S_0 and S_1. These two inputs select one of the four data inputs at any given time.

Now note from the truth table that $A = B = 0$ at two places: once when $C = 0$ and once when $C = 1$. Since output E is equal to C in these two instances, it is possible to say that $E = C$ whenever the select inputs are addressing D_0. Thus the D_0 input of the multiplexer can be connected to logic input C.

Whenever $A = 1$ and $B = 0$ (a condition that also occurs twice in the truth table), E is always equal to logic 0. Thus the D_1 input of the multiplexer (the one addressed by $A = 1$, $B = 0$) can be fixed at common, or logic 0.

The situation is slightly different whenever $A = 0$ and $B = 1$. Such a condition occurs twice in the truth table, with $E = 1$ when $C = 0$ and with $E = 0$ when $C = 1$. In other words, $E = \bar{C}$ whenever $A = 0$, $B = 1$. This situation is handled by simply connecting the multiplexer's D_2 input (the one addressed by $A = 0$, $B = 1$) to a $\bar{C}$ logic input.

This particular example illustrates the fact that any logic function can be solved with a multiplexer device having one less select line than variables in the logic expression. In the example just cited, it is possible to solve a 3-variable logic equation with a multiplexer having just two select lines. By the same token, an 8:1 multiplexer having three select lines can be used for solving logic equations having four input variables. And since a 16:1 multiplexer has four select lines, it is capable of handling logic expressions having as many as five independent input variables.

If it seems that the relatively simple application of a multiplexer to the example in Fig. 12-11 does not present a good case for its advantages, turn to the example in Fig. 12-12. The truth table in Fig. 12-12(a) was derived directly from the need for generating an image of a race can on the screen of a certain kind of TV game. This is, indeed, a complex logic function of five independent variables that would challenge the skill and patience of the most experienced engineer who still places undue faith in conventional mapping and circuit-mapping techniques.

The fact that there are five input variables makes it necessary to use a multiplexer device having at least four select lines, a 16:1 multiplexer, to be exact. It is a good bet that a technician could determine the D inputs to the multiplexer, wire the circuit and begin testing it before anyone else could get a good mapping/optimization procedure well underway.

And what's more, it is impossible to build a circuit to solve this truth table with

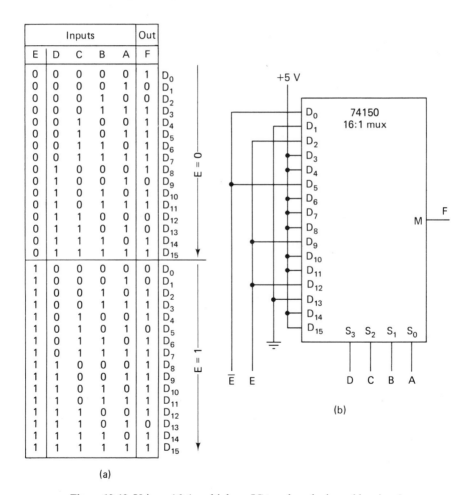

| Inputs | | | | | Out | |
E	D	C	B	A	F	
0	0	0	0	0	1	D_0
0	0	0	0	1	0	D_1
0	0	0	1	0	0	D_2
0	0	0	1	1	1	D_3
0	0	1	0	0	1	D_4
0	0	1	0	1	1	D_5
0	0	1	1	0	1	D_6
0	0	1	1	1	1	D_7
0	1	0	0	0	1	D_8
0	1	0	0	1	0	D_9
0	1	0	1	0	1	D_{10}
0	1	0	1	1	1	D_{11}
0	1	1	0	0	0	D_{12}
0	1	1	0	1	0	D_{13}
0	1	1	1	0	1	D_{14}
0	1	1	1	1	1	D_{15}
1	0	0	0	0	0	D_0
1	0	0	0	1	0	D_1
1	0	0	1	0	1	D_2
1	0	0	1	1	1	D_3
1	0	1	0	0	1	D_4
1	0	1	0	1	0	D_5
1	0	1	1	0	1	D_6
1	0	1	1	1	1	D_7
1	1	0	0	0	1	D_8
1	1	0	0	1	1	D_9
1	1	0	1	0	1	D_{10}
1	1	0	1	1	1	D_{11}
1	1	1	0	0	1	D_{12}
1	1	1	0	1	0	D_{13}
1	1	1	1	0	1	D_{14}
1	1	1	1	1	1	D_{15}

(a)

(b)

Figure 12-12 Using a 16:1 multiplexer IC to solve a logic problem involving five independent variables. (a) Truth table for the desired logic function. (b) Programming for the 16:1 multiplexer that solves the problem.

one conventional combinatorial logic package (NAND, NOR, INVERT gates and so on). Here, a very complex logic operation is implemented with a single 16:1 multiplexer IC.

This is not to say that *every* logic function should be solved with a multiplexer; but this multiplexer technique certainly becomes advantageous whenever there are a relatively large number of input variables and a relatively small number of outputs.

12-3 Demultiplexer Circuits

While the task of a multiplexer circuit is to select any one of a number of inputs at any given time, a demultiplexer performs the complementary operation—sending a 1-bit input to any one of a number of outputs. Figure 12-13 contrasts the selector-switch equivalents of a multiplexer and demultiplexer device.

Multiplexer   Demultiplexer

Figure 12-13 A comparison of switch-equivalents for multiplexers and demultiplexers. (a) A 4:1 multiplexer equivalent. (b) A 1:4 demultiplexer equivalent.

Figure 12-14 shows the NAND-gate equivalents of a 2-output and a 4-output demultiplexer. The 2-output demultiplexer in Fig. 12-14(a) has a single data input, D; a single select input, S; and two outputs, N_0 and N_1. Whenever S is set to logic 0, G_1 is enabled by virtue of the fact that the logic-0 S input is inverted to logic 1; and that action allows an inverted version of the D input to appear at N_0. As long as $S = 0$, however, NAND gate G_2 is effectively gated off, and its output is fixed at logic 1, regardless of the logic level appearing at input D. Setting S to logic 0 thus selects the N_0 output, delivering an inverted version of the D input to that point.

Setting input S to logic 1 reverses the situation, disabling NAND gate G_1 and enabling G_2 so that an inverted version of the D input appears at output N_1.

An inverted version of the D input thus appears at output N_0 or N_1, depending on the logic level at the select, S, input. See the truth table accompanying the diagram in Fig. 12-14(a).

The demultiplexer circuit in Fig. 12-14(b) works in a similar fashion, but it is capable of sending an inverted version of the D input to any one of four possible outputs, N_0 through N_3. The table in Fig. 12-14(b) summarizes the relationship

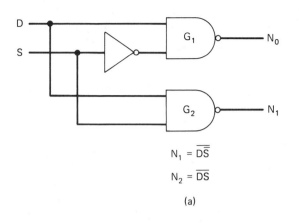

Inputs		Outputs	
S	D	N_0	N_1
0	0	1	1
0	1	0	1
1	0	1	1
1	1	1	0

$$N_1 = \overline{D\overline{S}}$$

$$N_2 = \overline{\overline{D}S}$$

(a)

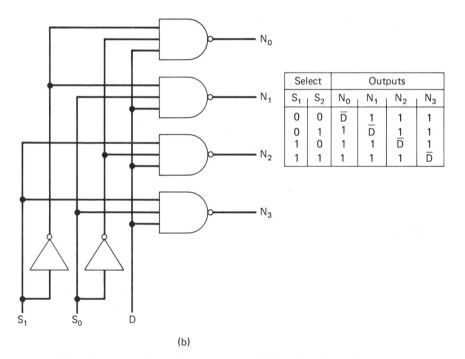

Select		Outputs			
S_1	S_2	N_0	N_1	N_2	N_3
0	0	$\overline{D}$	1	1	1
0	1	1	$\overline{D}$	1	1
1	0	1	1	$\overline{D}$	1
1	1	1	1	1	$\overline{D}$

(b)

Figure 12-14 Logic equivalents of demultiplexer functions. (a) A 1:2 demultiplexer. (b) A 1:4 demultiplexer or 2-line to 4-line decoder.

between the logic levels appearing at the two select inputs (S_0, S_1 and their complements) and the four data outputs.

It is no trivial fact that demultiplexer circuits such as these accept active-high data and select inputs but generate active-low outputs. All of the commercially available demultiplexer devices listed in Table 12-3 have active-low outputs.

Recall that data multiplexers can be applied in two entirely different kinds of operations: data multiplexing and executing relatively complex combinatorial logic functions. Demultiplexers are similarly used for two distinctly different kinds of operations, and it so happens that demultiplexers are specified differently in these two instances.

When used as a data demultiplexer (converting one line of serial or time-multiplexed data back into parallel data), the devices are specified as a ratio of the number of data inputs to data outputs. Used as a demultiplexer, the circuit in Fig. 12-14(a) would be classified as a 1:2 demultiplexer. And since the circuit in Fig. 12-14(b) has one data input and four outputs, it would be classified as a 1:4 demultiplexer.

Demultiplexers, however, can also be used as data selectors. If the D input in Fig. 12-14(b) is permanently fixed at logic 1, for example, the outputs would drop from logic 1 to logic 0 in sequence and one at a time as the select inputs are cycled through binary 0, 1, 2 and 3. The action here is identical to that of a BCD-to-decimal decoder circuit. The output range in Fig. 12-14(b), however, only covers four digits instead of ten.

When applied as a data selector, demultiplexers are specified by the ratio of the number of select lines to the number of data outputs. The circuit in Fig. 12-14(b), for example, has two select lines and four data output lines. It can thus be called a 2-line to 4-line decoder.

Note that the list of commercially available demultiplexer devices in Table 12-3 are specified as either 1:4 and 1:16 demultiplexers or, alternately, 2-line to 4-line and 4-line to 16-line decoders.

Table 12-3   Available TTL Demultiplexer/Decoder IC Devices

74155	Dual 1:4 demultiplexer (2-line to 4-line decoder)——TTL
74156	Dual 1:4 demultiplexer (2-line to 4-line decoder)——TTL, open collector
74154	1:16 demultiplexer (4-line to 16-line decoder)——TTL

Figure 12-15(a) shows the pinout and function designation of a 74155 or 74156 dual 4:1 demultiplexer IC device. The two TTL devices differ only in that the 74155 has a totem-pole output and the 74156 version has an open-collector output configuration.

The two sections of the 74155, '56 share a common set of select inputs, S_0 and S_1. The two sections of this dual demultiplexer, in other words, are selected in parallel. The sections have independent data inputs, strobe (G) inputs and data outputs.

The two function tables in Fig. 12-15 reflect an essential difference between the two halves of the 74155, '56 demultiplexer IC. Note that the A section has an active-high data input, D_a, and that the B section has an active-low data input, D_b. Other than that particular distinction, the two sections are identical: they share the same select inputs, they have independent but identically phased strobe (G) inputs; and they have active-low data outputs.

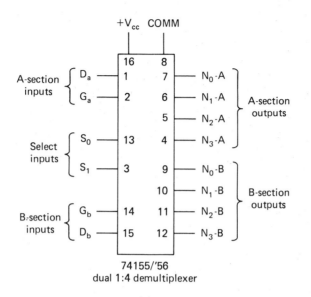

(a)

Inputs							
S_1	S_0	G_a	D_a	N_0-A	N_1-A	N_2-A	N_3-A
X	X	1	X	1	1	1	1
X	X	X	0	1	1	1	1
0	0	0	1	0	1	1	1
0	1	0	1	1	0	1	1
1	0	0	1	1	1	0	1
1	1	0	1	1	1	1	0

Active-high D input
Active-low outputs

(b)

Inputs							
S_1	S_0	G_b	D_b	N_0-B	N_1-B	N_2-B	N_3-B
X	X	1	X	1	1	1	1
X	X	X	1	1	1	1	1
0	0	0	0	0	1	1	1
0	1	0	0	1	0	1	1
1	0	0	0	1	1	0	1
1	1	0	0	1	1	1	0

Active-low D input
Active-low outputs

(c)

Figure 12-15   Terminology and pinout for a dual 1 : 4 demultiplexer or 2-line to 4-line decoder. (a) The 74155, '56 demultiplexer IC. (b) Function table for the A section. (c) Function table for the B section.

The reason for active-low inputs in the data-input section of the B-half of the IC will become clear when considering how this dual 1 : 4 demultiplexer is readily expanded to a single 1 : 8 demultiplexer or 3-line to 8-line decoder.

12-3.1 Expanding the 74155, '56 Functions

The 74155 dual 1 : 4 demultiplexer can be expanded to a single 1 : 8 demultiplexer (or from a dual 1-line to 4-line decoder to a single 3-line to 8-line decoder) by simply connecting together the two data inputs and the two strobe inputs. See the diagram and resulting function table in Fig. 12-16(a).

As long as the G input is at logic 1, the entire circuit is disabled, and all eight of its active-low outputs are at logic 1. Setting the G input to logic 0 enables the circuit, turning over control of the system to the S inputs.

There are three select inputs in this case, labeled S_0, S_1 and S_2. Inputs S_0 and S_1 go to their respective select inputs of the demultiplexer IC. The S_2 input, however, is connected to both the data inputs, D_a and D_b. Whenever $S_2 = 0$, the B section of the IC package is enabled, allowing the S_0 and S_1 bits to select one of the four B-section data outputs. Through the second half of the function table, however, $S_2 = 1$ and the A section of the IC are enabled. The fact that the data inputs to the two 4:1 sections of the 74155 are active-low in one instance and active-high in the other makes this kind of function expansion rather simple.

Figure 12-16(b) illustrates a second type of demultiplexer expansion. Here four 1:4 demultiplexers are connected in parallel such that they direct a 4-bit input word to one of four different destinations. The input word is made up of bits C_0, C_1, C_2 and C_3, with C_0 being the least-significant bit. The inverters at the C_0 and C_2 inputs are necessary because one-half of each of the 74155 demultiplexer sections has an active-low input. Using these inverters makes it possible to apply active-high input words.

Whenever the circuit's STROBE input is set to logic 1, it disables all four demultiplexer sections, forcing all of their outputs to logic 1. See the first line in the truth table accompanying the diagram in Fig. 12-16(b). Setting the G input to logic 0 enables all four sections, thus turning over circuit control to the select inputs, S_0 and S_1.

Now suppose $S_0 = S_1 = 0$. In this case, each demultiplexer section is being addressed such that its data input is directed to its N_0 output. Any 4-bit word appearing at the C inputs thus appears inverted at outputs C_{00}, C_{10}, C_{20} and C_{30}.

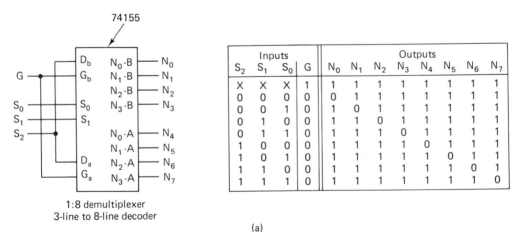

Inputs				Outputs							
S_2	S_1	S_0	G	N_0	N_1	N_2	N_3	N_4	N_5	N_6	N_7
X	X	X	1	1	1	1	1	1	1	1	1
0	0	0	0	0	1	1	1	1	1	1	1
0	0	1	0	1	0	1	1	1	1	1	1
0	1	0	0	1	1	0	1	1	1	1	1
0	1	1	0	1	1	1	0	1	1	1	1
1	0	0	0	1	1	1	1	0	1	1	1
1	0	1	0	1	1	1	1	1	0	1	1
1	1	0	0	1	1	1	1	1	1	0	1
1	1	1	0	1	1	1	1	1	1	1	0

1:8 demultiplexer
3-line to 8-line decoder

(a)

Figure 12-16 Expanding basic demultiplexer/decoder functions. (a) Two 1:4 demultiplexer sections expanded to do the job of a 1:8 demultiplexer.

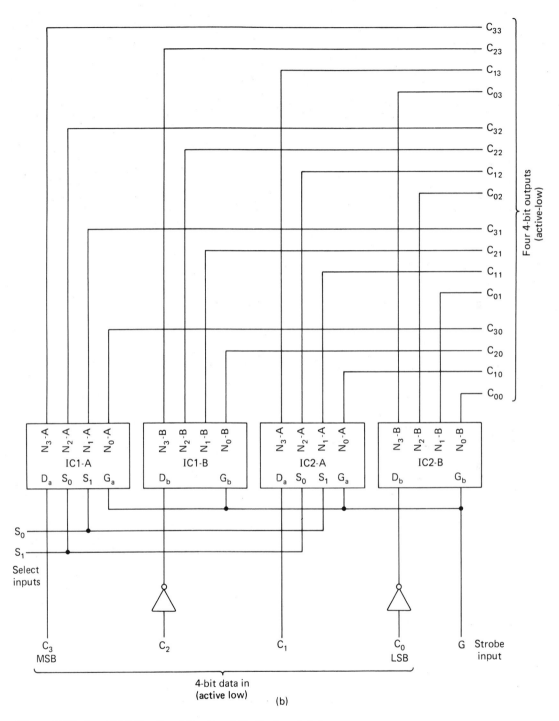

Figure 12-16 *(cont.)* (b) Circuit and function table for four 1:4 demultiplexers expanded to do the job of a quad 1:4 circuit.

Control inputs			Data outputs																
G	S_1	S_2	C_{33}	C_{23}	C_{13}	C_{03}	C_{32}	C_{22}	C_{12}	C_{02}	C_{31}	C_{21}	C_{11}	C_{01}	C_{30}	C_{20}	C_{10}	C_{00}
1	X	X	1	1	1	1	1	1	1	1	1	1	1	1	1	1	1	1
0	0	0	1	1	1	1	1	1	1	1	1	1	1	1	$\overline{C_3}$	$\overline{C_2}$	$\overline{C_1}$	$\overline{C_0}$
0	0	1	1	1	1	1	1	1	1	1	$\overline{C_3}$	$\overline{C_2}$	$\overline{C_1}$	$\overline{C_0}$	1	1	1	1
0	1	0	1	1	1	1	$\overline{C_3}$	$\overline{C_2}$	$\overline{C_1}$	$\overline{C_0}$	1	1	1	1	1	1	1	1
0	1	1	$\overline{C_3}$	$\overline{C_2}$	$\overline{C_1}$	$\overline{C_0}$	1	1	1	1	1	1	1	1	1	1	1	1

(c)

Figure 12-16 (*cont.*)

See the second line in the function table. All other outputs are fixed at logic 1, the deenergized condition for an active-low logic format.

Changing the SELECT inputs to another combination of 1's and 0's delivers an inverted version of the 4-bit input word to a different destination. Setting $S_0 = S_1 = 1$, for example, sends an inverted version of the C input word to outputs C_{03}, C_{13}, C_{23} and C_{33}. See this particular example illustrated in the last line of the function table in Fig. 12-16(b).

The basic idea of this circuit is to direct a single 4-bit word to any one of four different destinations. The scheme can be expanded to direct an 8-bit input word to any one of four different destinations by simply expanding the number of 1:4 demultiplexers from four to eight. The circuit in that case would work in a manner identical to an 8-deck, 4-position rotary switch.

12-3.2 Decoder versus Demultiplexer Functions

Whether a demultiplexer circuit is doing the job of a demultiplexer or decoder is often a matter of point of view rather than a matter of connecting the circuit one way for demultiplexing and in a different way for decoding.

It is generally easier for most people to see the expanded circuit in Fig. 12-16(a) as a 3-line to 8-line decoder than as a 1:8 demultiplexer function. Viewed as a decoder, it would make a fine binary-to-octal converter. Setting the G input to logic 1 lets the 3-bit binary input (select inputs S_0, S_1 and S_2) select one of the eight outputs. If S_0 is considered the least-significant input bit, N_0 could be the octal-0 output, N_1 could be the octal-1 output, N_2 could be the octal-2 output and so on.

Suppose, however, the circuit in Fig. 12-16(a) is to be used as a 1:8 demultiplexer. The idea is to convert an 8-bit string of serial (time-multiplexed) binary data into an 8-bit parallel output. To perform this demultiplexing function, the serial input data must be applied to the G input, and a modulo-8 counter must be connected to the three SELECT inputs. If it is assumed that there is a correspondence between the status of the counter and the position of the bit arriving at the G input, it follows that the bit arriving at the G input at any given instant will appear at the corresponding data output.

Whenever the first bit of the serial input word arrives at G, for example, the counter should be in a 000 condition that will deliver that bit (an inverted version of it actually) to the first-bit output position, N_0. Then when the second bit arrives at G, the counter increments to 001 to deliver an inverted version of that bit to the second-bit output, N_1.

The counter effectively scans the demultiplexer system, sending the data arriving at the G input to its corresponding parallel position at the output. Of course, the demultiplexer's counter must be synchronized in some fashion with the incoming serial data so that the first incoming bit always appears at the first data output, the second incoming bit appears at the second data output and so on.

Figure 12-17(a) is a block diagram of a simplified multiplexer/demultiplexer system. A modulo-4 counter addresses a 4 : 1 multiplexer circuit, effectively converting a 4-bit parallel input (bits B_0 through B_4) into a 4-bit serial data string.

The serial data is then applied to the data input of a 1 : 4 demultiplexer which is addressed by the same counter as the multiplexer circuit. The demultiplexer thus converts the incoming serial data into what appears to the eye to be parallel data at LEDs B_0 through B_3. Whatever 4-bit word is entered into the system at the multiplexer's data inputs appears at the demultiplexer's LED outputs. The two multiplexing and demultiplexing devices are synchronized in this particular instance by addressing them from the same counter.

The waveforms in Fig. 12-17(b) show how the system responds to data inputs 1001, or BCD 6. The visual impression at the LED outputs is that lamps B_1 and B_2 are lighted and lamps B_0 and B_3 are not lighted.

A system such as this one is useful for transmitting four bits of data over three communications lines: one for the serial data and two for 2-bit address from the counter. The advantage of such a scheme becomes more apparent, however, when sending larger word sizes, 16-bit words, for instance. A 16-bit communications link could be established using only five lines between them: one for the 16-bit serial data and four for the address from a 4-bit (modulo-16) counter.

Although the demultiplexer is scanning the outputs, showing the data one bit at a time and in sequence, running the counter at rates higher than about 100 Hz makes the LEDs appear to show a steady on or off state. If this scanning, or strobing, effect at the outputs of the demultiplexer cannot be tolerated for any reason, the outputs can be connected to data latches that will hold the logic levels fixed in spite of the continuous scanning effect.

12-4 Multiplexed Digit Displays

A multiplexed digital display assembly makes a fine example of how multiplexers and demultiplexers can be used together for controlling the flow of digital data through a system. Here is the situation: The output of some sort of digital circuit is generating four or more decades of BCD numbers, and it is necessary to display all the decades as efficiently as possible on an LED display assembly. Perhaps the

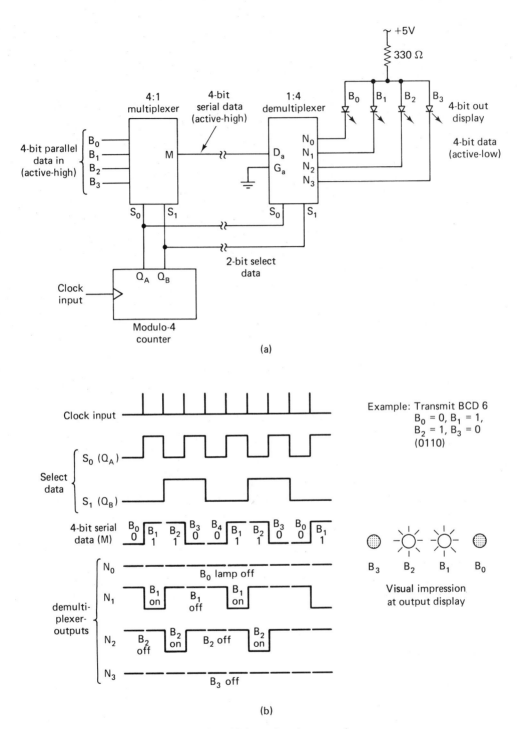

Figure 12-17 Using multiplexer and demultiplexer functions together. (a) Block diagram. (b) Waveforms and output visual impression when the inputs are 0110, or BCD 6.

system developing the BCD words is one of the 4-decade event or frequency counters described in Ch. 11. The 4-digit multiplexed display shown in Fig. 12-18 then replaces the nonmultiplexed display circuit shown in Fig. 11-2.

In the circuit in Fig. 12-18 the four decades of BCD input are represented by the four groups of inputs having designations A_1, B_1, C_1 and D_1, A_2, B_2, C_2 and D_2 and so on. The numeric suffix indicates the different BCD digits, where the suffix "1" is the least-significant digit to be displayed. The A, B, C and D prefixes indicate the bits within each BCD word, with the A bit being the least-significant bit.

The four incoming BCD words are sent to a quad 4 : 1 multiplexer assembly consisting of IC2-A, IC2-B, IC3-A and IC-3B. Multiplexer IC2-A multiplexes the least-significant bit of each BCD input; IC2-B handles the second bit; IC3-A handles the third (or C) bit of each word; and IC3-B multiplexes the most-significant bit of each incoming BCD word.

The four multiplexers are being scanned (addressed in sequence) by a 2-bit, modulo-4 binary counter, IC1. The corresponding select lines of all four multiplexers are connected together so that the counter addresses the same input of each multiplexer at the same time. The overall result is that the outputs of the multiplexers are always showing one of the input BCD words. Whenever the counter is generating a 00 output, for instance, the least-significant BCD digit appears at the M outputs; and when the count changes to 01, the second input digit appears at the multiplexers' M outputs.

The four input BCD numbers thus appear one at a time and in sequence at the input of a BCD-to-7 segment converter, IC4. This converter translates the 4-bit BCD numbers into the standard 7-segment numeric format for the 7-segment LED displays.

What appears at the output of the 7-segment converter is a 7-segment version of the input BCD numbers, appearing one at a time and in sequence.

Now note that the like segments of each display are all connected together: The S_a inputs to the displays are all connected together, as are the S_b, S_c, S_d, S_e, S_f and S_g terminals. All four LED displays thus see the same 7-segment code at any given moment. The problem at this point in the analysis is to see how the 7-segment coded digits can be demultiplexed so that only the appropriate digit is energized.

Return to the modulo-4 counter. While this counter is causing the multiplexers to scan the incoming BCD numbers, it is also addressing a digit-select demultiplexer, IC5. If this demultiplexer is viewed as a 2-line to 4-line decoder, it can be seen that its four outputs are normally at logic 1, dropping to logic 0 one at a time and in sequence as the counter increments the select inputs. The four digit-driver transistors, Q_1 through Q_4, are thus switched on one at a time and in sequence.

The emitters of these *PNP* transistors are connected to the V_{cc} terminals of the common-anode LED digit displays; and since these transistors are normally turned off by the normal logic-1 output from the demultiplexer, it follows that the 7-segment digit displays are normally turned off.

The demultiplexer circuit, however, is switching on the driver transistors one

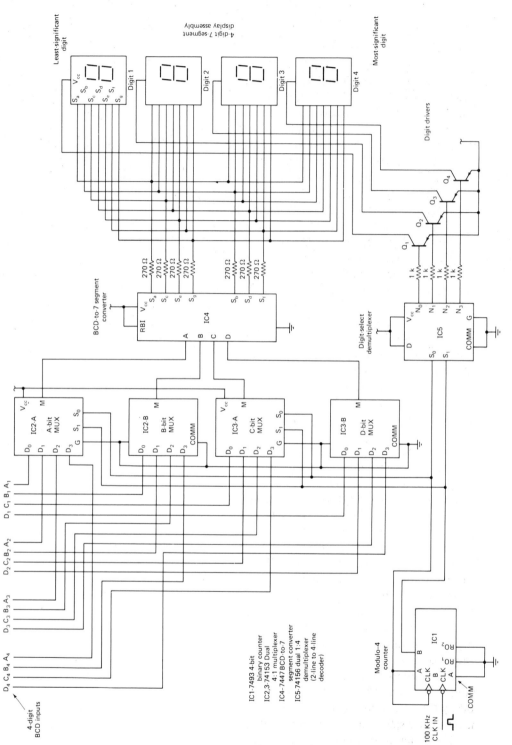

Figure 12-18 A multiplexed 4-digit display system.

at a time; and whenever one of these transistors is switched on, it supplies $+V_{cc}$ potential to the anodes of the corresponding 7-segment display unit.

The overall effect of the demultiplexer/digit-driver scheme is to energize the common-anode connections of the displays one at a time and in sequence. And since this demultiplexer system is being addressed by the same counter that addresses the multiplexer section of the circuit, it turns out that the two parts of the circuit are operating in perfect synch. As the multiplexer system scans the input BCD words and delivers them through the 7-segment converter to the segment inputs of the digit displays, the demultiplexer system scans the V_{cc} terminals of the digit displays.

Suppose, for example, the counter is generating address 00. The multiplexers are thus set to deliver the least-significant BCD decade through the 7-segment converter and to the segment connections of all four digit displays. Only one of the digit-display devices has $+V_{cc}$ power applied to its common-anode connection, however; and since the counter is addressing output 00 (N_0) in the demultiplexer, it follows that Q_1 is switched on to apply V_{cc} power to the least-significant digit display unit, the unit that is supposed to respond to the least-significant BCD input.

As the counter switches over to output 01, the multiplexer assembly and 7-segment decoder deliver a 7-segment version of the second incoming digit to the displays; and at the same time the demultiplexer circuit energizes digit 2 with $+V_{cc}$ power.

The counter continuously cycles both the multiplexer and demultiplexer systems, energizing the displays one at a time and in sequence. The demultiplexer is always energizing the same digit position the multiplexer is selecting from the BCD inputs.

The 4-digit multiplexed display system just described is, indeed, rather complicated compared to the nonmultiplexed display shown in Fig. 11-2. One of the real advantages of a multiplexed display is that it requires only one BCD-to-7 segment decoder, no matter how many digits are in the system. A nonmultiplexed display, however, requires a BCD-to-7 segment converter for each digit in the display.

Note that the multiplexed display in Fig. 12-18 uses five ICs and four driver transistors, whereas the nonmultiplexed 4-digit equivalent in Fig. 11-2 requires only four ICs. As far as IC-count efficiency is concerned, the nonmultiplexed version is somewhat better. In fact, the only real advantage of using a multiplexed 4-digit display is that the display assembly itself (the four 7-segment display units) is available as a single unit having all the corresponding segment terminals internally connected together.

The real advantage of multiplexed displays becomes more apparent in display systems handling larger numbers of digits. As shown in Table 12-4, an 8-digit multiplexed display requires seven ICs: four 8:1 multiplexers (one for each bit in the BCD format), a single 7-segment converter, a modulo-8 counter package and a 3-line to 8-line demultiplexer. A corresponding nonmultiplexed 8-digit display requires eight IC packages: eight 7-segment converters. There is thus a slight gain in design efficiency, or at least a break-even condition when considering that the digit-driver transistors in the multiplexed display can be replaced with a single proprietary digit-driver IC package.

The design efficiency of multiplexed displays is especially apparent in the 12-digit display assembly cited in Table 12-4. The multiplexed version calls for seven ICs again: four 16:1 multiplexers, a single 7-segment converter, a modulo-12 counter and a demultiplexer connected as a 4-line to 12-line decoder. A non-multiplexed display having a 12-digit output requires twelve 7-segment converter ICs.

Table 12-4 A COMPARISON OF DESIGN AND CURRENT-DRAIN EFFICIENCY
OF MULTIPLEXED AND NONMULTIPLEXED DIGITAL DISPLAYS

Type of display	Number of digits	Multiplexer ICs	Demultiplexer ICs	Counter ICs	7-segment converter ICs	Total number of IC devices	Maximum display current (70 mA/digit)
Multiplexed	4	4	1	1	1	7	70 mA
non-mux		0	0	0	4	4	280 mA
Multiplexed	8	4	1	1	1	7	70 mA
non-mux		0	0	0	8	8	560 mA
Multiplexed	12	4	1	1	1	7	70 mA
non-mux		0	0	0	12	12	840 mA
Multiplexed	16	4	1	1	1	7	70 mA
non-mux		0	0	0	16	16	1.12 mA

Table does not include digit drivers for multiplexed displays. Drivers can be transistors or proprietary digit-driver ICs

Although a 4-digit display might be more effectively constructed around a non-multiplexed scheme, the multiplexed versions become relatively simple in systems calling for larger numbers of display units.

Another distinct advantage of multiplexed displays is that they have a far lower average power dissipation then their nonmultiplexed counterparts. Suppose, for example, that each segment in the display assembly draws 10 mA of current from the power supply. The worst-case condition is one in which the display is showing all 8's—each digit drawing 7×10 mA, or 70 mA. If a 4-digit nonmultiplexed display is showing all 8's, the total current drain for the display alone is 4×70 mA, or 280 mA.

A multiplexed display, however, energizes only one digit at a time; therefore, the maximum current drain for the display unit never exceeds that of a single digit. A multiplexed display built around 70 mA display units will never draw more than 70 mA from the power supply, no matter how many digits are in the display.

Under any condition, then, a multiplexed LED display is always more efficient from a power-dissipation point of view than any nonmultiplexed display. This fact is especially relevant in pocket calculators that must operate several hours at a time from a single battery charge.

Exercises

1. Use a logic-equation proof to justify the output equation expressed for the AOI circuit in Fig. 12-5(c).

2. Use a logic-equation proof to justify the output equation expressed for the AOI circuit in Fig. 12-5(d).

3. What is the simplest logic expression for the output of a 4-wide, 2-input AOI if all inputs are inverted (active-low)?

4. Draw a 2-wide, 2-input AOI circuit that generates the EQUALITY function expressed in the truth table in Fig. E-12.4. Logic inverters may be used only when necessary.

Inputs		Output
A	B	C
0	0	1
0	1	0
1	0	0
1	1	1

Figure E-12.4

5. How many select lines are required for a 10 : 1 multiplexer?

6. Show how to program a 4 : 1 multiplexer to satisfy the truth table in Fig. E-12.6.

Inputs			Output
C	B	A	E
0	0	0	1
0	0	1	0
0	1	0	1
0	1	1	1
1	0	0	0
1	0	1	0
1	1	0	1
1	1	1	1

Figure E-12.6

7. How can a 7442 BCD-to-decimal decoder be compared to a 1 : 10 demultiplexer?

8. If the circuit in Fig. 12-10 is considered a quad 8 : 1 multiplexer, what is the name of the circuit in Fig. 12-16(b) (expressed in the same context)?

9. Are the multiplexer and demultiplexer circuits in Figs. 12-10 and 12-16(b) compatible?

10. Name one distinct advantage of a multiplexed display system over a nonmultiplexed display, regardless of the number of digits in that display.

SHIFT REGISTERS

In the context of digital electronics, a *register* is any sort of simple memory circuit. It is generally made up of an arrangement of flip-flop elements, each capable of accepting, storing and reading out a data bit in response to commands from an external control circuit.

The 7475 TTL quad D latch is one example of a 4-bit data register. It is not generally classified as a shift register, however. The memory circuits described in Ch. 14 are also registers; but again, they are not shift registers because they are incapable of shifting their stored data anywhere but in and out.

A *shift register* is a flip-flop memory device that not only accepts, stores and reads out binary data, but it is also capable of shifting that data in a systematic fashion through the memory system.

13-1 Serial-In, Serial/Parallel-Out Shift Registers

Figure 13-1 shows two ordinary J-K master–slave flip-flops connected as a 2-bit serial-in, serial/parallel-out shift register. Note that the two flip-flops are clocked in parallel. The fact that the flip-flop elements are clocked in parallel is one of the distinguishing features of a shift-register circuit, but it is not the only distinguishing feature. Recall that synchronous counters are also clocked in parallel.

Also note that the J and K inputs are connected to complementary data sources. The data input at S_R, for instance, is connected directly to the J input of FF-A,

318

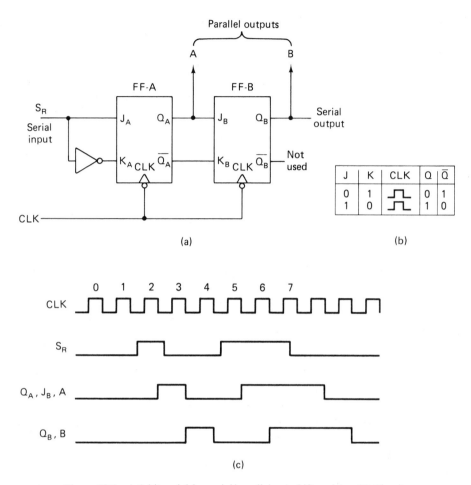

Figure 13-1   A 2-bit serial-in, serial/parallel-out shift register. (a) Circuit diagram using J-K master–slave flip-flops. (b) Truth table for the synchronous J-K operating modes. (c) Sample waveforms.

while an inverted version of that same input data goes to input K. And, in a similar fashion, the J and K inputs of FF-B are seeing complemented data from the Q and $\bar{Q}$ outputs of FF-A.

If there were a third flip-flop in the circuit, its CLK input would be connected to the common CLK bus and its J and K inputs would be tied to Q_B and $\overline{Q_B}$, respectively.

The flip-flops in a shift register *must* be master–slave flip-flops. In this particular example, J and K data are fed into the master sections on the positive-going edge of the CLK pulse, and then they are fed to the slave section and outputs on the subsequent negative-going edge of the CLK pulse. The table in Fig. 13-1(b) shows how the outputs of each flip-flop respond to complemented inputs at J and K.

Since the PRESET and CLEAR inputs are not indicated on the drawing, it can be properly assumed these are the only valid operating states for these flip-flops.

Now suppose the outputs of both flip-flops, the Q_A and Q_B outputs, have somehow been set to logic 0. If the S_R input is likewise at logic 0 when the 0th clock pulse occurs as shown in the waveforms in Fig. 13-1(c), it follows that both flip-flops will continue showing logic-0 outputs after that clock pulse is completed. In this particular instance, J_A and J_B equal logic 0, while K_A and K_B equal logic 1, the conditions required for setting their outputs to $Q_A = Q_B = 0$, $K_A = K_B = 1$ after completing one clock pulse.

If the S_R input remains at logic 0 through the first clock pulse, the outputs remain unchanged again: $Q_A = Q_B = 0$, $\overline{Q_A} = \overline{Q_B} = 1$.

As shown in the waveforms, however, the S_R input is switched to logic 1 between the first and second clock pulses. When the positive-going edge of the second clock pulse occurs, the master section of FF-A is "reprogrammed" to show an output of $Q_A = 1$, $\overline{Q_A} = 0$. This change at the J and K inputs of FF-A does not affect the Q outputs of FF-A, however, until the negative-going edge of the CLK pulse occurs. Between the first and second CLK pulses, then, the J and K inputs of FF-B are still being "programmed" to read $J_B = 0$, $K_B = 1$.

As a result of changing the S_R input from logic 0 to logic 1 between the first and second clock pulses, Q_A switches to logic 1 and $\overline{Q_A}$ changes to logic 0 on the negative-going edge of the second clock pulse. Output A is then equal to logic 1 and output B is still at logic 0.

Any change in the S_R input logic level appears at the output of FF-A only after one complete clock pulse occurs.

Now let the S_R input drop back to logic 0 between the second and third clock pulses. Two important changes thus take place on the positive-going edge of the third clock pulse. First, the master section of FF-A is "programmed" to return the A output to logic 0 when the negative-going edge of the third clock pulse occurs: Second, equally important is the fact that the master section of FF-B is now being programmed to change output B to logic 1, that is, FF-B is being programmed this way because $Q_A = 1$ and $\overline{Q_A} = 0$ the moment the positive-going edge of the third clock pulse takes place.

When the third CLK pulse drops to logic 0, then, output A follows the S_R input by returning to logic 0, but output B takes on the previous state of FF-A, letting its B output go to logic 1.

Any change in the S_R input logic level appears at the output of FF-A after one complete CLK pulse occurs; and that same change appears at the output of FF-B after *two* CLK pulses occur. If there were a third flip-flop element in the circuit in Fig. 13-1(a), its output would respond to any change in the S_R input only after three clock pulses occurred.

This particular shift register thus shifts any input data at S_R first to FF-A and then to FF-B. Each change occurs only on the negative-going edge of the CLK pulse. Compare the waveforms for S_R, Q_A and Q_B in Fig. 13-1(c). Note that the S_R

waveform is duplicated at Q_A and Q_B, but it is shifted in each case by one clock-pulse interval.

This is considered a serial-input shift register because the data are entered into the system one bit at a time in a serial sequence. If the circuit's output is taken from the B connection, whatever change takes place at the serial input appears at this serial output exactly two clock pulses later. This would be an example of a 2-bit, serial-in, serial-out shift register.

If, however, the circuit's outputs are taken from both A and B, it would be considered a serial-in, parallel-out shift register. It would be a serial-in shift register because the input data is presented in a serial fashion, one bit at a time. But taking the outputs from both A and B at the same time creates a 2-bit parallel-out format. A sequence of two input bits would appear simultaneously at outputs A and B after two complete clock pulses. In a sense, a serial-in, parallel-out shift register is a serial-to-parallel data converter.

Figure 13-2 is a practical demonstration circuit for a 4-bit serial-in, parallel-out shift register. The flip-flop units are 7476 dual J-K flip-flops. The PRESET and

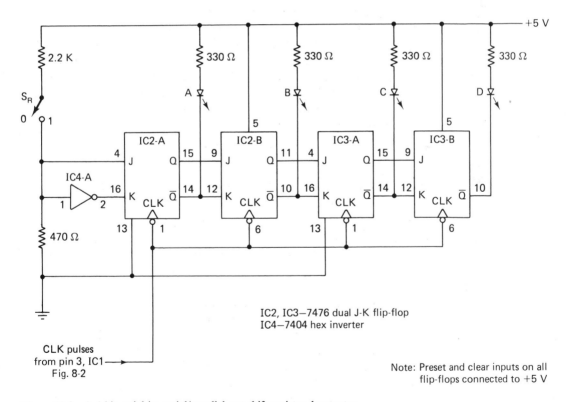

Figure 13-2   A 4-bit serial-in, serial/parallel-out shift register demonstration circuit.

CLEAR inputs (not shown on the diagram) ought to be disabled by connecting them to +5 V.

The circuit can be clocked by means of a 555-type astable multivibrator, such as the one illustrated in Fig. 8-2. The clocking frequency should be held to one pulse every 2 s or so, giving the experimenter ample time to change the S_R input level between successive clock pulses.

Any logic level entered via the S_R switch will propagate down the line of flip-flops, entering the A output on the first clock pulse and finally reaching the D output after four successive clock pulses. That bit of data is then lost—"dropped off the end"—after the fifth clock pulse occurs.

This same circuit can be used as a serial-in, serial-out shift register by omitting all LEDs except the one at the D output.

13-2 Parallel-In Shift Registers

In most respects, the circuit in Fig. 13-3 is identical to the two serial-in shift registers described in the previous section. Note that the flip-flops are clocked and intercon-nected between their Q and J-K inputs in exactly the same fashion.

The essential difference between the circuit in Fig. 13-3 and the serial-in shift registers is the fact that this particular circuit has provisions for asynchronously loading the flip-flops from a set of "preset" inputs, P_A through P_D. Whenever the LOAD bus is set to logic 1, the NAND gates deliver complemented logic levels from the P inputs to the PRESET and CLEAR inputs of each flip-flop. This action overrides the normal synchronous J-K action of a serial-in shift register, and it forces the logic levels at the P inputs to appear at the corresponding parallel data outputs as long as the LOAD bus is held at logic 1.

This is an example of parallel loading a shift register. It is sometimes referred to as *jam loading*. It is an asynchronous operation that stops the serial-shifting effect altogether. See the bottom line in the function table in Fig. 13-3(b).

Returning the LOAD bus to logic 0 effectively turns off all of the NAND gates, forcing their outputs to logic 1, regardless of the logic levels appearing at the P inputs. This action returns the J-K flip-flops to their normal serial-shifting mode, and the 4-bit data is shifted to the right, followed by whatever logic level happens to be at the S_R input at the time.

The serial input, S_R, is normally held at logic 0 when the system is used for parallel loading. Any 4-bit word entered into the shift register from the P inputs thus appears at the parallel outputs, A through D, as long as the LOAD bus is at logic 1. Dropping the logic level at the LOAD bus to 0 enables the serial-shift feature, and the loaded bits appear at the D output in sequence, proceeding from the D-bit through the A-bit.

One LOAD-bus pulse is adequate for loading the system; but after that, at

Parallel outputs

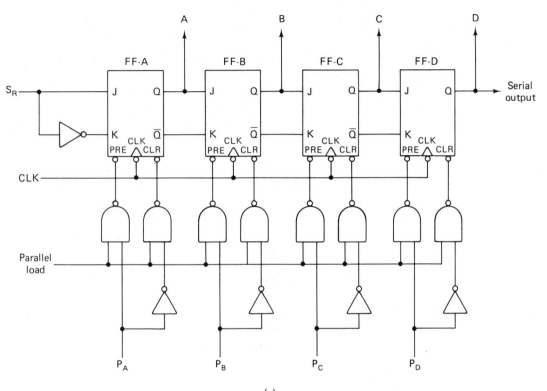

(a)

Load	S_R	CLK	Mode
0	1	⎍	Serial load 1's
0	0	⎍	Serial load 0's
1	X	X	Parallel load

(b)

Figure 13-3   A 4-bit serial/parallel-in, serial/parallel-out shift register. (a) Circuit diagram using J-K flip-flops, NAND gates and INVERT functions. (b) Function table.

least three CLK pulses are necessary for serially shifting that data from the D output. The system, in effect, can work as a parallel-to-serial data converter. The data is loaded from the P inputs via a positive LOAD pulse and then it is shifted from the D output by pulses on the CLK bus.

13-3 Right/Left Serial Shift Registers

The serial-shifting effects described thus far in this chapter have all been in a direction from the least-significant to the most-significant bit position. (Designating the A-bit position as the least-significant position is a carryover from binary-counter terminology.) These are all examples of right-shift registers.

A left-shift register, by contrast, simply moves the data in the opposite direction: Data is serially entered at the most-significant bit position and proceeds through the system toward the least-significant position.

Whether a serial-shift operation is considered right- or left-shift depends on the position of the least-significant bit, not on how the circuit is drawn. It so happens that the data in the figures shown thus far moves from left to right, but a cautious technician should not rely on how a draftsman might draw the figure. The circuits in Figs. 13-1, 13-2 and 13-3 are, indeed, right-shift registers; but they would still be right-shift registers if they were drawn in reverse, making the data move from right to left as far as the drawing is concerned. As long as the data first enters the A-bit (least-significant bit) position, it is still a right-shift register.

Figures 13-4(a) and 13-4(b) illustrate the difference between right- and left-shift registers. The 2-bit shift register in Fig. 13-4(a) is a true right-shift register. The circuit in Fig. 13-4(b), however, is a left-shift register: Serial data from the S_{LI} (shift-left input) is fed to the most-significant bit position and proceeds to the least-significant bit position. Both circuits operate according to the same general shift-register principles, but data shifts to the right in the first case and to the left in the second.

A closer comparison of the circuits in Fig. 13-4 shows the following logic relationships between the inputs and outputs of the J-K flip-flops:

1. When right shifting, $J_A = S_{RI}$, $J_B = Q_A$ and S_{RO} (shift-right output) $= Q_B$. Inputs K_A and K_B are merely complements of their respective J inputs.
2. When left shifting, $J_A = Q_B$, $J_B = S_{LI}$ and S_{LO} (shift-left output) $= Q_A$.

It is possible to derive some simple logic equations relating the circuit inputs and outputs to the desired direction of shifting. Suppose we let L designate a left-shift operation and R designate a right-shift operation. It is then possible to say that $J_A = R \cdot S_{RI} + L \cdot Q_B$. Connection J_A, in other words, is connected to S_{RI} or Q_B, depending on whether the shifting action is to be to the right or to the left.

In a similar fashion, it can be seen that $J_B = R \cdot Q_A + L \cdot S_{LI}$. J_B is thus connected to either Q_A or S_{LI}, depending on whether the serial shifting is to be to the right or to the left.

The K inputs in either case are merely complements of their respective J inputs. S_{RO} (shift-right output) is always taken from Q_B and S_{LO} (shift-left output) always appears at Q_A. The flip-flops are clocked in exactly the same fashion, regardless of the desired direction of shift.

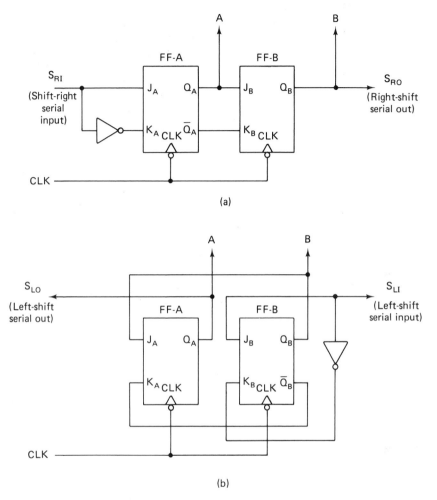

Figure 13-4 A comparison of right- and left-shift serial shift registers.
(a) Right-shift register. (b) Left-shift register.

Figure 13-5 shows how the logic equations for the J inputs can be implemented by means of a couple of AOI gates and inverters. These logic circuits satisfy the conditions for selecting right- or left-serial shifting. In this particular circuit, setting the SS bus (shift select) to logic 1 sets up the conditions for right shifting—entering data at the S_{RI} connection and letting it proceed to the S_{RO} terminal. Setting the SS bus to logic 0, however, satisfies the requirements for left shifting—entering data serially at the S_{LI} input and letting it proceed to the S_{LO} output.

A logic-equation analysis of the AOI circuits in Fig. 13-5 ought to clear up any mystery about the right/left shifting action. The basic idea can be extended to

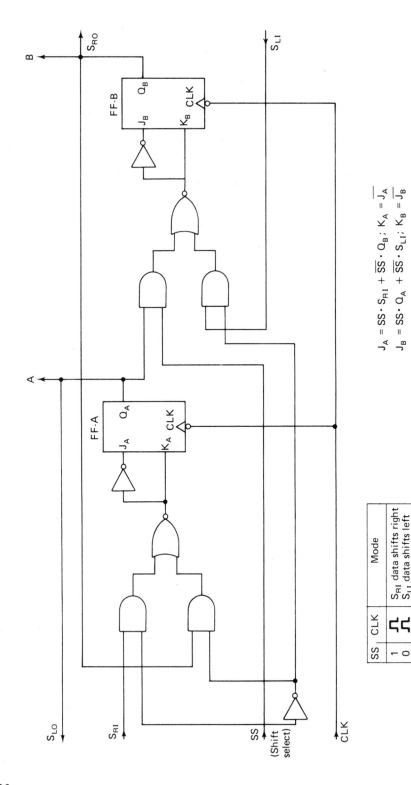

SS	CLK	Mode
1	⊓	S_{RI} data shifts right
0	⊓	S_{LI} data shifts left

$$J_A = SS \cdot S_{RI} + \overline{SS} \cdot Q_B; \; K_A = \overline{J_A}$$
$$J_B = SS \cdot Q_A + \overline{SS} \cdot S_{LI}; \; K_B = \overline{J_B}$$

Figure 13-5 Logic scheme for combining left- and right-shift features into a single circuit.

4-bit shift registers by connecting a similar kind of AOI circuit between each flip-flop in the system.

13-4   The "Universal" Shift Register

A universal shift register is one that combines all of the features described thus far in this chapter. By means of a set of programming inputs, a universal shift register can be operated in any one of the following modes:

1. Shift-right, serial/parallel-out
2. Shift-left, serial/parallel-out
3. Parallel-in, parallel-out
4. Parallel-in, shift-right serial out
5. Parallel-in, shift-left serial out

Universal shift registers are rather popular because they can perform the function of any of the simpler shift-register devices. The only justification for providing simpler shift registers is the simple fact that they are generally less expensive than their universal counterparts.

The main operating features of universal shift registers are described in greater detail in the following sections.

13-5   Commercially Available Shift-Register Devices

Table 13-1 summarizes the basic shift register ICs on the market today. They are classified according to their input and output configurations, and then they are subclassified according to the number of bits they can handle and certain kinds of control functions.

Bear in mind that any shift register having parallel outputs can also be used as a serial-out register. The serial output in such instances is always taken from the most-significant parallel output terminal.

The shift register with the simplest input/output configuration is the 7491A 8-bit serial-in, serial-out device. As indicated in Fig. 13-6(a), serial inputs appearing at SRA and SRB are clocked into the register on the negative-going edge of each CLK pulse. Eight clock pulses later that data appears at the serial outputs Q_H and $\overline{Q_H}$.

In this particular instance, the two serial inputs are AND-ed together. The system thus normally propagates 0's through the register. A logic 1 can be entered only by setting $SRA = SRB = 1$. That logic-1 level will appear exactly eight clock pulses later at Q_H, along with its 0 complement at $\overline{Q_H}$. The only way to clear the shift register in Fig. 13-6(a) is by setting one or both of the serial inputs to logic 0 and clocking the device at least eight times.

The shift register in Fig. 13-6(b) is typical of an 8-bit serial-in, parallel-out device. Like the 7491A register just described, serial data is entered into this

Table 13-1 Common Shift-Register Devices

Serial-In, Serial-Out	

7491A	8-bit serial-in, serial-out——TTL
4006	18-stage programmable serial-in, serial-out——CMOS

Serial-In, Parallel-Out	

74164	8-bit serial-in, parallel-out with CLEAR——TTL
4015	Dual 4-bit serial-in, parallel-out with RESET——CMOS

Serial/Parallel-In, Serial-Out	

74165	8-bit serial/parallel-in, serial-out with CLOCK INHIBIT——TTL
74166	8-bit serial/parallel-in, serial-out with CLOCK INHIBIT and CLEAR——TTL

Serial/Parallel-In, Parallel-Out	

74195	4-bit serial/parallel-in, parallel-out with CLEAR——TTL
74199	8-bit serial/parallel-in, parallel-out with CLOCK INHIBIT and CLEAR——TTL

Universal Shift Registers	

74194	4-bit universal with CLEAR——TTL
74198	8-bit universal with CLEAR——TTL

74164 IC by means of a pair of AND-ed serial inputs, SRA and SRB. Data is shifted to the right on the positive-going edge of each CLK pulse.

The 74164, however, features parallel outputs and an asynchronous CLEAR. All eight elements of the register are immediately cleared to 0 the instant the CLR input is pulled down to logic 0. And while this clearing operation is taking place, the shifting operations are completely disabled.

Suppose the register in Fig. 13-6(b) has been cleared and the CLR input is returned to logic 1. Now data can be serially entered through inputs SRA and SRB. Further suppose both of these inputs are at logic 1. That means the first positive-going edge of the CLK pulse will enter a logic 1 into the Q_A position. The next CLK pulse after that will carry that 1-bit to the Q_B position, and the CLK pulse following that will carry the bit to the Q_C position and so on. After eight complete CLK pulses, the logic 1 entered at the serial inputs will finally arrive at Q_H, the most-significant bit position. If the system is then clocked again, that original bit is lost.

One of the primary applications of a serial-in, parallel-out shift register such as the 74164 is for serial-to-parallel data conversion. Suppose data is arriving one bit at a time and in sequence at the SRA/SRB inputs. Each time the register is clocked, a different bit is entered into the system; and after eight clock pulses a

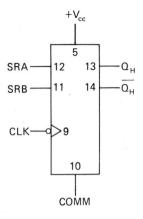

$Q_H = SRA \cdot SRB$ after
$8 \downarrow$ CLK pluses

7491A—8-bit serial-in,
serial-out
shift register

(a)

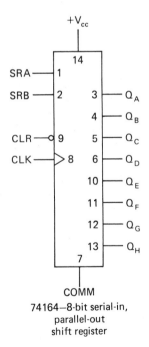

74164—8-bit serial-in,
parallel-out
shift register

(b)

CLR	SRA	SRB	CLK	Mode
0	X	X	X	All outputs to 0
1	0	X	↑	Enter 0 at Q_A
1	X	0	↑	Enter 0 at Q_A
1	1	1	↑	Enter 1 at Q_A

Note: Shift-right serial input
= SRA · SRB

Figure 13-6 Two types of serial-in shift registers. (a) An 8-bit serial-in, serial-out register. (b) An 8-bit serial-in, parallel-out register.

full eight bits of serial data appear in a parallel format at the Q outputs, Q_A through Q_H.

The shift register in Fig. 13-7(a) is typical of most serial/parallel-in, serial-out ICs. Note that the device has a single serial-data input, S_{RI}, and a set of eight parallel-data inputs, P_A through P_H. Data can be entered serially at the S_{RI} input by first setting SS (SHIFT SELECT) to logic 1, pulling the CLK INHIB (CLOCK INHIBIT) terminal to logic 0, and then clocking the CLK terminal. The positive-going edge of each CLK pulse then carries the data from S_{RI} one step further through the register, allowing it to appear at Q_H after completing eight CLK pulses. Used in this serial-shift mode, the 74165 works almost exactly like its simpler 8-bit serial-in, serial-out counterpart, the 7491A in Fig. 13-6(a).

The 74165 can be set to its parallel-entry mode by dropping the SS input to logic 0. Any 8-bit word appearing at the P inputs is immediately and asynchronously entered into the system. Output Q_H responds immediately to the data bit entered at P_H. The only way to recover the parallel-entered data is by returning the SS input to logic 1 and clocking the device serially. The parallel-entered data then appears serially: one bit at a time and in a sequence from P_H to P_A at the Q_H output. At least seven CLK pulses are required to feed the entire 8-bit word to the serial output. (Only seven CLK pulses are required because the P_H bit appears at Q_H the instant the parallel data is loaded.)

Parallel loading is thus asynchronous for the 74165, but serial shifting is fully synchronous.

There are two methods for clearing all the flip-flops in the 74165: A logic 0 can be applied to the serial input, followed by at least eight CLK pulses; or a parallel string of eight 0's can be parallel loaded. In either case, the register is cleared—loaded with 0's. A somewhat more elaborate version of this 74165 shift register, the 74166, has a CLR input terminal that allows all the flip-flop sections to be cleared by setting the CLR input to logic 0. The clearing operation in this instance is asynchronous.

The 74165 shift register illustrated in Fig. 13-7(a) also includes a CLK INHIB terminal. Setting this control terminal to logic 1 (while $SS = 1$) stops the serial-shifting action once the next positive-going edge of the CLK pulse occurs. In a sense, this is a "memory" function.

To appreciate the usefulness of the CLK INHIB function, consider the fact that the 74165 shift register can be used as either a serial-data storage register or a parallel-to-serial converter. As a serial-data storage register, the circuit is used in its serial input mode. An eight-bit string of serial data appearing at the S_{RI} input is entered one bit at a time and in sequence through the system. Once all eight bits are entered, it is most often necessary to stop entering data for some period of time (for a period of time representing the 8-bit storage time). After the eighth bit is entered through S_{RI}, the CLK INHIB control input can be pulled up to logic 1, overriding the CLK input and holding the stored serial data for any desired length of time. Retrieving the stored data is then a simple matter of returning the CLK INHIB input to logic 0. Normal right-serial shifting then resumes, and the

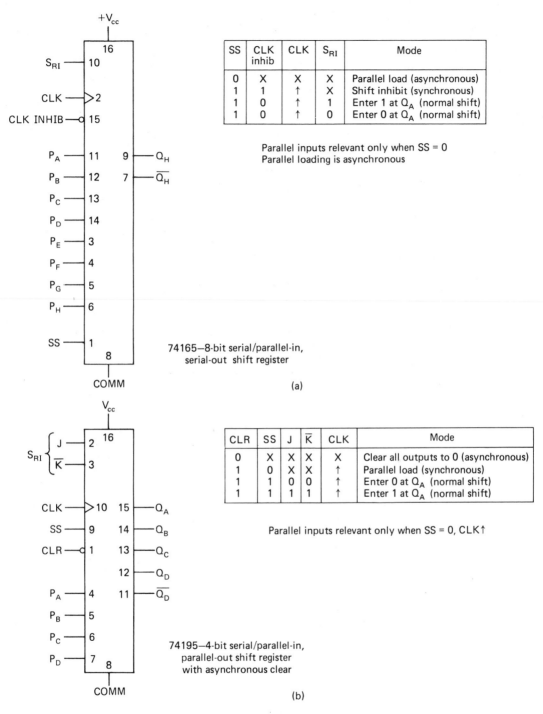

SS	CLK inhib	CLK	S_{RI}	Mode
0	X	X	X	Parallel load (asynchronous)
1	1	↑	X	Shift inhibit (synchronous)
1	0	↑	1	Enter 1 at Q_A (normal shift)
1	0	↑	0	Enter 0 at Q_A (normal shift)

Parallel inputs relevant only when SS = 0
Parallel loading is asynchronous

74165—8-bit serial/parallel-in,
serial-out shift register

(a)

CLR	SS	J	$\overline{K}$	CLK	Mode
0	X	X	X	X	Clear all outputs to 0 (asynchronous)
1	0	X	X	↑	Parallel load (synchronous)
1	1	0	0	↑	Enter 0 at Q_A (normal shift)
1	1	1	1	↑	Enter 1 at Q_A (normal shift)

Parallel inputs relevant only when SS = 0, CLK↑

74195—4-bit serial/parallel-in,
parallel-out shift register
with asynchronous clear

(b)

Figure 13-7 Two types of serial/parallel-in shift registers. (a) An 8-bit serial/parallel-in, serial-out register. (b) A 4-bit serial/parallel-in, parallel out register.

eight bits of stored data appear at the Q_H output, in sequence and one bit at a time.

When using the system as a parallel-to-serial converter, CLK INHIB can be set to logic 1 to override any CLK activity, and then parallel data at the P inputs can be asynchronously loaded into the register by setting SS to logic 0. Even after returning SS to logic 1, the data remains in the register as long as CLK INHIB is at logic 1. Changes at the P inputs have no effect on the register at all; the parallel-stored data is protected. That data can be retrieved at any later time by returning the CLK INHIB input to logic 0. Each positive-going edge at the CLK input then pushes the stored data out of the register in a serial fashion from Q_H.

The 74195 4-bit serial/parallel-in, parallel-out shift register illustrated in Fig. 13-7(b) is typical of this class of shift-register devices. According to the function table accompanying the diagram, all four outputs can be asynchronously cleared to 0 by setting the CLR input to logic 0. Parallel data entry is accomplished by setting the SS input to logic 0 and waiting for the next positive-going edge of the CLK pulse. Parallel data entry in this particular case is synchronous; it is controlled by the CLK input.

Parallel data at the P inputs thus appears at the Q outputs only when $SS = 0$ and after the subsequent CLK pulse occurs. The data can then be retrieved in a parallel format at that time, or it can be retrieved in a serial format by setting SS to logic 1 and taking the output from Q_D.

Remember that any shift register specified as a parallel-out register can also be used as a serial-out register by simply taking data from the most-significant bit position.

Data can be entered into the 74195 serially through the J and $\overline{K}$ inputs. As long as $SS = 1$, 1's and 0's can be propagated through the register while the J-$\overline{K}$ inputs are both fixed at logic 1 or logic 0, respectively. The J and $\overline{K}$ inputs are most often connected together to ensure proper entry of serial data.

In any event, the data is shifted serially to the right, in the direction of Q_A to Q_D, on the positive-going edge of each CLK pulse (assuming, of course, CLR $= SS = 1$).

The 74195 shift register can thus be used for serial-to-parallel data storage and conversion, serial-to-serial data storage, parallel-to-parallel data storage, and parallel-to-serial data storage and conversion.

The diagram and function table in Fig. 13-8 represent a typical universal, bidirectional shift register. It combines all the features of simpler shift registers, including serial/parallel input, serial/parallel output and one additional feature: the option of shifting data serially to the right or left.

The control inputs to this 74194 universal shift register include CLK, CLR and a pair of mode-programming terminals, SS_0 and SS_1. Clearing in this instance is asynchronous, pulling all outputs down to logic 0 whenever the CLR input is set to 0. The clearing action overrides any other ongoing activity.

Clocking takes place on the positive-going edge of the CLK waveform, parallel loading the register or shifting data serially to the right or left, depending on the logic levels at SS_0 and SS_1.

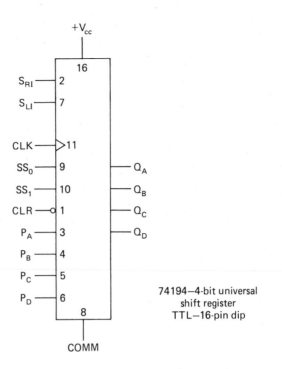

Figure 13-8   A 4-bit bidirectional universal shift register.

CLR	SS_0	SS_1	S_{RI}	S_{LI}	CLK	Mode
0	X	X	X	X	X	All outputs to 0 (asynchronous)
1	0	0	X	X	X	Shift inhibit (asynchronous)
1	1	0	0	X	↑	Enter 0 at Q_A (right shift)
1	1	0	1	X	↑	Enter 1 at Q_A (right shift)
1	0	1	X	0	↑	Enter 0 at Q_D (left shift)
1	0	1	X	1	↑	Enter 1 at Q_D (left shift)
1	1	1	X	X	↑	Parallel load (synchronous)

Notes: (1) Parallel inputs relevant only when
$SS_0 = SS_1 = 1$, CLK ↑
(2) Right-shift serial out from Q_D
Left-shift serial out from Q_A

Setting $SS_0 = SS_1 = 0$ puts the register into its shift-inhibit mode. Shift-inhibit is anynchronous in this case, immediately stopping any serial-shift action, independent of the status of the CLK input.

Setting $SS_0 = SS_1 = 1$ allows parallel loading from the P inputs when the next positive-going edge of an incoming CLK pulse occurs. Parallel loading is thus synchronous.

Setting the SS_0 and SS_1 inputs so that they are complements of one another allows serial-data shifting through the register. If $SS_0 = 1$ while $SS_1 = 0$, data is shifted to the right. That data might have been entered in parallel via the P inputs or serially from the SRI (SHIFT RIGHT INPUT) terminal. As long as the register is set to this right-shift mode, the SLI (SHIFT LEFT INPUT) terminal is totally disabled. It is impossible to enter serial data from the SLI terminal while the system is shifting to the right.

On the other hand, setting $SS_0 = 0$ and $SS_1 = 1$ places the register into its left-shift mode. Data can then be entered serially at the SLI terminal, and each positive-going edge of the incoming CLK pulse will shift the serial data in a direction from Q_D to Q_A.

Being a universal shift register, the 4-bit 74194 device can be used for a large number of different data-shifting and storage applications. It is possible, for example, to clock in four bits of data serially, store that data for any desired length of time and then retrieve it serially. This is a serial-storage function. This particular function, however, can be altered to change the order in which the data is entered or retrieved. Suppose, for instance, the serial data is entered from the S_{RI} input, that is, it is shifted into the system from Q_A to Q_D. That same data, however, can be retrieved in reverse order by setting up the system for left shifting and taking the serial data from the Q_A output. Four-bit data can be entered most-significant-bit-first, stored for any length of time and then retrieved least-significant-bit-first. An alternative is to serially enter the data with a left shift, store it and then retrieve it from Q_D with a right-shift action.

The same device can be used for serial-to-parallel data storage and conversion. Again, the universal feature of this register allows the serial data to be entered in either direction. A right-shift-only register must enter the most-significant bit first; but this device allows the option of entering the least-significant bit first.

Of course, simple parallel data storage is possible with the 74194, making it work very much like a triggered 4-bit data latch.

And, finally, this device can be used for parallel-to-serial data storage and conversion. Four-bit data can be entered via the P inputs, stored in the device for any length of time by setting $SS_0 = SS_1 = 0$ and then retrieved serially from Q_D or Q_A by allowing right- or left-shift, respectively.

13-6 A Shift Register Demonstration Circuit

The wide range of shift register operations possible with the 74194 IC device makes it an ideal demonstration circuit. Figure 13-9 shows how this particular shift register can be connected to a set of control switches and a source of active-low clock pulses. The outputs in this case are LEDs connected to the register's Q outputs.

The CLK pulse is manually generated each time the user depresses the CLK push button. Depressing that button initiates an 11-ms positive pulse at the output connection of the 555-type monostable multivibrator. This pulse is then inverted

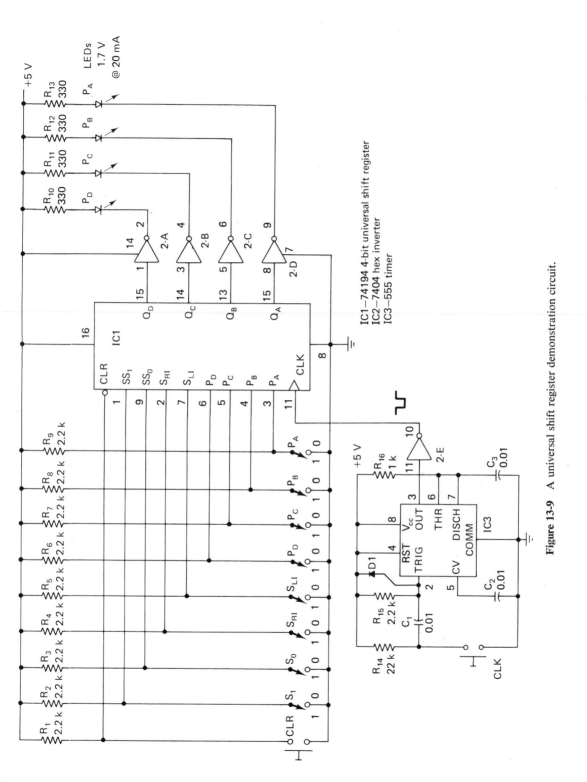

Figure 13-9 A universal shift register demonstration circuit.

IC1—74194 4-bit universal shift register
IC2—7404 hex inverter
IC3—555 timer

by means of inverter 2-E and then it is applied to the CLK input of the shift register.

Since the shift register responds only to the positive-going edge of any CLK pulse, the monostable acts as an 11-ms delay circuit. To be more specific, it is a switch-debouncing circuit that delivers only one CLK pulse to the shift register each time the CLK button is depressed, even when the CLK switch contacts bounce several times within that 11-ms interval.

The other switches in the system do not have to be debounced in this fashion. Depressing the CLR push button, for instance, might cause a noisy, switch-bouncing signal at the CLR input of the shift register, but that sort of noise does not cause any noticeable problems as far as the experimenter is concerned.

All of the switch inputs are pulled up to logic 1 through various pull-up resistors. The inputs, in other words, are normally high. Closing any of the switches pulls their respective shift-register terminals down to logic 0.

The outputs from the shift register are active-high; but as described in Ch. 5, TTL circuits can be properly interfaced with LEDs only by an active-low logic format. Thus the need for logic inverters between the four Q outputs of the shift register and the corresponding display LEDs. Of course, the LEDs light up whenever their Q outputs from the shift register are at logic 1.

The following descriptions show all possible shift-register operations. Most of these operations can be performed with simpler shift registers, and it is up to the experimenter to decide which of the shift registers summarized in Table 13-1 could be used for each of the basic functions.

13-6.1 Serial-In, Store, Serial-Out

One of the simplest shift-register routines is shifting data into it serially, allowing it to be stored in the register for any desired length of time and then shifting it out serially.

A bidirectional serial-in, serial-out shift register offers four options: (1) right-shift in, right-shift out; (2) left-shift in, left-shift out; (3) right-shift in, left-shift out; and (4) left-shift in, right-shift out.

A serial operation described as right-shift in, right-shift out can be alternately described as *MSB in first, MSB out first*. To do this particular operation with the demonstrator circuit in Fig. 13-9, first program the unit for right shifting by setting $S_0 = 1$, $S_1 = 0$. Depress the CLR push button to clear the display to all 0's, and then set the desired MSB logic level at the S_{RI} switch. Depress the CLK push button one time to enter that bit into the Q_A section of the register. Enter three more bits, one at a time, through the S_{RI} switch, momentarily depressing the CLK push button after each entry.

The 4-bit word you entered serially should now be stored in the register, with the first-entered bit at the P_D output lamp.

To retrieve the data serially, set the S_{RI} switch to logic 0 and depress the CLK push button at least three times. The serially entered 4-bit word should then appear at the P_D lamp, showing the second, third and fourth bits in that sequence.

In this particular phase of the demonstration, the first bit to be entered was the first one to be retrieved from the P_D output.

It is also possible to do a first-in, first-out serial operation using left shifting. Merely repeat the steps just described for right shifting, but change the direction of shift by setting $S_0 = 0$ and $S_1 = 1$. The serial data input is the S_{LI} switch in this case, and the P_A lamp is the serial data output. Nevertheless, the first-entered bit is the first one to appear at the P_A output when unloading the register.

Observing a first-in, last-out serial operation is a matter of entering the data serially in one direction and then serially unloading the register in the opposite direction.

For instance, load the register serially and to the right by setting $S_0 = 1$, $S_1 = 0$ and entering the data at S_{RI}. After the register is thus loaded, change the programming switches to $S_0 = 0$ and $S_1 = 1$. Operating the CLK push button then unloads the register serially at lamp P_A. The first bit entered is thus the last one out.

13-6.2 Parallel-In, Parallel-Out

Purely parallel-in/out operations make a shift register act as a data latch, a temporary storage place for a data word. The circuit in Fig. 13-9 has a 4-bit capacity; therefore, it is quite suitable for use as a 4-bit data latch.

To demonstrate the latch feature (parallel-in, parallel-out), clear the register by depressing the CLR push button and then set up the circuit for parallel entry by setting $S_0 = S_1 = 1$. Any 4-bit word at the P inputs should not appear at the parallel outputs (LEDs P_A through P_D) at this time, however, because parallel loading with the 74194 is synchronous; the register must be clocked.

Next set some 4-bit binary word at the parallel inputs by using switches P_A through P_D. Momentarily depress the CLK push button and note the response at the parallel outputs. Whatever word was entered at the parallel inputs should then appear at the parallel outputs.

The 4-bit word at the inputs can then be changed without affecting the outputs at all. The register is thus latched and totally immune to any ongoing activity at the inputs, at least until the circuit is clocked again.

Any 4-bit word can be stored in the register by clocking the circuit. Note that it is not necessary to clear the register before clocking in a new word. This circuit is capable of writing over any previously stored data; and that is not a trivial feature when considering larger memory registers in the following chapter.

13-6.3 Serial-In, Parallel-Out

The circuit in Fig. 13-9 can be used as a serial-to-parallel converter by first entering the data serially and then reading it from the parallel outputs. Such circuits are, for example, used in pocket calculators in which any multi-digit number is entered serially via the keyboard and displayed in a parallel fashion at the display assembly. Such converters are also found at the receiving end of a multiplexed data-transmission link. Multi-bit data arrives along a single channel in a serial form and is loaded into the register where it is finally read out in parallel form.

The serial data in this demonstration circuit can be entered either from the left or right, depending on whether the first-entered bit is to appear in the least- or most-significant bit position.

If the first bit in the serial input string is to be the least-significant bit, set up the circuit for left-shift serial entry, setting $S_0 = 0$, $S_1 = 1$. Enter the serial data (LSB first) one bit at a time at the S_{LI} terminal, depressing the CLK push button after setting the position of the S_{LI} switch. After thus entering four bits, the LSB should appear at the P_A output lamp, and the last-entered bit should be at lamp P_D.

Once the serial data is loaded into the register, it can be latched there by setting $S_0 = S_1 = 0$. Any further clocking or attempt to enter serial data will not affect the parallel-stored data.

Of course, the serial data can be entered with a right-shift input. In this instance, the first-entered serial bit will end up at the P_D output after entering four serial bits in succession. This is accomplished by setting $S_0 = 1$, $S_1 = 0$ and clocking in the serial data from the S_{RI} input switch.

13-6.4 Parallel-In, Store, Serial-Out

A 4-bit word can be clocked into the register via the parallel inputs, stored in that register for any desired length of time and then read out serially with either a left or right shift. This would be an example of parallel-to-serial data conversion. The scheme differs from a multiplexer system in that the data can be stored in the register until the serial-shifting action takes place. Data storage of this sort is impossible with a multiplexer parallel-to-serial conversion scheme.

Set up the circuit in Fig. 13-9 for parallel entry by setting $S_0 = S_1 = 1$. Enter the 4-bit word at the parallel input switches P_A through P_D and depress the CLK push button. The data should then appear at the data output lamps. Setting $S_0 = S_1 = 0$ then latches the data, overriding any subsequent clock activity or changes in the serial or parallel data inputs. The data is thus preserved for any length of time, unless, of course, the power supply voltage is interrupted for any reason.

The stored data can be retrieved serially from either the P_D or P_A output, depending on the direction of serial shift. To retrieve the data from the P_A data output, the circuit must be adjusted for left shifting, setting $S_0 = 0$, $S_1 = 1$. Each time the CLK push button is depressed, one bit of the stored data should appear at the P_A output. The sequence is from the A to the D bit.

Parallel-stored data can be retrieved serially, beginning with the D bit, by unloading the register with a right shift ($S_0 = 1$, $S_1 = 0$) and reading the output from LED P_D.

13-7 Keyboard Entry of Decimal Data

The applications of shift registers cited thus far in this chapter have been limited to data storage and to serial-to-parallel and parallel-to-serial conversion. These particular applications demonstrate the operation of single shift-register IC devices. Shift registers, however, find their most powerful applications when used in groups.

The example described in this section uses a number of shift-register devices for entering multi-digit numbers into a digital system via a typical keyboard assembly.

The point of data entry in Fig. 13-10 is a set of ten key switches labeled with decimal numerals 0 through 9. A decimal-to-BCD encoder circuit immediately reduces this ten-line format to a simpler 4-line BCD format.

Any time one of the keys is depressed, a 555-type monostable multivibrator generates a 10-ms CLK $\emptyset1$ pulse that clocks the BCD word into a parallel-in, parallel-out shift register, SR-1. Data is clocked into this particular register on the positive-going edge of the CLK $\emptyset1$ pulse. Thus there is a 10-ms delay from the time any keyboard key is depressed and the BCD data is entered into that register. The purpose of the delay is to erase any keyboard bouncing effect. (See Sec. 7-2.1 for details and typical circuits.)

The positive-going edge of the CLK $\emptyset1$ pulse not only enters BCD data into SR-1, it also initiates a second 10-ms monostable pulse, CLK $\emptyset2$. The monostable timing interval for the CLK $\emptyset2$ pulse isn't as critical as the debouncing interval of CLK $\emptyset$; its primary function is to allow some time for the SR-1 register to settle down before feeding its parallel outputs to a set of four serial, right-shift registers.

When the positive-going edge of the CLK $\emptyset2$ pulse occurs, it clocks the data in SR-1 into the Q_A positions of registers SR-2 through SR-5. Note that the LSB of the latched BCD word from SR-1 is directed to SR-2, the LSB serial shift register. The second bit from SR-1 goes to the serial input of SR-3, the C-bit goes to SR-4, and the MSB of the BCD word goes to the serial input of SR-5.

Depressing any one of the ten keys on the keyboard assembly thus sets off a 2-phase clock operation, first entering a BCD version of the key designation into SR-1, followed by splitting up the BCD word among four serial shift registers, SR-2 through SR-5.

Depressing another key on the keyboard assembly repeats the action, entering a new BCD number into SR-1 and then shifting it serially into the four larger registers. The original word is shifted down the line, appearing at the Q_B outputs of the larger registers.

Each time the keyboard is used, all previously entered numbers are shifted one location to the right in the serial registers, and the new one appears at the Q_A positions.

This particular keyboard-entry circuit has a capacity for accumulating up to eight BCD words from the keyboard. If the user attempts to enter nine numbers from the keyboard, the first-entered number will be lost from the Q_H positions of registers SR-2 through SR-5.

The system in Fig. 13-10 includes an 8-digit multiplexed decimal display scheme. Generally speaking, this decimal display system accepts up to eight BCD numbers and displays them at an 8-digit, 7-segment display assembly. (See Sec. 12-4 for a more detailed description of multiplexed 7-segment display circuits.)

Now considering the overall operation of this system, suppose the user wants to display the number 5678. People normally enter decimal numbers into a system with the most-significant digit first, 5 in this case. The user can first clear the

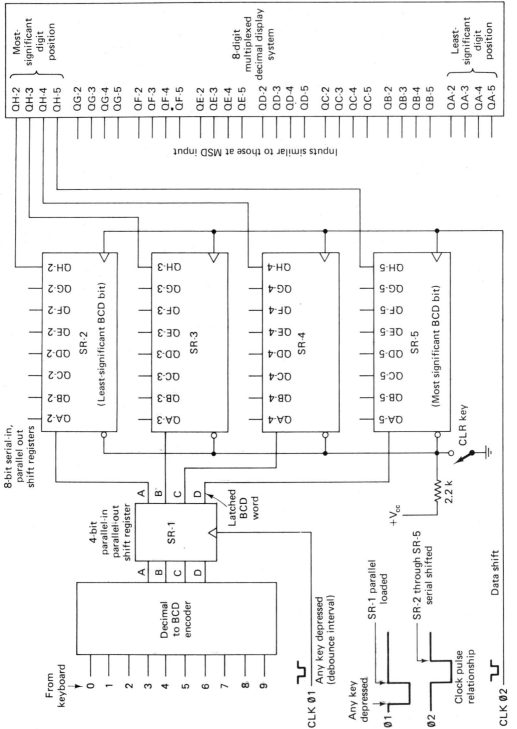

Figure 13-10 An 8-digit keyboard entry system using shift registers.

display by depressing the CLR key on the keyboard and then striking the "5" key. A BCD version of decimal 5 is then entered in a parallel fasion into SR-1; and shortly after that, the word is split up between the Q_A inputs of serial registers SR-2 through SR-5. At that same instant, the display system displays decimal numeral 5 in its least-significant bit (right-hand) position. The BCD version of decimal 5 is stored at the Q_A outputs of the serial registers, and those outputs are connected to the LSD input position of the display assembly, inputs Q_{A-2} through Q_{A-5}.

The next numeral to be entered from the keyboard in this particular example is 6. When the user strikes the "6" key, the 2-phase clock scheme first loads SR-1 with BCD 6 and then shifts it into the serial registers. Now the 5 is stored in the Q_B positions, while 6 is in the Q_A positions. The display thus shows 56.

The remaining two numbers in this example, 7 and 8, are entered in that sequence via the keyboard; and each time a key is struck, previously entered numbers move one decimal place down the line toward the MSD position in the display.

As shown here, the circuit in Fig. 13-10 merely displays any decimal number (up to eight digits) on a 7-segment display assembly. The outputs of the serial shift registers could also go to a memory system arithmetic accumulator, the preset inputs of digital counters, or any other sort of digital mechanism that calls for a numeric input.

In any event, the user can see the digits he or she is entering into the system as he or she enters them. If the user makes any errors, correcting the situation is a simple matter of striking the CLR key and starting all over again. Once the decimal entry is complete and correct, the user can depress another key that will enter the data from the serial registers in a parallel fashion into another part of the working system. If this is part of a calculator system, for example, the data entered into the serial registers can be parallel loaded into another set of registers upon depressing one of the function keys such as $+$, $\times$, $\div$ and so on.

Remember that this is merely one common example of how a number of shift registers can be interconnected within a digital system.

Exercises

1. What is the critical difference between the appearance of a synchronous binary counter and a serial-shift register built from J-K master–slave flip-flops?

2. What is the meaning of the term *jam entry* in the context of shift registers?

3. What is the basic definition of a *universal shift register*?

4. List five possible shifting formats for a universal shift register.

5. Suppose an 8-bit serial shift register is loaded with data. If the serial input is set to logic 0, how many times should it be clocked to fill the register with 0's?

6. In what way is a parallel-in, parallel-out shift register like a data latch?

7. What changes in the programming for a serial shift register are necessary if it is to be used in a first-in, last-out data scheme?

8. Generally speaking, how can two 8-bit serial shift registers be interconnected (cascaded) to perform the function of a 16-bit serial shift register?

9. Suppose a serial shift register is in the process of shifting a single logic-1 level. If the serial output is connected back to the serial input during the shifting time, what happens to that 1 bit as it leaves the last position in the register?

10. Using a single 556 dual timer IC, sketch a circuit that will generate the CLK ∅1 and ∅2 pulses shown in Fig. 13-10.

BASIC MEMORY SYSTEMS

Shift-register devices can, indeed, be used as temporary storage places for digital data. Data can be written into them in a serial or parallel fashion, stored there for any desired length of time and then read out in a serial or parallel fashion.

The only problem with shift registers as data-storage devices is that they have such a limited capacity. Shift-register ICs are available with bit capacities on the order of 4 or 8 bits; and the relatively common task of temporarily storing 256 bits of digital data would call for as many as 86 shift-register ICs! Obviously, shift registers are out of the question when it comes to storing moderate-to-large amounts of digital data.

This chapter deals with the basic principles and applications of semiconductor memory devices. Although the standard TTL family offers a very limited selection of memory ICs [a RAM (random-access memory) with the capacity for accepting, storing and reading out sixteen 4-bit words and a ROM (read-only memory) capable of reading out 32 pre-programmed 8-bit words], the principles they embody can be extended to the much wider range of MOS/LSI memory devices on the market today.

While studying this chapter, bear in mind the fact that semiconductor memories can be (and most often are) assembled in a building-block fashion to create more complex and higher-capacity memory systems. Virtually all TTL/MSI and MOS/LSI memory systems are assembled as described here.

14-1 A Basic Memory Cell Model

Memory ICs are all built around a system of basic 1-bit memory cells. A memory cell is an addressable flip-flop element that can be set to a 1 or 0 state, allowed to hold that state for any desired length of time and then display that stored logic level on demand.

There are a number of different possible ways to represent the essential characteristics of a basic 1-bit memory cell. For our purpose here, however, the basic memory cell is represented by a common level-D flip-flop. As illustrated in Fig. 14-1(a), a level-D flip-flop has a Q output that follows its D input as long as the G terminal is fixed at logic 1. The moment the G input is pulled down to logic 0, however, the Q output retains the D input level that existed just before the transition from 1 to 0 at input G. Section 6-4 describes the characteristics of level-D flip-flops in greater detail.

This flip-flop element is only one of many identical elements in a memory system; therefore, a basic memory cell must also include some provisions for accessing one particular element out of the group. As shown in Fig. 14-1(b), the operation of the D flip-flop element can be selected and controlled by means of some inputs to AND logic gates.

The memory select and control circuit has four inputs and a single output. The inputs include D_k (a data input terminal), X_n (a memory-cell select input), and a pair of true and complemented $R/\overline{W}$ (read-or-write) inputs. The sole S_n output is the cell's data output terminal.

Since this particular cell is actually one of many identical cells in some memory IC, there must be some provision for selecting that one cell out of the group. In this particular instance, the cell is selected by setting the X_n input to logic 1. According to the function table in Fig. 14-1(b), setting X_n to logic 1 enables the cell for either writing or reading activity. As long as $X_n = 0$, however, all input and output activity is blocked, and the cell is in a HOLD mode where its stored content is protected.

After selecting this particular cell by setting X_n to logic 1, the next step is to choose either a writing or reading operation. If the user wishes to write a logic 1 or logic 0 into the memory, the $R/\overline{W}$ input is set to logic 0 (and $\overline{R/\overline{W}}$ to logic 1). Setting $R/\overline{W}$ to logic 0 effectively gates off AND gate A_3, maintaining a 0 output from S_n. Setting the complemented logic-1 level at $\overline{R/\overline{W}}$, however, enables AND gate A_2 and, ultimately, A_1 as well.

Now both inputs to A_2 are at logic 1, thus enabling the G and D inputs of the flip-flop element. The flip-flop's G input is set at logic 1 so that any data appearing at D_k is directed to the Q output. Under these memory-writing conditions, Q can be set to logic 0 if $D_k = 0$ or to logic 1 if $D_k = 1$. See the second and third lines in the function table in Fig. 14-1(b).

After the WRITE operation is completed, the cell can be disabled by setting X_n to logic 0. The stored logic bit remains at the Q output of the flip-flop, but it

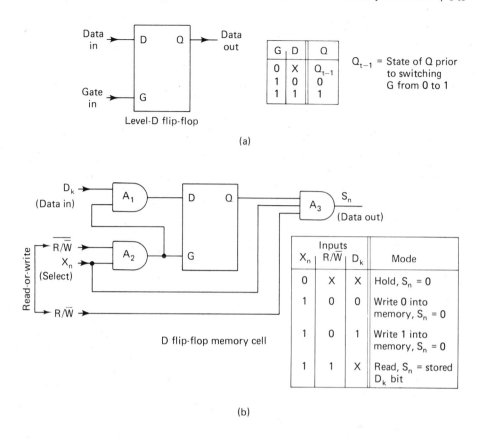

(a)

(b)

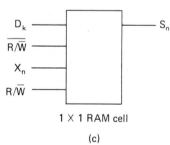

1 X 1 RAM cell

(c)

Figure 14-1   A basic memory cell. (a) D flip-flop diagram and truth table. (b) Adding gates to allow enabling and selection of READ or WRITE modes. (c) Simplified block diagram of a memory cell.

cannot appear at the S_n data output unless the cell is enabled and set for the READ operation ($X_n = 1$, $R/\overline{W} = 1$).

It is important to note that the READ operation is nondestructive. That is, data can be read at the S_n output without disturbing the stored content of the flip-flop element. Once a logic level is stored in the flip-flop via a WRITE operation, it can be read out any number of times by means of the READ operations. The only way to alter the content of this little register is by writing in a new bit of data.

Figure 14-1(c) represents the basic memory cell in a block-diagram format.

14-1.1 Serial Expansion of Memory Cells

It is possible to assemble any number of basic memory cells in such a way that any one of them is selectable for reading or writing operations. The basic idea is to connect together inputs D_k, $R/\overline{W}$ and $\overline{R/\overline{W}}$ and OR together the S_n outputs. All cells are then set to their READ or WRITE modes at the same time, but they are selected separately. If the system is set for the WRITE mode, the logic level at the D_k inputs affects only the selected memory cell; and if the system is set for the READ mode, all S_n outputs, except the one being selected, are at logic 0. The OR-ed output thus responds only to the S_n output of the selected memory cell.

Figure 14-2(a) shows two basic memory cells interconnected to make up a 2×1 RAM, a random-access memory capable of storing and reading out two 1-bit digital words.

Whenever the A_0 input is set to logic 0, memory cell MC-1 is disabled; but because of the logic inverter between the A_0 input and the X_0 select terminal of MC-0, that cell is enabled. Data can thus be written into MC-0 or read out of it, depending on the logic level at the $R/\overline{W}$ input.

Setting A_0 to logic 1 enables MC-1 and disables MC-0, making it possible to write or read data at MC-1.

The system can be further expanded to a 4×1 RAM as shown in Fig. 14-2(b). In this case, there are four basic memory cells that are selected one at a time by means of a 2-line to 4-line address decoder. The decoder shown here could be a $1:4$ demultiplexer having active-high outputs. MC-0 is thus selected whenever $A_1 = A_0 = 0$, MC-1 is selected whenever $A_1 = 0$ and $A_0 = 1$, and so on down to MC-3 which is selected whenever $A_1 = A_0 = 1$. See the address selection table accompanying the diagram in Fig. 14-2(b).

Data can thus be written into any one of the four memory cells by setting the desired logic level to be stored at the D_i input and setting $R/\overline{W}$ to logic 0. The data stored in one of the memory cells in this fashion can be inspected at any later time by addressing that cell via the A_0 and A_1 inputs while $R/\overline{W}$ is fixed at logic 1, the READ mode.

Further expanding the memory is a matter of enlarging the size of the address decoder, adding more memory cells and OR-ing their outputs. The 1024×1 RAM described later in this chapter includes 1024 separate memory cells, a 10-line to 1024-line address decoder and a fantastic 1024-input wired-OR output.

Serially expanded $n \times 1$ RAMs, however, require only one data input and one $R/\overline{W}$ terminal, no matter how large n might be.

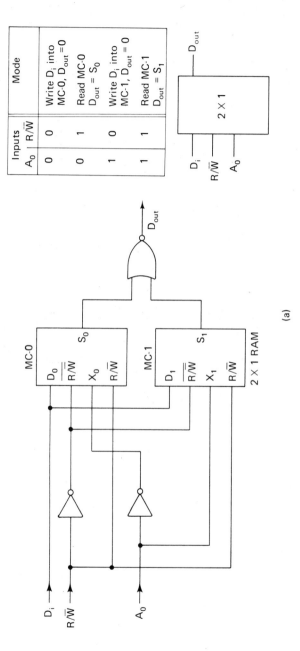

Inputs		Mode
A_0	$R/\overline{W}$	
0	0	Write D_i into MC-0, $D_{out} = 0$
0	1	Read MC-0 $D_{out} = S_0$
1	0	Write D_i into MC-1, $D_{out} = 0$
1	1	Read MC-1 $D_{out} = S_1$

(a)

Figure 14-2 Serial expansion of basic memory cells. (a) 2×1 format.

347

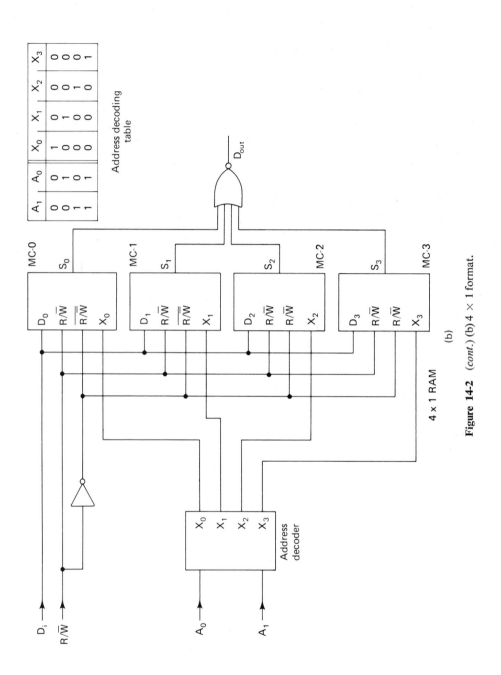

A_1	A_0	X_0	X_1	X_2	X_3
0	0	1	0	0	0
0	1	0	1	0	0
1	0	0	0	1	0
1	1	0	0	0	1

Address decoding
table

4 x 1 RAM

(b)

Figure 14-2 (*cont.*) (b) 4 × 1 format.

348

14-1.2 Parallel Expansion of Memory Cells

The serially expanded memory unit just described has only a single data input and one output. The system, in other words, can accept or read out only one bit at a time. A parallel-expanded memory can work with more than one bit at a time.

The circuit in Fig. 14-3 represents a 1×4 RAM, a random-access memory capable of accepting, storing and reading out just one 4-bit word. This circuit could actually be classified as a 4-bit data latch: There are four separate data inputs (D_1 through D_4), four separate outputs (S_1 through S_4), a single enable bus (EN) and one read/write control ($R/\overline{W}$).

The system is totally disabled as long as $EN = 0$; the outputs are all set at logic 0, and the stored contents of the register are preserved by disabling the data inputs to the memory cells. Setting the $R/\overline{W}$ input to 0 and enabling the system by setting $EN = 1$, it is possible to WRITE the parallel data at the four D inputs into their respective cells. The data outputs remain at logic 0 throughout the writing process, however.

Once the data is entered, the EN input can be returned to logic 1 to preserve the stored data, in spite of any changes at the D inputs and the $R/\overline{W}$ bus. The stored information can be recovered at any later time without altering the contents of the cells by first setting $R/\overline{W}$ to logic 1 and enabling the system. The stored data thus appears at the four S outputs.

This 1×4 RAM system can be expanded to accommodate any number of parallel inputs. A $1 \times n$ RAM, for example, would include n memory cells that could accept, store and read out one n-bit word.

The $1 \times n$ RAM concept is presented here only for the purpose of showing the evolution of more complex memory devices. One-word memories do not really exist as available IC devices. A $1 \times n$ RAM is properly classified as a D-type data latch; and even then, such latches are available with no more than eight parallel inputs and outputs.

14-1.3 Serial/Parallel Expansion of Memory Cells

Some of the most useful memory systems in the digital business today combine the features of serial- and parallel-expanded memory cells. Figure 14-4 shows how four 4×1 RAM assemblies can be parallel-expanded to make up a 4×4 RAM, a random-access memory that can handle four 4-bit words.

The 4×1 RAM assemblies in this case are identical to the one illustrated in Fig. 14-2(b). Each one includes four serially connected memory cells that are addressed by inputs A_0 and A_1. The four separate 4×1 units are then expanded in a parallel fashion to accept four parallel-data inputs.

The building-block nature of memory-cell systems ought to be apparent at this point in the discussion.

A 4-bit word can be written into this 4×4 memory by entering the word at inputs D_1 through D_4, addressing one of the four possible locations via the A_0 and A_1 inputs and then setting $R/\overline{W}$ to logic 0. If it happens that $A_0 = A_1 = 0$,

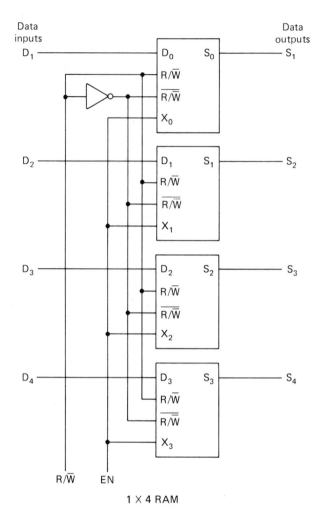

Data inputs
Data outputs

Figure 14-3   Parallel expansion of basic memory cells to form a 1 × 4 RAM.

each one of the four input bits will be written into the X_0 memory cell of its respective 4 × 1 RAM assembly. The data outputs, D_{01} through D_{04}, remain at logic 0 as long as the system is in this WRITE mode, however.

Three additional 4-bit words can be written into cells X_1, X_2 and X_3 by addressing each of them at A_0 and A_1. [Refer to Fig. 14-2(b) for a definition of X_0 through X_4.]

Serial and parallel expansion can be carried out to create RAM systems having any desired bit size and word capacity. The 7489 RAM IC described in the following section is one example of serial/parallel memory expansion. The result in that instance is a 16 × 4 memory, a memory that is capable of storing up to 16 different 4-bit words.

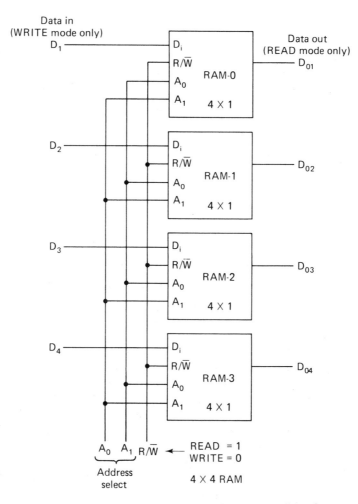

Figure 14-4   Serial/parallel expansion of basic memory cells to form a 4 × 4 RAM.

14-2   The 7489 TTL X 4 RAM

The 7489 TTL integrated circuit is a 16 × 4 RAM. In other words, it is a random-access memory having 64 separate memory cells arranged in a 16 × 4 parallel/serial fashion. It is a MSI (medium-scale integrated) device that can accept, store and readout 16 different 4-bit words.

Figure 14-5 shows the pinout and function table for the 7489 TTL RAM. Note there are four data inputs (D_1 through D_4) and four corresponding data outputs (S_1 through S_4). Since the memory has 16 word-storage locations, it follows that it must have four binary address lines, inputs A_0 through A_3, in this case.

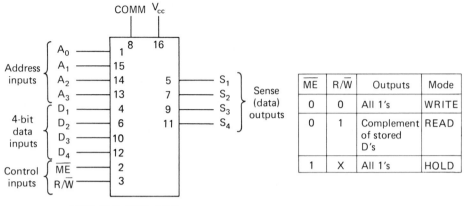

Figure 14-5   The 7489 16 × 4 TTL RAM: pinout and function table.

There is an $R/\overline{W}$ control terminal that determines whether data will be written into the memory via the data inputs or read out via the data outputs. According to the function table in Fig. 14-5, data is written into the memory whenever $R/\overline{W}$ is fixed at logic 0; and the complement of any previously written data is read out when $R/\overline{W}$ is equal to logic 1.

The $\overline{ME}$ terminal serves the function of a chip-select or memory-enable control input. According to the function table, setting $\overline{ME}$ to logic 0 enables the chip for either writing or reading operations, depending on the logic level at $R/\overline{W}$. Setting $\overline{ME}$ to logic 1, however, effectively turns off the IC, putting it into a hold mode where the stored data is protected from any accidental writing operations and the outputs are all set to logic 1.

This memory is actually a rather simple device from the user's point of view. Writing a 4-bit word into this memory is a matter of first disabling the chip by setting $\overline{ME}$ to logic 1, setting the desired address location and data to be stored at the A and D inputs, setting $R/\overline{W}$ to logic 0 and finally enabling the chip by setting $\overline{ME}$ to logic 0. The data specified at the D inputs is then written into the address specified at the address inputs.

Reading that data out at any later time is a matter of disabling the chip, addressing that same location via the A inputs, setting $R/\overline{W}$ to logic 1 and then enabling the chip. An inverted version of the previously stored data then appears at the S outputs.

It is important to bear in mind that any data read out of the 7489 is a complemented version of any data stored in it. The S outputs, in fact, have an open-collector configuration that allows rather simple memory-expansion techniques. Expansion techniques are described later in this section; but whether the 7489 is expanded for larger bit sizes or not, the open-collector outputs call for wiring pull-up resistors between each S output and $+V_{cc}$. Without these external resistors, logic-1 outputs are not defined properly.

14-2.1 A Simple Test and Demonstration Circuit for the 7489 RAM

Figure 14-6 shows a circuit for testing and demonstrating the operation of this 16×4 random-access memory IC. The data and address inputs are connected to SPST toggle switches and 2.2-k pull-up resistors. As long as any of these data and address switches are open, they deliver a logic-1 level to their respective input terminals. Closing these switches changes their respective input logic level to logic 0.

The READ/WRITE toggle switch determines whether the system is being used for reading or writing operations. As long as this particular switch is open, the $R/\overline{W}$ terminal sees a logic-1 level that sets up the memory for reading operations. Closing that switch switches the system to its WRITE mode.

The ENAB push button, connected to the IC's $\overline{ME}$ terminal, keeps the chip normally disabled, thus protecting any previously stored data and fixing the S outputs at logic 1. Data can be written into or read out of the memory only while this push button is depressed, but the ENAB button should never be depressed while changing the address. Generally speaking, any memory device ought to be disabled during addressing intervals. Changing the address while a memory is enabled in its READ mode creates a jumble of data at the outputs that might confuse the systems operating from it; and changing the address while the memory is enabled in its WRITE mode causes data to be written in a lot of different (and perhaps undetermined) locations in the memory.

The data outputs are connected through LEDs and limiting resistors to $+V_{cc}$. This particular output scheme takes advantage of the device's active-low, open-collector output configuration. The LED/resistor combination provides the pull-up circuit required for each of the data outputs.

To write 16 words into this memory, first address location 0000 at the ADDRESS INPUTS, make certain the READ/WRITE switch is in its WRITE position, set the desired 4-bit word to be stored at the DATA INPUTS and then momentarily depress the ENAB push button. The outputs do not respond at all during this phase of the operation; the LEDs remain switched off at any time the circuit is disabled or in its WRITE mode.

To enter the second word, address its location at the ADDRESS INPUTS—perhaps address 0001—and set the new word to be stored at the DATA INPUTS. Depressing the ENAB push button then stores that data at the designated address.

Continue this process until all 16 address locations have data stored in them. Since this is a random-access memory, the data can be stored in any desired address sequence.

Once the desired data is loaded into the memory in this fashion, it can be read out by setting the READ/WRITE switch to READ, addressing any of the locations in the memory and depressing the ENAB push button. Each time the ENAB button is depressed, the stored 4-bit word appears at the LED readout assembly.

The data can be thus stored and read out at any later time, just as long as the

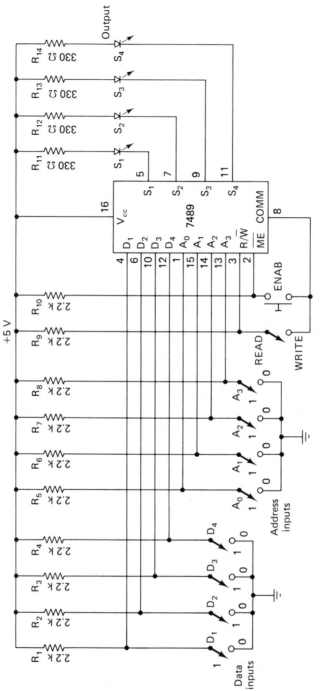

Figure 14-6 A test and demonstration circuit for the 7489 TTL RAM.

354

$+V_{cc}$ supply is not interrupted. Any stored data will be hopelessly scrambled in any RAM if the power supply voltage fails for even a brief instant.

The circuit in Fig. 14-6 can be used for testing the operation of the 7489 RAM by first writing logic 0's into all address locations and then systematically reading all those locations. A logic-1 appearing at any point during the readout phase of this test is a good indication that the memory system is defective. If it so happens that the circuit properly stores logic 0's, load the memory with logic 1's, and then read all the address locations. This time, every location should read out a logic 1. If a logic-0 level appears, the IC is probably defective.

It is important to test the chip by entering all 0's one time and then all 1's the next. A defect in the chip might fix one of the outputs at either state; and the only way to check the system thoroughly is by running all 0's at one time and all 1's at another.

14-2.2 Expanding the 7489 Memory System

Any memory system can be expanded in much the same fashion elementary memory cells are expanded to make up a useful memory chip. Two 7489 16 × 4 RAMs, for example, can be paralleled to make up a 16 × 8 memory, a memory that can deal with sixteen 8-bit words. The same two chips, however, can be re-organized to build a 32 × 4; and then, of course, four of these chips can be combined in a series/parallel fashion to make up a 32 × 8 memory.

In fact, the bit size of a memory system built around 7489 building blocks can be expanded in multiples of four to any desired word size; and by the same token, the number of words the system can handle can be expanded in multiples of 16. Thirty-two 7489's can be assembled, for instance, to a rather common 256 × 8 scratch-pad memory for smaller computer systems.

Figure 14-7 shows the technique for paralleling two 7489's to create a 16 × 8 RAM system. Note that the address and control inputs are all connected in parallel but that the data inputs and outputs are separate. Since the address inputs are connected in parallel, addressing one particular location in one section automatically addresses the same location in the other. Both ICs are operated in exactly the same mode because their $\overline{ME}$ and $R/\overline{W}$ terminals are similarly connected in parallel.

IC1, however, handles the four lower-order bits of the 8-bit input word, while IC2 takes care of the four higher-order bits. Thus any 8-bit word stored in one particular address location in the ICs appears inverted at the eight S outputs whenever that location is addressed with the system in its READ mode.

Although the word size could be expanded indefinitely in this manner, practical digital systems seldom call for word sizes larger than 16 bits. A 16-bit system built around the 7489 RAM IC would call for four paralleled chips of this type.

Serial expansion of memory ICs expands the word-storage capacity rather than the size of each word to be stored. Figure 14-8 shows how a pair of 7489's can be interconnected to make up a 32 × 4 memory scheme. In this case, the four data inputs and outputs are paralleled and the $R/\overline{W}$ terminals are connected together.

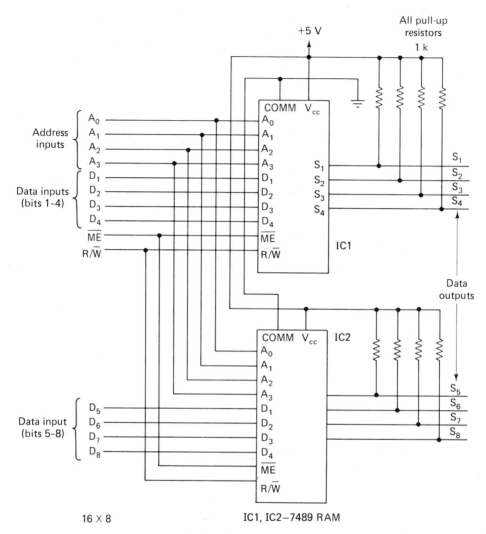

Figure 14-7 Parallel expansion of two 16×4 RAMs to produce a 16×8 RAM.

A memory having 32 addressable word locations calls for five address lines, however; and the 7489 has only four of them. The fifth line (A_4) is used for enabling the two ICs one at a time. The four lower-order address bits are shared between the two ICs, but A_4 is connected directly to the $\overline{ME}$ input of IC1 and through an inverter to the $\overline{ME}$ input of IC2.

Whenever $A_4 = 0$, then, memory IC1 is enabled and IC2 is disabled and cannot participate in any ongoing writing or reading activity. Changing A_1 to logic 1, however, disables IC1 and enables IC2; and under this set of circumstances, IC2 becomes the active memory element and IC1 is effectively switched off.

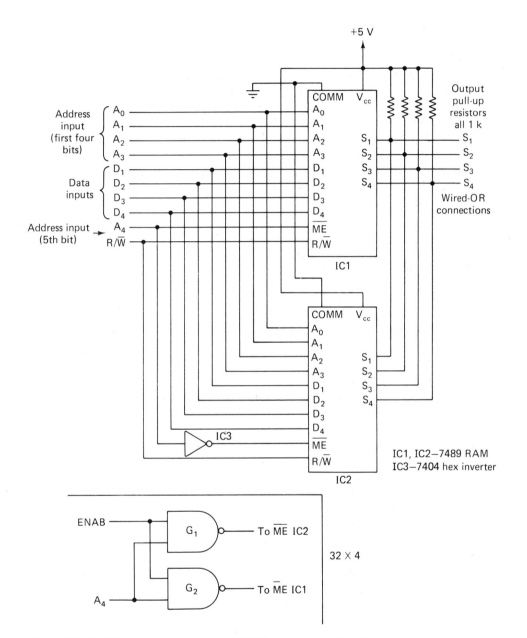

Figure 14-8 Serial expansion of two 16 × 4 RAMs to produce a 32 × 4 RAM.

As long as the user is addressing the binary equivalent of address 0 through 15, IC1 is active. Addressing locations 16 through 31, however, enables IC2 and switches off IC1. Each IC handles sixteen 4-bit words; and by enabling only one chip at a time, it is possible to work with thirty-two 4-bit words.

It is especially important to note that the S outputs are connected directly together. This cannot be done with totem-pole TTL ICs; but the 7489 features an open-collector output scheme that permits this kind of "wired-OR" connection.

Suppose the system is set for its READ mode ($R/\overline{W} = 1$) and $A_4 = 0$. In this case, IC1 is enabled and in its READ mode. IC2 is disabled, however, and all four of its data outputs are at logic 1. A logic 0 output from IC1 will pull down that memory output, though, in spite of the logic 1 appearing from IC2.

Recall that a memory system ought to be addressed only while it is totally disabled. The circuit in Fig. 14-8 has no provisions for disabling both ICs at the same time while the user is changing the address. Using this system properly would then require another logic gate to the $\overline{ME}$ inputs of both ICs. See the insert in Fig. 14-8. Whenever ENAB (ENABLE) = 1, both NAND gates are effectively switched on, and an inverted version of A_4 appears at the gates' output; IC1 is thus enabled whenever $A_4 = 0$ and IC2 is enabled whenever $A_4 = 1$. This is perfectly consistent with the scheme just described for the 32 × 4 memory system. Setting the ENAB terminal to logic 0 effectively turns off both NAND gates, thus forcing them to send a logic-1 level to both memory ICs to disable them. Inverter IC3 must be eliminated when the enabling scheme shown in the insert is used.

The block diagram in Fig. 14-9 illustrates the operation of a 256 × 4 RAM assembly that is built up from sixteen 7489 16 × 4 RAMs. The important feature here is the technique used for enabling one of the 16 × 4 RAMs, depending on the 8-bit address word.

For the sake of simplicity, the 16 basic RAM elements aren't shown in this diagram. Their input and output designations are shown, however. In this instance, M_0 indicates the first 16 × 4 RAM device, M_1 indicates the second and so on through M_{15}.

Note that the system's READ/$\overline{WRITE}$ input is connected to the $R/\overline{W}$ terminal on all 16 RAM devices in the memory assembly. The significance of this feature is that all 16 are set to either their read or write modes at the same time.

Then see that the D_1 data input terminals of all 16 RAMs are connected together, as are the D_2, D_3 and D_4 inputs. In a similar fashion, the four lower-order address inputs are connected in parallel: A_0 to the A_0 terminals of all 16 RAMs, A_1 to all 16 RAMs and so on through A_3. All four address inputs on the 16 RAM elements are thus committed to the four lower-order bits of the overall addressing scheme.

The point of special interest concerns the treatment of the four higher-order address bits, A_4 through A_7. These four address bits are connected to the select inputs of a standard 4-line to 16-line decoder (1 : 16 demultiplexer). Each of the 16 active-low outputs from this demultiplexer goes to one of the 16 × 4 RAM devices, the $\overline{ME}$ inputs of these devices. So whenever this demultiplexer is enabled by

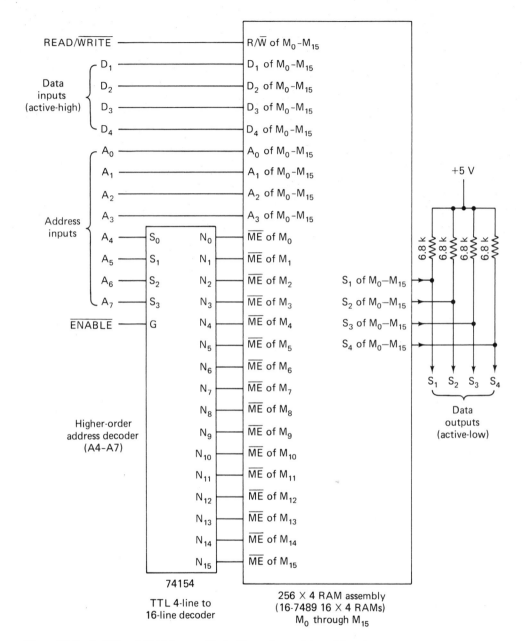

Figure 14-9 A 256 × 4 RAM assembly built up from sixteen 16 × 4 RAMs and a 4-line to 16-line decoder.

applying a logic 0 at the $\overline{\text{ENABLE}}$ terminal of the system, one (and only one) of the 16 basic memory elements is enabled. The other 15 are disabled.

Suppose, for example, the user addresses 00001111. In this case, the four lower-order bits are 1111; and since these bits are connected directly to the address inputs of all 16 basic RAM devices, all 16 of them are being addressed at their last word position, position 16. The four higher-order bits in this example are 0000. The demultiplexer circuit responds to this situation by pulling its N_0 output to logic 0, leaving the other 15 outputs at logic 1. So although the four lower-order address bits are simultaneously addressing the sixteenth position in each RAM device in the assembly, memory element M_0 is the only one that is properly enabled by the decoder circuit. Entering address 00001111 thus accesses the sixteenth-word position in the M_0 memory element. All other memory elements are disabled. The user can then write or read data at that position, depending on the logic level at the READ/$\overline{\text{WRITE}}$ input terminal.

Whenever the user sets the $\overline{\text{ENABLE}}$ input to logic 1, all 16 outputs from the demultiplexer go to logic 1, thereby disabling all the memory elements, regardless of the input address, input data and setting of the READ/WRITE terminal. This overall disabling feature is most important for disabling the entire system while changing the address inputs.

This technique for expanding sixteen 16×4 RAMs to build up a 256×4 RAM system might look rather complicated as it appears in Fig. 14-9. There is, however, a very definite and orderly pattern to the system. If this particular circuit were to be built up on a printed-circuit board, it would include 16 RAM ICs plus a 74154 demultiplexer IC. The board would appear very complex because of the large number of interconnecting lines; but again, the pattern often becomes clear after studying the circuit for a moment.

The memory assembly, as such, is often the least complicated part of a digital memory system. There are a lot of parts and interconnecting wiring, but the clear-cut pattern makes the design, testing and troubleshooting operations relatively easy. The most complicated part of a memory system in most instances is the addressing circuitry. The memory scheme illustrated in Fig. 14-9 could appear in any number of different kinds of digital systems. It is the addressing scheme, however, that sets one digital memory format apart from another.

14-3 Basic Memory Address and Control Circuits

One of the most critical features of any RAM addressing circuit is to make certain the memory cannot be in its WRITE mode while any change of address is taking place. Any addressing scheme requires some settling time; and if the WRITE mode happens to be enabled during that settling time, any data at the circuit's data inputs will be written into an indeterminate number of places in the memory. Scattering data throughout the memory in this manner is, of course, highly undesirable.

The circuit in Fig. 14-10 is a basic memory address register that automatically

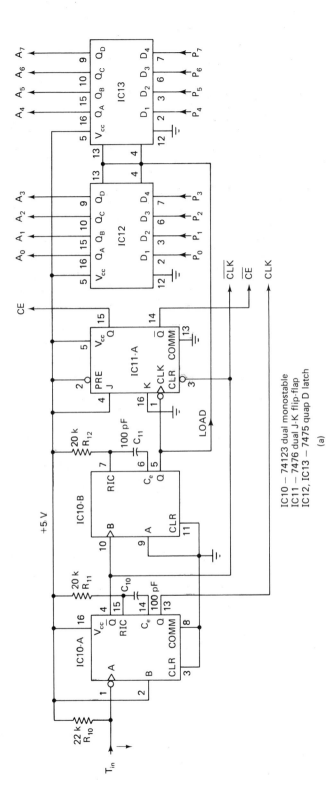

IC10 – 74123 dual monostable
IC11 – 7476 dual J-K flip-flap
IC12, IC13 – 7475 quap D latch

(a)

Figure 14-10 An 8-bit address register for a 256 × 4 **RAM** assembly. (a) Schematic diagram.

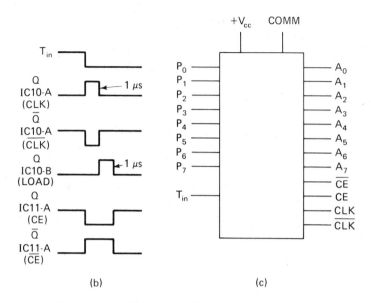

(b)   (c)

Figure 14-10 (*cont.*) (b) Waveforms for the 2-phase address interval and RAM disabling. (c) Simplified block diagram.

ensures that the memory chip is disabled during the addressing intervals. And, as illustrated in Fig. 14-11, this address register can be interfaced directly with the 256×4 RAM assembly described in the previous section of this chapter.

The outputs of the memory address register in Fig. 14-10 include a set of eight address connections, A_0 through A_7. These terminals are to be connected directly to the corresponding address input terminals of the RAM assembly. Any 8-bit address word fixed at the output of the address register thus determines the address location of the RAM.

The address outputs from the circuit in Fig. 14-10 come from a pair of 7475 quad D latches. In effect, these latches isolate the RAM assembly from any ongoing change-of-address activity. During a normal addressing cycle the latches hold the output address word fixed until the new address is set up; and only then are the latches enabled so they can respond to the new address information.

The address information can come from a set of eight address switches as illustrated in the block diagram in Fig. 14-11. These switches are connected to the data inputs of the quad latches, IC12 and IC13. The user can thus enter any desired 8-bit address via the address switches. The new address does not reach the RAM assembly, however, until the latches see a positive-going latch pulse; and it is IC10-B in Fig. 14-10 that is responsible for generating this latch pulse at the proper time.

IC10-A and IC10-B are halves of a 74123 dual monostable multivibrator IC package. The first monostable section, IC10-A, is wired to generate a 1-μs pulse whenever the T_{in} connection sees a negative-going edge of a trigger waveform.

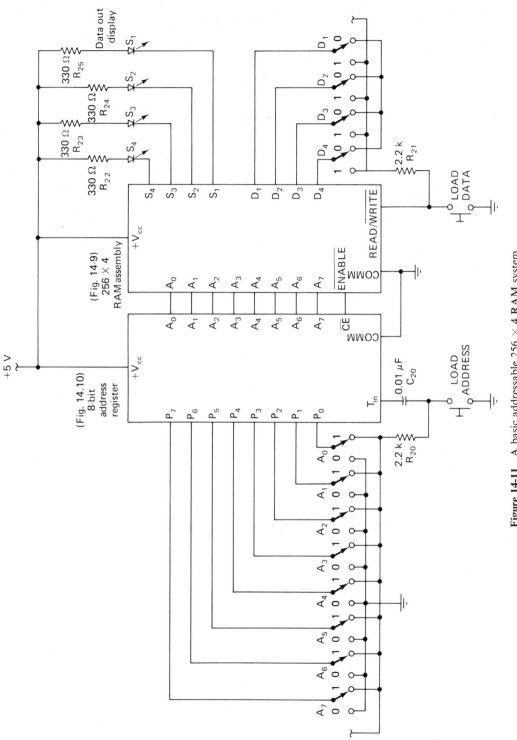

Figure 14-11 A basic addressable 256 × 4 RAM system.

Compare the first three waveforms in Fig. 14-10(b). Neither of the outputs from IC10-A is connected to the load terminals of the latch circuits, however.

An active-low version of the 1-μs pulse from IC10-A goes to the B input of the second monostable timer, IC10-B. This monostable device is wired so that it generates a 1-μs pulse in response to a positive-going edge of an input waveform.

When the operations of IC10-A and IC10-B are considered together, it happens that triggering the input of IC10-A yields a sequence of two 1-μs pulses: first one from IC10-A and then the other from IC10-B. It is the second of these two pulse waveforms that loads a new address into the address latches, IC12 and IC13. Note the connection between the Q output of IC10-B and the LOAD terminals of the latches (pins 4 and 13).

As shown in the waveforms in Fig. 14-10(b), a negative-going edge at T_{in} will load the address latches only after the first 1-μs delay is completed. And besides generating this delay-before-latching interval, the output of IC10-A serves a couple of other important functions.

Recall that any RAM system should be disabled during an addressing interval, and then note how the $\bar{Q}$ output of IC10-A is connected to the CLR input of a J-K flip-flop, IC11-A. Whenever the first monostable multivibrator generates its 1-μs pulse, then, it clears the flip-flop, making certain its $\bar{Q}$ output (pin 14 of IC11-A) is set to logic 1.

The $\bar{Q}$ output of IC11-A then remains at logic 1 until the negative-going edge of a pulse waveform appears at the CLK input (pin 1 of IC11-A). And when that clock waveform occurs, it sets the $\bar{Q}$ output to logic 0 again because the J and K inputs of the flip-flop are permanently fixed at logic 1 and logic 0, respectively.

Finally, note that the CLK pulse for IC11-A is taken from the Q output of IC10-B. This means that the $\bar{Q}$ output of the flip-flop will be clocked to logic 0 at the end of the second 1-μs timing pulse.

Thus the $\bar{Q}$ output of IC11-A is set to logic 1 at the beginning of the first of two 1-μs pulses, and it is clocked back to logic 0 at the end of the second pulse. The output of IC11-A, in other words, goes to logic 1 and remains there as long as the addressing interval lasts. The $\bar{Q}$ output of the flip-flop goes to the circuit's $\overline{CE}$ output terminal; and as illustrated in Fig. 14-11, $\overline{CE}$ is connected to the $\overline{ENABLE}$ connection of the RAM assembly.

When all of this is put together, it turns out that the RAM assembly is disabled from the moment IC10-A is triggered until the time the second 1-μs interval (the latch-loading interval) is completed. See the relationships between the $\overline{CE}$ pulse, T_{in} and the Q outputs of IC10-A and IC10-B in Fig. 14-10(b).

The block diagram in Fig. 14-11 represents a complete RAM system. The system is addressed by first entering the desired address at switches A_0 through A_7 and then depressing the LOAD ADDRESS push button. Depressing the LOAD ADDRESS button momentarily pulls the T_{in} connection of the address register circuit to logic 0, thereby initiating its 2-phase timing operation. The second of the two pulses loads the latches with the address information set at the address switches.

The RAM assembly is disabled, however, until the 1-μs latching interval is completed.

If the user wishes to write data into the memory, all he or she has to do is set the desired 4-bit data word at the four D switches (D_1 through D_4) and then depress the LOAD DATA switch. This particular switch is connected to the READ/$\overline{\text{WRITE}}$ terminal of the memory assembly; and as described in the previous section of this chapter, pulling that terminal down to logic 0 places the memory assembly into its WRITE mode. The data is thus written into the memory.

The moment the user releases the LOAD DATA push button, the RAM assembly returns to its READ mode, and the data just stored in the memory appears displayed at the output LEDs.

Entering further data into the memory system is then a matter of selecting the desired address location with switches A_0 through A_7, depressing the LOAD ADDRESS push button, setting the data to be stored at data switches D_1 through D_4 and finally depressing the LOAD DATA push button.

Since system is normally in its READ mode, the reading operations are slightly simpler. All the user must do to read the contents of the memory is set the desired address location at the address switches and depress the LOAD ADDRESS push button. Two microseconds later (after the 2-phase addressing operation is over) the previously stored data will appear at the output LEDs.

As useful as the RAM system outlined in Fig. 14-11 might be under many different kinds of circumstances, running through 256 different address locations can be an awkward and time-consuming process. Many digital applications call for addressing schemes that are more convenient to use and faster. The circuit in Fig. 14-12 represents yet another add-on assembly, one that can be interfaced with the system just described. The result is a RAM system that is far easier to use. Figure 14-13 shows how the three assemblies can be interconnected.

The outputs of the address control circuit in Fig. 14-12 are eight address lines (A_0 through A_7) and a trigger pulse, $\overline{\text{T}}$. The address lines go to the P inputs of the address register assembly, and the $\overline{\text{T}}$ connection goes to the T_{in} terminal of that same circuit. The address control circuit, in other words, provides the address register assembly with new address words and a pulse that initiates the 2-phase address timing cycle.

Note that one of the inputs to the address control circuit is a CLK connection from the address register. This is the first of the two 1-μs address timing pulses.

Adding the circuit in Fig. 14-12 provides some convenient addressing options. It is possible, for instance, to load the address in much the same fashion described for the system in Fig. 14-11. In this case, the user sets the desired address at a set of eight address switches and depresses an ADR push button.

It is also possible to increment or decrement the address by depressing that same ADR push button. The memory can thus be addressed through a sequence of address locations without having to address each location manually. Each time

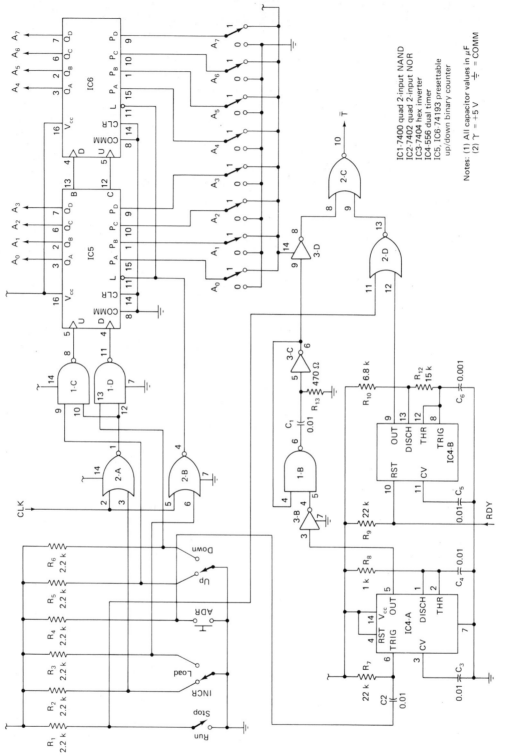

Figure 14-12 An address control assembly.

Notes: (1) All capacitor values in μF
(2) ⊤ = +5 V   ⏚ = COMM

IC1-7400 quad 2-input NAND
IC2-7402 quad 2-input NOR
IC3-7404 hex inverter
IC4-556 dual timer
IC5, IC6-74193 presettable
up/down binary counter

366

the ADR button is depressed, the address automatically increases or decreases one location, depending on the setting of the UP/DOWN switch.

And, finally, it is possible to scan the entire sequence of 256 address locations at a 40-kHz rate. This feature is especially useful for clearing all data from the RAM. The idea is to load 0's from the data switches while the memory is being scanned. In about 6.4 ms the user can completely clear the memory.

The heart of this address control assembly is a pair of 74193 4-bit binary, presettable up/down counters. These counters generate an 8-bit address word that is eventually loaded into the latch circuits of the address register and that 8-bit word ultimately reaches the address inputs of the RAM assembly.

The counters (IC5 and IC6) are either clocked or loaded by means of the CLK pulse from the address latch assembly. Recall that the address sequence involved two 1-μs pulses. The second pulse is used mainly for loading the latch circuits in the address register. It is the first of these address pulses, however, that controls the counters in the address control system.

The address is thus determined during the first of the two address pulses, and then that address is loaded into the RAM at the end of the second pulse. The address-changing operations at both the address control and register assemblies are clearly isolated from the RAM's data-handling operations.

A new address can be loaded into the address register and RAM assembly only after the T_{in} connection of the register sees the negative-going edge of a pulse waveform. This particular waveform in Fig. 14-13 comes from the $\overline{T}$ terminal of the address control circuit (Fig. 14-12); so it follows that an address-changing operation begins whenever the logic level at $\overline{T}$ falls from logic 1 to logic 0.

To see how the $\overline{T}$ pulse is generated, find the ADR push button in the address control assembly. The "hot" side of this push button is normally held at logic 1 by the 2.2-k pull-up resistor, R_4. And in a similar fashion, the TRIG input of a 555-type monostable multivibrator, IC4-A, is held at logic 1 through pull-up resistor R_7. As long as the ADR switch remains open, IC4-A remains in its non-timing, quiescent state.

Depressing the ADR push button, however, momentarily pulls the TRIG input of IC4-A down to logic 0, thereby initiating a 10-ms pulse at its OUT connection. The primary purpose of this timer is to mask any switch-bouncing effects that are bound to occur whenever the ADR switch is depressed.

The pulse from the OUT connection of IC4-A is thus a clean 10-ms pulse that is initiated the moment ADR is depressed. That positive pulse is inverted by IC3-B and applied to a pulse-shortening circuit comprised of IC1-B, IC3-C, C_1 and R_{13}. What emerges from IC3-C is a negative-going pulse that lasts about 5 μs. That particular pulse is coincident with the depression of the ADR switch.

The negative, 5-μs pulse from IC3-C is inverted by IC3-D and applied to one input of a 2-input NOR gate. If it is assumed for the time being that the other input of this NOR gate is fixed at logic 0, the pulse emerges from IC2-C as a 5-μs negative pulse at the $\overline{T}$ output connection.

Depressing the ADR switch thus ultimately causes an active-low, 5-μs pulse

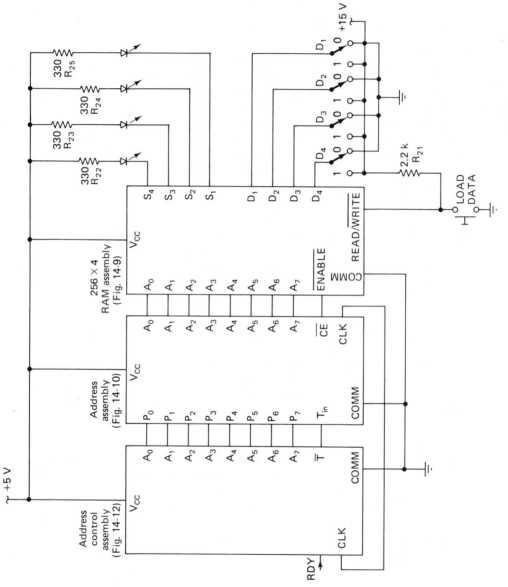

Figure 14-13 Block diagram of a complete 256 × 4 RAM System.

to appear at $\overline{T}$; and as described previously, that is the condition required for initiating the 2-phase address-changing sequence in the address register assembly.

The first of the two address-changing pulses in the address register is designated the CLK pulse. Whenever this CLK pulse arrives at the CLK input of the address control, it is applied to an input of two NOR gates, IC2-A and IC2-B. This pulse will pass through one of these two NOR gates, depending on the setting of the INCR/LOAD switch.

If the INCR/LOAD switch is set to the LOAD position, the incoming CLK pulse will pass through IC2-B and appear as an active-low pulse at the LOAD (L) inputs of the two counter circuits, IC5 and IC6. This is the pulse requirement for asynchronously loading these presettable counters.

Any 8-bit word at the address switches will be loaded into the counters; and when the second address-changing pulse occurs in the address register assembly, the content of the counters is loaded into the address latches.

Depressing the ADR push button while the INCR/LOAD switch is in its LOAD position lets the user manually address any desired location set on the address switches.

Depressing the ADR push button while the INCR/LOAD switch is in its INCR position, however, automatically increments or decrements the RAM address. In this case, depressing the ADR push button ultimately generates a CLK pulse at the input of the address control assembly; and since IC2-A is gated on by having the INCR/LOAD switch in its INCR position, an active-low CLK pulse appears at the inputs of two NAND gates, IC1-C and IC1-D. That pulse will be inverted by one of these two NAND gates and applied to the UP CLOCK (U) or DOWN CLOCK (D) pin of counter IC5. If the UP/DOWN switch happens to be in its DOWN position, a positive-going CLK pulse appears at terminal U of counter IC5; and the leading edge of that pulse increments the entire counter assembly by one integer.

If the UP/DOWN switch is in its DOWN position, the incoming CLK pulse eventually arrives at the D input of counter IC5, and the count is decremented one integer.

Whether this CLK pulse increments or decrements the count, the new address appears at the output of the address control assembly a few nanoseconds after the user depresses the ADR push button. The second address-changing pulse in the address register assembly then loads the new address into the RAM system.

The address thus increments or decrements one count (or address location) each time the user depresses the ADR push button. The INCR/LOAD switch must be in its INCR position to operate the address system in this particular mode, however.

The incrementing/decrementing feature makes it relatively convenient for the user to read or write data at the RAM assembly in a long sequence of address locations.

All of these operating modes described thus far assume the RUN/STOP switch is in its STOP position. If the user changes this switch to the RUN position, the logic-0 level from the "hot" side of that switch allows the output of a free-funning, 40-kHz astable multivibrator (IC4-B) to pass through NOR gate IC2-D. What appears at $\overline{T}$ in this case is a 40-kHz string of active-low pulses. Each of these pulses initiates the 2-phase address-changing operation in the address register.

The CLK pulses at the CLK input appear at this same 40-kHz rate; and if the INCR/LOAD switch is in its INCR position, these pulses will appear at the output of IC2-A. The waveform is then directed to either the U or D input of the counter circuit, IC5, depending on the setting of the UP/DOWN switch.

If the system is in its RUN, INCREMENT mode, it will advance the address at a 40-kHz rate if the UP/DOWN switch is in its UP position. Setting the UP/DOWN switch to its DOWN position forces the address to decrement at the 40-kHz clock rate.

The whole system can be modified further to create even more sophisticated and useful operating modes. The RDY (READY) input to the address control circuit in Fig. 14-12, for example, can be pulled down to logic 0 by some outside mechanism, thereby inhibiting the free-running action of the counters. The moment this terminal returns to logic 1, however, the multivibrator begins running once again. Such operations are often used to control memory operations in calculator and computer systems.

The user might also be given the option of loading the counters from an external source of 8-bit data. In this case, it would be possible to perform loop and branching operations, operations calling for storing an address location as data in the RAM system.

14-4 RAM Switching Parameters

RAM devices are mainly specified according to their bit storage capacities and data I/O formats. There are, however, two critical switching parameters that limit the speed of operation: memory access time and minimum write pulse width.

Memory access time is generally defined as the propagation delay interval between the time the memory is either address enabled in the READ mode and the previously stored data appears at the data outputs. Basically, memory access time is the response time of a RAM operating in its READ mode.

In a practical sense, the access time of any memory device places an upper limit on its readout frequency. In the case of the 7489 TTL RAM, the access time is on the order of 80 ns. The maximum allowable readout frequency is thus about 12.5 MHz. Attempting to address the memory any faster than 12.5 MHz does not allow the outputs to become well established before the next address command comes along.

The write pulse interval of a RAM device is the amount of time the input data must be held stable in order to guarantee reliable writing into the memory. Just as the access time limits the upper readout frequency, the write pulse time limits the writing frequency of the system.

The minimum write pulse width for the 7489 RAM is about 50 μs; so it follows that WRITE operations must take place at a frequency less then 20 MHz.

As a rule of thumb, it can be said that the 7489 TTL RAM is a 10-MHz device; it can both read and write at a 10-MHz rate with no access time or write pulse time problems. The RAM system described in Sec. 14-3 operates well below this 10-MHz level.

14-5 ROM Devices

The primary feature of a RAM device is that it is possible to read and write data with it. A ROM (Read-Only-Memory), however, is capable of performing only readout operations; the programming is internal and permanent.

The 7488 IC device is a TTL ROM organized with a 32 × 8 data output format. It is, in other words, capable of reading out thirty-two 8-bit data words. In addition to having eight data outputs, the 7488 has five address inputs to accommodate the 32 different address locations.

The internal ROM programming is specified by the design engineer, but it is the manufacturer's responsibility to implement the prescribed programming during the last phase of the IC manufacturing process. In essence, the 7488 ROM is totally useless until that final programming operation is completed by the manufacturer.

All TTL ROM devices are thus custom ICs; and the practical implication is that they can be very expensive. The manufacturer requests a mask set-up charge that can run into the thousands of dollars. The only way a user can justify that initial charge is by ordering large quantities of identically programmed ROMs and amortize the cost by selling or using all of them.

Several members of the TTL logic family are actually made from 7488 ROMs that have be programmed to perform the required operations. Most priority encoders and some of the code converters are actually 7488 ROMs the manufacturer has programmed to perform the specified functions.

14-5.1 Programmable ROMs (PROMs)

As the old saying goes, we are all human and we all make mistakes. Engineers make mistakes, too; and an engineer who makes a mistake in specifying the programming for a custom ROM can waste a lot of time and money. When specifying the programming for a 7488 ROM, the engineer could only hope the program is perfect. If there is anything wrong with it, he or she will most likely detect the error only after paying the set-up charge and receiving the first prototype models from the manufacturer.

Because a single error in specifying a ROM program can be so expensive and time consuming, there rose a need for a special kind of ROM that could be programmed in the field (or at the engineer's own location). Such a ROM, commonly known as a *programmable ROM*, can be internally programmed by fusing some minute connections in the device. The fusing operation calls for addressing the location and then running a few milliamperes of current through a special program-location and then running a few milliamperes of current through a special pro-gramming connection. The outputs at each address are thus set permanently to the desired patterns of 1's and 0's.

If there are any errors in the programming, all that is lost is the programming time and the few dollars that a PROM costs. Correcting the error is then a matter of running the program into a fresh PROM.

Once the engineer is satisfied with the performance of the PROM, he or she can send it to the manufacturer who will, in turn, begin mass producing mask-programmed duplicates at a few dollars each.

Electronics enthusiasts have taken advantage of the large quantity of PROM devices now available. It is altogether possible to build a ROM-programming device and use it for programming a one-of-a-kind PROM IC. Anyone who has the right knowledge can now create a reliable ROM device, a permanent memory that can be programmed to perform an unlimited variety of sophisticated logic operations.

14-5.2 Erasable PROMs

The advent of PROMs was a boon to both engineers and electronics hobbyists. Many of these people didn't like the idea of having to discard a PROM that was either improperly programmed or had outlived its usefulness. Thus the need for EPROMs (Erasable PROMs).

An EPROM is programmed the same way PROMs are. The advantage of an EPROM is that the program can be erased at any time. A quartz window in the EPROM chip is normally covered with an opaque "door." Erasing the program-ming is a simple matter of opening the window to ultraviolet light for an hour or so. After that, the device can be programmed again.

Exercises

1. List the number of address inputs, the number of data inputs and outputs and the number of 7489 RAM ICs required for the following RAM assemblies: (a) 8×8; (b) 8×32; (c) 256×8; (d) 1024×8.

2. What is the essential difference between a 256×8 RAM and a 256×8 ROM?

3. Why is it important to disable a RAM assembly while changing its address?

4. State at least one characteristic of a ROM that makes it superior in some respects to a comparable RAM.

15

EXCLUSIVE-OR, EQUALITY
AND COMPARATOR FUNCTIONS

All of the circuits and systems described to this point have dealt with major areas of digital electronics such as combinatorial logic, binary counting, display systems and memories. There is one more major classification of digital devices and systems that remains to be discussed: arithmetic systems.

Although the devices discussed in this chapter are not arithmetic devices as such, they bridge the gap between ordinary logic functions and binary arithmetic. Then, too, these circuits have some useful applications in their own right.

15-1 EXCLUSIVE-OR Functions

An OR function is characterized by producing a logic-1 output whenever any one or all of its inputs are set to logic 1. By way of comparison, an EXCLUSIVE-OR function is one that shows a logic-1 output whenever any one or more, but not all, inputs are at logic 1.

Figure 15-1 summarizes the nature of a standard 2-input EXCLUSIVE-OR gate. The symbol and truth table are very similar to that of a 2-input OR gate. The symbol, however, has an arc drawn parallel to the OR-gate input arc; and the truth table shows a logic-1 output whenever either—but not both—inputs are at logic 1.

In a manner of speaking, an EXCLUSIVE-OR gate is an OR gate that *excludes* the condition in which all inputs are high.

The logic equations in Fig. 15-1 are three distinctly different, but valid, ways

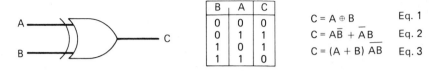

B	A	C
0	0	0
0	1	1
1	0	1
1	1	0

$$C = A \oplus B \qquad \text{Eq. 1}$$
$$C = A\bar{B} + \bar{A}B \qquad \text{Eq. 2}$$
$$C = (A + B)\,\overline{AB} \qquad \text{Eq. 3}$$

Figure 15-1 EXCLUSIVE-OR function: symbol, truth table and logic equations.

to express the EXCLUSIVE-OR function. Equation 1 is the simplest of the three, but it calls for introducing a new Boolean operator: an OR operator with a circle around it. This is a rather convenient way to express the EXCLUSIVE-OR function, but there are no Boolean theorems that relate that particular operator with the standard AND and OR operators.

Equations 2 and 3 provide more useful ways to express the EXCLUSIVE-OR function; and every engineer and technician should be prepared to recognize them whenever they appear in a design or circuit analysis situation.

Equation 2 is actually taken directly from the EXCLUSIVE-OR truth table. Output C in this instance is equal to logic 1 whenever $A = 1$ AND $B = 0$, OR $A = 0$ AND $B = 1$. Equation 3 is then derived from Eq. 2. The derivation in this instance is left as an exercise at the end of this chapter.

Design situations often call for executing logic functions of the forms listed in Eqs. 2 and 3, but there are a few other features of the EXCLUSIVE-OR function that are responsible for its popularity.

Notice, for example, that an EXCLUSIVE-OR gate yields a logic-1 output whenever the two input logic levels are different. Whenever the inputs are equal (both logic 0 or both logic 1), the output is 0. The output goes to logic 1 only when the inputs are unequal. Bear in mind this feature when considering the equality, comparator and parity circuit described later in this chapter.

Another useful feature of the EXCLUSIVE-OR function concerns the way it can be used as an inverting or non-inverting gate. Figure 15-2(a) shows an EXCLUSIVE-OR gate having one input permanently connected to a logic-1 source. According to the truth table in Fig. 15-1, fixing one of the inputs to logic 1 in this fashion makes the output look like an inverted version of the A input. If, however, the B input is permanently connected to COMM, that input is essentially fixed at logic 0; and the truth table in Fig. 15-1 shows that the A input will appear non-inverted at the device's C output.

An EXCLUSIVE-OR gate can therefore operate as either an inverter or non-inverter, depending on the logic level at the second input terminal.

Figure 15-2(b) shows how four EXCLUSIVE-OR gates can be connected to form a 4-bit invert/non-invert circuit. According to the function table accompanying that circuit, the I outputs are equal (non-inverted versions) of the four D inputs whenever the S terminal is at logic 0. Changing the S input to logic 1, however, makes the I outputs show inverted versions of their respective D inputs.

The circuit in Fig. 15-2(c) is a simple circuit that demonstrates an interesting

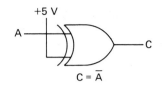

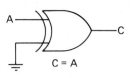

$$C = \overline{A}$$

$$C = A$$

(a)

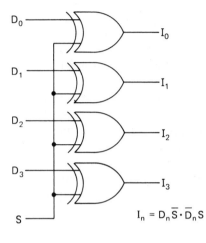

S	I_0	I_1	I_2	I_3
0	D_0	D_1	D_2	D_3
1	$\overline{D_0}$	$\overline{D_1}$	$\overline{D_2}$	$\overline{D_3}$

$$I_n = D_n \overline{S} \cdot \overline{D_n} S$$

(b)

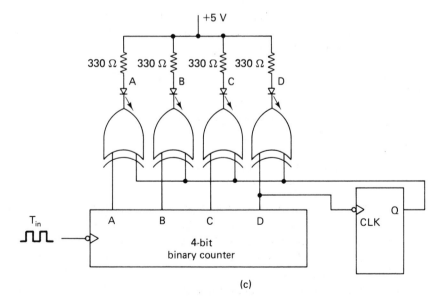

(c)

Figure 15-2 Some simple applications of EXCLUSIVE-OR gates. (a) As a logic inverter or non-inverter. (b) As an invert/non-invert selector. (c) In a counter circuit that alternately counts up and down.

375

application of the invert/non-invert feature of EXCLUSIVE-OR gates. A standard 4-bit binary counter generates its sequence of 16 different outputs in response to clocking pulses from T_{in}. Each time the up counter resets to 0, the counter's D output shows a negative-going edge that toggles a flip-flop.

The toggling action of the flip-flop alternately sets the EXCLUSIVE-OR gates for inverting or non-inverting each time the counter resets to 0. The LEDs thus respond to inverted or non-inverted versions of the counter's outputs.

Recall that a counter will appear to count up or down, depending on whether the outputs are inverted or non-inverted. In this particular case, the LEDs will show an up-counting effect whenever the Q output of the flip-flop is at logic 1, and they will display a down-counting effect whenever the Q output of the flip-flop is at logic 0.

Since the output of the flip-flop changes state each time the counter makes a transition from 1111 to 0000, it follows that the display will show an up-counting effect (0000 through 1111) and then switch to a down-counting effect (1111 down through 0000).

The counter IC in this case can be a 7493 4-bit binary up counter, the flop-flop can be one section of a 7476 dual J-K flip-flop (programmed for a T-type mode) and the EXCLUSIVE-OR gates can come from a 7486 quad 2-input EXCLUSIVE-OR IC package.

15-2 EQUALITY Functions

One application of an EXCLUSIVE-OR gate is to detect the inequality of two inputs. If the output of such a gate is inverted, the circuit will show a logic 1 output whenever the two inputs are equal—both logic 1 or both logic 0. See the simple circuit and truth table in Fig. 15-3.

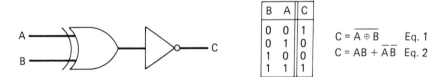

B	A	C
0	0	1
0	1	0
1	0	0
1	1	1

$$C = \overline{A \oplus B} \qquad \text{Eq. 1}$$
$$C = AB + \overline{A}\,\overline{B} \qquad \text{Eq. 2}$$

Figure 15-3 EXCLUSIVE-NOR or EQUALITY function: symbol, truth table and logic equations.

This inverted EXCLUSIVE-OR function can be called either an EXCLUSIVE-NOR or EQUALITY function. In fact, the EXCLUSIVE-NOR function can be properly represented by an EXCLUSIVE-OR symbol having an inverting "bubble" at its output. EXCLUSIVE-NOR or EQUALITY gates, however, are not available in the standard TTL family.

Perhaps the most popular application of EXCLUSIVE-NOR or EQUALITY functions is in circuits that must be able to detect a certain BCD or binary number from a counter system. Suppose, for example, a piece of automated machinery is supposed to carry out a number of different operations according to a timed sequence. The timer in this case can be a binary counting circuit operating at a 1-Hz clock rate. The idea, then, is to program the machine to do certain functions whenever the counter reaches some specified sets of numbers.

The outputs of the counter can be connected to the EXCLUSIVE-NOR function and compared with some other numbers prescribed by the user. Whenever the count reaches the numbers prescribed by the user, the system outputs a logic-1 level that initiates the desired response. See the circuit in Fig. 15-4.

The four EXCLUSIVE-OR gates in Fig. 15-4 compare the BCD outputs of a BCD up counter with the status of the programming switches. The "A" number from the counter is made up of bits A_0 through A_3, where A_0 is the least-significant bit. The programming inputs, "B," are similarly made up of four bits designated here as B_0 through B_3. B_0 is the least-significant bit in the program word.

The outputs of the EXCLUSIVE-OR gates are connected to a set of open-collector inverters. The inverting feature combines with the EXCLUSIVE-OR function to create the desired EXCLUSIVE-OR or EQUALITY function.

If any of the paired inputs to the EXCLUSIVE-OR gates are complements of one another, the output of that EXCLUSIVE-OR gate is set to logic 1. The inverter following that gate, however, inverts the level to logic 0, thereby pulling down the open-collector outputs to 0. In short, if any of the four pairs of inputs are unequal, the overall output of the circuit is at logic 0.

Whenever BCD inputs A and B are exactly equal, $A_0 = B_0$, $A_1 = B_1$, $A_2 = B_2$ and $A_3 = B_3$; and the outputs of all four inverter circuits are pulled up to logic 1 by the 2.2-k pull-up resistor. The output of this circuit goes to logic 1, then, only when the A word is exactly equal to the B word. Otherwise, the output is fixed at logic 0.

This equality-detecting scheme can be expanded to include any number of bits per word, and it can similarly be expanded to deal with any number of BCD decades.

Suppose an experimenter wants to add an alarm feature to a digital clock circuit that has BCD outputs available. Most digital clocks display a $3\frac{1}{2}$-digit output,—a full digit for 1's minutes, 10's minutes and 1's hours and a half digit (numeral 1 only) for 10's hours.

As shown in the block diagram in Fig. 15-5, the BCD outputs of such a clock can be connected to EQUALITY functions where they are compared with the setting of some simple toggle switches. The output of the EQUALITY function is normally at logic 0, but it rises to logic 1 the moment the time of day matches the time set on the switches. That logic-1 pulse lasts only 1 min, but it can be used to toggle a flip-flop and gate on a tone alarm. The alarm can then be switched off by clearing the flip-flop with a push-button control.

Of course, the same scheme can be used for energizing a relay to apply 120-VAC utility power to a coffee pot, a radio and tape recorder or any other such appliance that is to be switched on at some prescribed time of day.

EQUALITY functions can also be used in control instruments that must set off an alarm whenever some critical parameter reaches a certain level. Suppose a

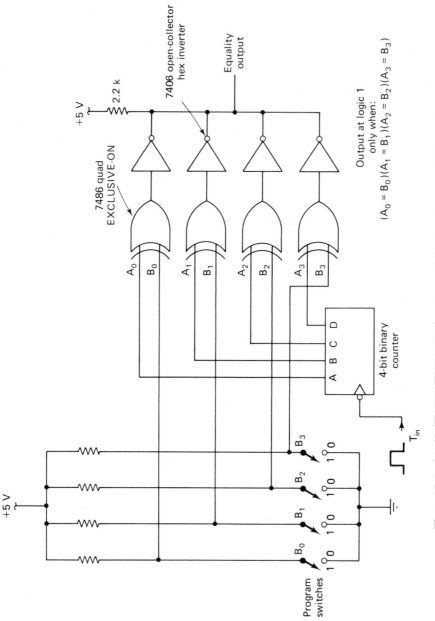

Figure 15-4 A 4-bit EXCLUSIVE-NOR function used for detecting one of 16 counts from a 4-bit binary counter.

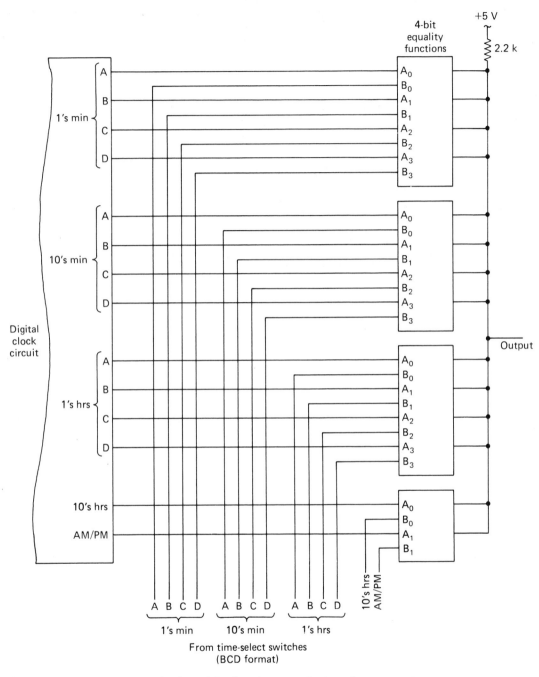

Figure 15-5 An EQUALITY circuit used for detecting a certain time of day.

digital voltmeter is monitoring the analog voltage output from a temperature-sensing device. The digital output of the voltmeter can be compared with logic levels from a set of programming switches. Whenever the temperature reaches the prescribed level, the output of the EQUALITY-detecting scheme will change logic state to sound an alarm or shut down the temperature-controlling operation.

Figure 15-6 shows how an EQUALITY circuit can be used as part of a stop-count counting system. The output of the EQUALITY function in this instance is returned to the active-low enable input of an 8-bit binary counter. Whenever the output of the counter reaches the 8-bit number prescribed by the "B" inputs to the EQUALITY circuit, the counting automatically stops and holds at that count.

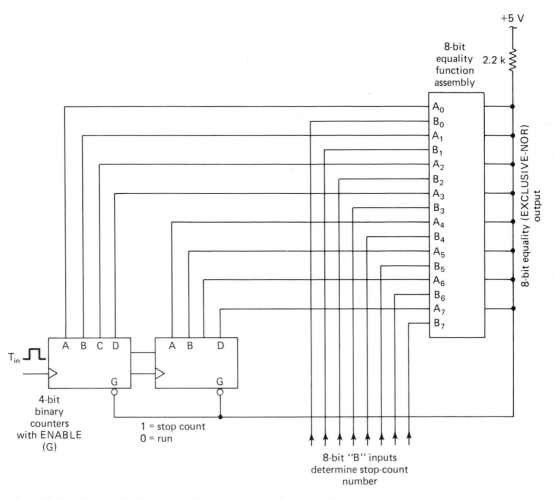

Figure 15-6 Using an 8-bit EQUALITY function to stop the count of an 8-bit counter.

The output of the EQUALITY circuit disables the counter whenever the A and B numbers are identical.

OR-ing together the outputs of a number of different EQUALITY functions allows an operation to be initiated at a number of different points in the counting operation. The system illustrated in Fig. 15-7, however, is one that can be programmed to initiate four entirely separate functions from the same source of clock information.

As presented in Fig. 15-7, the circuit is scaled down to handle 4-bit words, but it can be expanded to include any word size, including the $3\frac{1}{2}$ BCD digits that characterize most digital electronic time clocks.

Blocks EF_0 through EF_3 are 4-bit EQUALITY or EXCLUSIVE-NOR functions. Each one represents a different function, F_0 through F_3. Whenever any one of these function outputs switches to logic 1, it initiates a corresponding operation of some sort. This particular circuit can handle up to four separate functions in this fashion.

The "A" word input is applied to the A inputs of each of the four EQUALITY functions. This word comes from a master counter circuit, the circuit that does the main counting or timing operations. This "A" word is compared with corresponding 4-bit data from a set of 4-bit latches. It is the data from these latches that makes up the "B" word to each of the EQUALITY functions.

The purpose of the latch circuits, DL_0 through DL_3, is to act as memory devices for the "B" words the user enters into them. Entering this data is a matter of first setting the desired function count on the PROGRAM switches, B_0 through B_3, and then depressing the desired LOAD/SELECT push button.

If, for instance, the user wants function output F_2 to show a logic 1 whenever the master counter reaches 1101, he or she sets 1101 on the PROGRAM switches and then depresses the FS_2 push button. Depressing that particular push button enables latch DL_2 so that it will accept and store the data at the PROGRAM switches. Whenever the count at the A inputs reaches 1101, then, output F_2 goes to logic 1 as long as that count exists.

15-3   Magnitude Comparators

The EQUALITY functions described in the foregoing section are capable of detecting numeric equality between two binary numbers. They are incapable of telling whether one of the two numbers is greater or less than the other; that is the job of a magnitude comparator circuit.

Stated simply, a magnitude comparator compares the values of two binary numbers (usually designated numbers A and B) and generates a logic-1 level at one of three outputs. The outputs indicate $A = B$, $A > B$ (read "A greater than B") and $A < B$ (read "A less than B").

Table 15-1 illustrates a magnitude comparison operation as it applies to two 2-bit numbers. Their decimal equivalents are shown for convenience in the two right-hand columns. Binary inputs A_1 and A_0 represent the two bits of the A num-

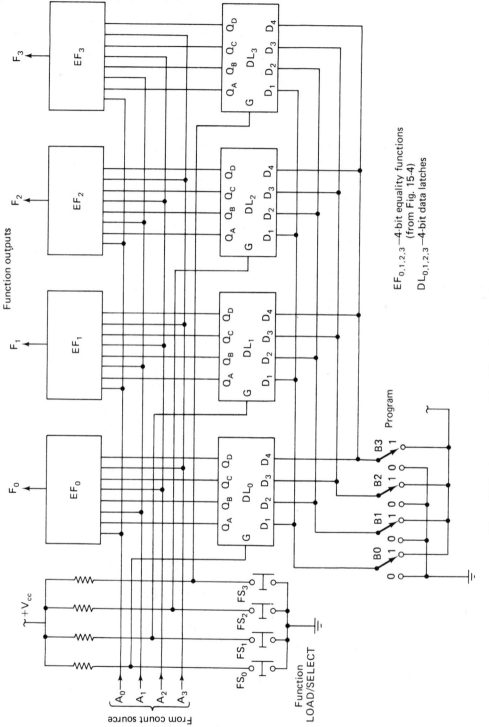

Figure 15-7 A programmable, 4-function count detector.

$EF_{0,1,2,3}$—4-bit equality functions (from Fig. 15-4)
$DL_{0,1,2,3}$—4-bit data latches

Table 15-1   Two-Bit Magnitude Comparator Truth Table

Decimal Equiv. A	B	Binary inputs A A_1	A_0	B B_1	B_0	Outputs A = B	A > B	A < B
0	0	0	0	0	0	1	0	0
0	1	0	0	0	1	0	0	1
0	2	0	0	1	0	0	0	1
0	3	0	0	1	1	0	0	1
1	0	0	1	0	0	0	1	0
1	1	0	1	0	1	1	0	0
1	2	0	1	1	0	0	0	1
1	3	0	1	1	1	0	0	1
2	0	1	0	0	0	0	1	0
2	1	1	0	0	1	0	1	0
2	2	1	0	1	0	1	0	0
2	3	1	0	0	1	0	0	1
3	0	1	1	0	0	0	1	0
3	1	1	1	0	1	0	1	0
3	2	1	1	1	0	0	1	0
3	3	1	1	1	1	1	0	0

ber, with A_0 being the least-significant bit. Similarly, columns B_1 and B_0 show the binary values of number B.

The outputs indicate whether or not the two numbers are numerically equal, that is, whether A is greater than B or whether A is less than B. Note, for instance that the $A = B$ column shows 1's only when $A = B = 0, 1, 2$ or 3. The $A > B$ column shows a logic 1 only when number A is greater than B and the $A < B$ column shows logic 1's only when A is less than B.

It is possible to devise several different kinds of logic circuits that yield the truth table in Fig. 15-1. The idea is to come up with a circuit having four inputs (A_0, A_1, B_0 and B_1) and three outputs ($A = B$, $A > B$ and $A < B$). Two-bit magnitude comparators have limited usefulness, however, because so few comparison applications call for comparing simple 2-bit words.

A 4-bit magnitude comparator, one that compares a pair of 4-bit words, is very useful. Such comparators, for instance, can compare the values of two BCD numbers; and the idea can be expanded to any number of decades of BCD numbers by cascading 4-bit comparators.

A conventional, straightforward approach to building the logic circuitry for a 4-bit magnitude comparator, however, usually results in a highly cumbersome array of ICs. The design of magnitude comparators is thus based on an algorithm, a pattern of bit comparison that can be repeated any number of times until the desired result is achieved.

15-3.1 The Comparator Algorithm

The table in Fig. 15-8 is a guide to understanding this comparator algorithm. First note that any two bits are considered equal if they are both at logic 1 or both at logic 0. In other words, any two bits A_n and B_n are equal if $A_n, B_n = 0$ or $A_n, B_n = 1$. The only way A_n can be greater than B_n is if $A = 1$ and $B = 0$; and by the same token, A_n is less than B_n only if $A = 0$ while $B = 1$.

The inputs to a 4-bit magnitude comparator are two 4-bit numbers designated A and B. The bits making up each number are bits 3, 2, 1 and 0, with the 0-bit being the LSB. This breakdown of the two numbers is illustrated in Fig. 15-8(a).

The first column in the table in Fig. 15-8(b) shows all possible magnitude relationships between the MSBs, A_3 and B_3. In the first line, for instance, A_3 is shown to be greater than B_3. In the second line A_3 is shown to be less than B_3; and in

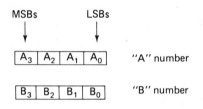

"A" number

"B" number

(a)

A and B inputs				Outputs		
A_3, B_3	A_2, B_2	A_1, B_1	A_0, B_0	$A > B$	$A < B$	$A = B$
$A_3 > B_3$	X	X	X	1	0	0
$A_3 < B_3$	X	X	X	0	1	0
$A_3 = B_3$	$A_2 > B_2$	X	X	1	0	0
$A_3 = B_3$	$A_2 < B_2$	X	X	0	1	0
$A_3 = B_3$	$A_2 = B_2$	$A_1 > B_1$	X	1	0	0
$A_3 = B_3$	$A_2 = B_2$	$A_1 < B_1$	X	0	1	0
$A_3 = B_3$	$A_2 = B_2$	$A_1 = B_1$	$A_0 > B_0$	1	0	0
$A_3 = B_3$	$A_2 = B_2$	$A_1 = B_1$	$A_0 < B_0$	0	1	0
$A_3 = B_3$	$A_2 = B_2$	$A_1 = B_1$	$A_0 = B_0$	0	0	1

$A_n = B_n$ if $A_n, B_n = 0$ or $A_n, B_n = 1$
$A_n > B_n$ if $A_n = 1, B_n = 0$
$A_n < B_n$ if $A_n = 0, B_n = 1$

(b)

Figure 15-8   Basic 4-bit magnitude comparator functions.

the remaining lines of the first column $A_3 = B_3$. The X's in the remaining input columns indicate irrelevant comparisons.

Whenever A_3 and B_3 are not equal, none of the other bits is relevant. There is, in other words, adequate information to determine whether number A is greater or less than number B. If it happens that A_3 is greater than bit B_3, there is enough information to determine that number A is greater than number B. Thus a logic 1 appears in the first line of the $A > B$ output column. If A_3 is less than B_3, however, number A must be smaller than B; and the second line of the $A < B$ output column shows a logic 1.

Now suppose the MSBs, A_3 and B_3, are equal. Whenever that is the case, the relative magnitudes of A_2 and B_2 become relevant. If it is assumed that A_2 is not equal to B_2, the relative magnitude of numbers A and B can be decided on the basis of A_2 and B_2.

If it turns out that $A_3 = B_3$ and $A_2 = B_2$, the algorithm calls for comparing bits A_1 and B_1. Bits A_0 and B_0 aren't relevant in this case, unless A_1 happens to equal B_1. And if the three more-significant bits of both input numbers are equal, the relative magnitudes of numbers A and B rest entirely on a comparison of the A_0 and B_0 bits. The last line in the table shows the output that occurs when the corresponding bits in both numbers are all equal: $A = B$.

In summary, the magnitude-comparing algorithm always begins with the MSBs and then proceeds toward the LSBs. If an inequality exists anywhere along the line, the bit having a value of 1 belongs to the larger of the two input numbers. If the entire operation proceeds through the LSBs without encountering an inequality, the two input numbers are considered equal.

15-3.2 A Practical 4-Bit Magnitude Comparator

The 7485 TTL magnitude comparator IC is illustrated in Fig. 15-9. Its inputs include four bits for the A number, four for the B number and three additional inputs for cascading the operation. The outputs are simply the three comparison expressions, $A > B$, $A < B$ and $A = B$.

If the cascading inputs in Fig. 15-9(b) are ignored, the table is identical to the basic 4-bit comparator table in Fig. 15-8(b). When used as a simple 4-bit magnitude comparator, however, the $(A = B)_i$ input must be connected to logic 1. Otherwise, the circuit could not output the $A = B$ condition.

Cascading 7485's to build 8-, 12- or 16-bit magnitude comparators is a matter of connecting the outputs of one stage to the corresponding cascading inputs of the next stage. The four most-significant bits of the two large numbers are applied to the first magnitude comparator. These four bits are compared, and if they are equal, the comparison job shifts to the second comparator IC down the line.

Magnitude comparators are frequently used in computer-type circuits that feature conditional programming steps. The operation of such programs depends on the relative values of two different numbers. If they happen to be equal, the program executes one kind of operation. If the numbers are different, the program

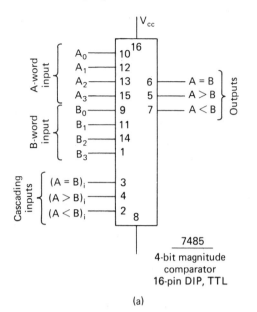

(a)

Comparing inputs				Cascading inputs			Outputs		
A_3, B_3	A_2, B_2	A_1, B_1	A_0, B_0	$(A > B)_i$	$(A < B)_i$	$(A = B)_i$	$A > B$	$A < B$	$A = B$
$A_3 > B_3$	X	X	X	X	X	X	1	0	0
$A_3 < B_3$	X	X	X	X	X	X	0	1	0
$A_3 = B_3$	$A_2 > B_2$	X	X	X	X	X	1	0	0
$A_3 = B_3$	$A_2 < B_2$	X	X	X	X	X	0	1	0
$A_3 = B_3$	$A_2 = B_2$	$A_1 > B_1$	X	X	X	X	1	0	0
$A_3 = B_3$	$A_2 = B_2$	$A_1 < B_1$	X	X	X	X	0	1	0
$A_3 = B_3$	$A_2 = B_2$	$A_1 = B_1$	$A_0 > B_0$	X	X	X	1	0	0
$A_3 = B_3$	$A_2 = B_2$	$A_1 = B_1$	$A_0 < B_0$	X	X	X	0	1	0
$A_3 = B_3$	$A_2 = B_2$	$A_1 = B_1$	$A_0 = B_0$	1	0	0	1	0	0
$A_3 = B_3$	$A_2 = B_2$	$A_1 = B_1$	$A_0 = B_0$	0	1	0	0	1	0
$A_3 = B_3$	$A_2 = B_2$	$A_1 = B_1$	$A_0 = B_0$	0	0	1	0	0	1

(b)

Figure 15-9 The 7485 4-bit magnitude comparator IC. (a) IC pinout. (b) Function table.

can perform one of two other kinds of operations, depending on whether the "*A*" number is greater or less than the "*B*" number.

Magnitude comparators also find their way into industrial-type servomechanisms that control some critical parameter such as motor revolutions per minute, temperature, pressure, liquid level and so on. A digital system monitors the parameter to be controlled, generating a binary number that is usually proportional to the actual operating level. That number is sent to a magnitude comparator where it is compared with another number representing the desired operating control point.

If the actual operating point is above or below the desired operating point, the magnitude comparator generates a "greater-than" or "less-than" output that ultimately tells the servomechanism to make a corrective effort, an effort to correct the discrepancy between the desired and actual operating points.

Of course, the corrective effort of the system is reduced to zero when the actual and desired operating points are identical and the magnitude comparator generates an "equality" output.

15-4 Parity Generators and Detectors

Transmitting digital data from one place to another brings up the distinct possibility of that data's being garbled by electrical noise from some outside source. The transmitting medium might be simply a cable carrying the data 20 ft or so across the room, or it might be a space communications link between the earth and the outer reaches of the solar system. In either case, static interference can effectively "erase" or add unwanted bits of data to the transmitted information.

Such errors occurring during the transmission of data can be minimized by keeping the data link as electrically quiet as possible, but such errors cannot be eliminated altogether. Therefore, there must be some reliable technique for detecting whether or not a word of transmitted data has been garbled. If, indeed, the received data is different from the transmitted data, the receiver can request a repeat of the data—and continue requesting and receiving that same word of data until it is correct.

The technique for detecting errors in transmitted data relies on a special comparator operation that compares the number of 1 bits in the transmitted word with the number of 1 bits received. To do this, the transmitter system must generate an additional bit of data, a parity bit.

The parity bit tells the receiver system whether the transmitted binary word contained an even or odd number of 1 bits. It doesn't say how many 1 bits are in the transmitted word; it just says whether there were an odd or even number of them.

The table in Fig. 15-10(a) shows ten 4-bit words representing the standard BCD counting format. This set of 4-bit words merely illustrates one particular format, which can actually be any binary code of any word length.

If the system is using an even-parity format, it sends the four original bits in each word, plus a fifth bit that always adjusts the total number of 1's transmitted to an even number. A parity generator for this table is shown in Fig. 15-10(b).

If the system is sending the number 0001, for instance, there is an odd number of 1's in the number; and the parity bit is logic 1, thereby bringing the total number of 1's transmitted to 2, an even number of 1's. If, however, the system is sending the word 0011, it already contains an even number of 1's, and the parity bit is set to 0.

Comparing the 4-bit word entered into the parity generator with the output of an even-parity generator circuit, the transmitted word always contains an even number of 1's, including the parity bit, P.

An odd-parity generator works exactly the same way, but it transmits a word that always has an odd number of 1's, including the parity bit. An odd-parity

Inputs				Output	
BCD word				Parity bit	
D	C	B	A	Even	Odd
0	0	0	0	0	1
0	0	0	1	1	0
0	0	1	0	0	1
0	0	1	1	1	0
0	1	0	0	0	1
0	1	0	1	1	0
0	1	1	0	0	1
0	1	1	1	1	0
1	0	0	0	0	1
1	0	0	1	1	0

(a)

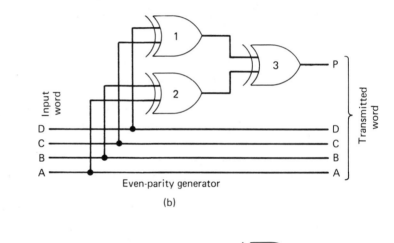

(b)

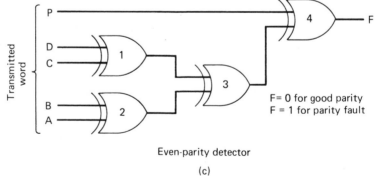

Even-parity detector

(c)

Figure 15-10 Parity functions. (a) Odd and even parity table for a 4-bit BCD word format. (b) An even-bit parity generator circuit. (c) An even-bit parity detector circuit.

generator is not shown in Fig. 15-10, but the truth table shows that it is simply an even-parity generator with an inverted parity bit.

The parity detector is located at the receiving end of the communications link. Its job is to determine whether a 1 bit has been added or lost during transmission. If there is a parity error, there will be a discrepancy between the level of the parity bit and the "evenness" or "oddness" of the transmitted data.

The parity detector in Fig. 15-10(c) actually generates a parity bit of its own at the output of EXCLUSIVE-OR gate 3. Since this is an even-parity generator circuit, it will generate a logic-1 level if there is an odd number of 1's in the received 4-bit word or it will generate a logic-0 level if there is an even number of 1's in the received word. This parity output from gate 3 is, in turn, compared with the parity bit received from the transmission system. If they agree (if they are equal), the F output of gate 4 is at logic 1, thereby indicating good parity between the transmitted and received data. If the F output goes to logic 1, however, that means the parity bit from gate 3 in Fig. 15-10(c) is different from that generated at the transmitter end of the data link. In other words, whenever $F = 1$, the data has been garbled in transit; and usually the receiver system will request a repeat of that word.

Of course, it is possible to defeat the whole parity scheme by garbling one data bit and the parity bit during transmission, thus yielding a false indication of good parity. Suppose, as an example, the system is transmitting word 0111. If it is an even-parity system, $P = 1$ to make an even number (4) of 1's. The transmitted word is then 10111, where the parity bit appears at the MSB end of the word. If the word is garbled such that the system recieves 00101 (the parity bit and "B" bit have both been changed to 0's), the parity detector will be fooled into seeing good parity because the received word does indeed include an even number of 1's. The statistical probability of such an error occurring is infinitesimally small, however; and the scheme works well within the tolerance of the most high-performance data transmission systems available today.

Exercises

1. Use algebraic methods to prove that $(A + B)\overline{AB} = A\overline{B} + \overline{A}B$. Use a truth-table analysis to demonstrate that $(A + B)\overline{AB} = A \oplus B$.

2. Use a logic-equation analysis and a truth-table demonstration to determine whether the circuit in Fig. E-15.2 is an EXCLUSIVE-OR or an EQUALITY function.

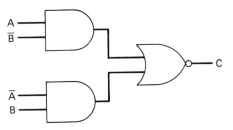

Figure E-15.2

3. The following binary numbers are written in standard form (with the LSB on the right). Use the magnitude comparator algorithm in each case to show whether $A < B$, $A > B$ or $A = B$. (a) $A = 11011$, $B = 11001$; (b) $A = 11001001$, $B = 11001011$; (c) $A = 101110110$, $B = 101110111$; (d) $A = 011011$, $B = 011010$.

4. Indicate the *even*-parity bit (1 or 0) for each of the following binary numbers: (a) 1101; (b) 10110; (c) 1110110; (d) 00000.

5. Show by a logic-equation or truth-table analysis that the circuit in Fig. 15-10(b) will be an odd-parity generator if each of the EXCLUSIVE-OR gates is replaced with EQUALITY functions.

16

BASIC ARITHMETIC FUNCTIONS

The advent of relatively low-cost, MOS/LSI microprocessors has made TTL and CMOS arithmetic circuits all but obsolete. Arithmetic functions once painstakingly assembled from a fairly large selection of TTL arithmetic ICs, registers and memories are now replaced with a few microprocessor-oriented chips.

It is important to study some of the basic arithmetic operations on the TTL level, however, because much of the nomenclature from such systems has carried over into modern microprocessor technology. And without at least a passing acquaintance with TTL-oriented arithmetic, some of the origins of microprocessor terms and procedures can be rather obscure.

16-1 Fundamentals of Binary Addition

Figure 16-1 shows a truth table, circuit diagram and functional block diagram for the simplest kind of binary adder circuit. This circuit simply adds two 1-bit numbers (A and B) and yields a sum (S) and a carry bit (C). Note from the truth table that 0 plus 0 equals a sum of 0 and a carry of 0. This particular operation is perfectly in line with ordinary decimal arithmetic.

The sum of 1 plus 0 is equal to 1, no matter whether the A_0 or B_0 number is 1. The carry in these two instances, however, is still 0. Adding 1 and 1 yields a sum of 0 and a carry of 1. This is actually the point of departure for any similarity between decimal and binary addition. Consider the problem of adding binary 1 and 1 in

B	A	S	C
0	0	0	0
0	1	1	0
1	0	1	0
1	1	0	1

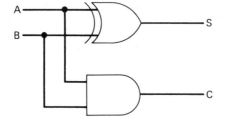

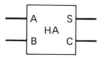

Figure 16-1 The basic half-adder circuit: truth table, gate-circuit equivalent and block diagram.

this fashion:

$$\begin{array}{rl} 01 & \text{``}A\text{'' number} \\ +01 & \text{``}B\text{'' number} \\ \hline 10 & \text{``}A\text{'' plus ``}B\text{''} \end{array}$$

In this instance, 01 plus 01 equals 10. Or translating it to decimal equivalents, $1 + 1 = 2$.

Adder circuits can be cascaded to handle numbers of any desired bit length. The basic idea is to sum the two least-significant bits, generate a sum bit and send the carry bit to the next adder stage down the line. Although the circuit in Fig. 16-1 is wholly adequate for adding the two least-significant bits, it cannot be used for adding higher-order bits. Why not? It has no provisions for including a carry-in bit from a previous, lesser-significant bit.

The adder circuit in Fig. 16-2 has provisions for handling a carry bit from a previous stage. The truth table shows that this kind of adder actually sums three bits at one time: an *A* bit, a *B* bit and a carry-in bit from the previous stage. The circuit generates a sum of the three bits as well as a carry-out bit for the next adder down the line.

Note from the block diagram in Fig. 16-2 that this particular adder is made up of two of the simpler adder circuits. The *A* and *B* bits are summed at one adder stage, and the resulting sum is then added to the carry-in bit. Since this circuit is made up of two of the simpler adders, it is called a *full adder*. The circuit in Fig. 16-1 is called a *half adder*.

The block diagram in Fig. 16-3 is that of a 4-bit adder. Two 4-bit numbers are entered at the *A* and *B* inputs, with the LSBs going to the A_0 and B_0 inputs of the half-adder circuit. The half adder generates the sum of the LSBs at S_0 and, at the same time, generates a carry bit for the second-bit full adder.

Similarly, the output of the first full adder is both a sum (S_1) and a carry bit to the next stage. The sum-and-carry operations are carried out through the entire line of adders until the MSB adder generates its sum (S_3) and carry (C_3) outputs.

Perhaps the best way to analyze the operation of this 4-bit adder is to try adding two 4-bit numbers and tracing the effect through the entire circuit.

C_i	B	A	S	C
0	0	0	0	0
0	0	1	1	0
0	1	0	1	0
0	1	1	0	1
1	0	0	1	0
1	0	1	0	1
1	1	0	0	1
1	1	1	1	1

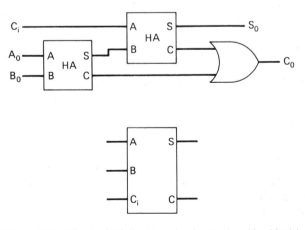

Figure 16-2 The basic full-adder circuit: truth table, block/gate-circuit diagram and full block diagram.

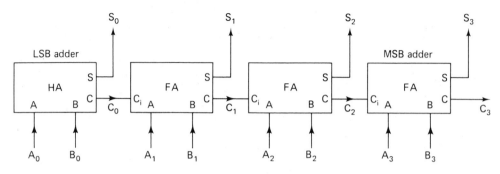

Figure 16-3   A 4-bit binary adder circuit block diagram.

Suppose the A number is 0011 (binary 3) and the B number is 0110 (binary 6). Of course, we know the sum of these two numbers should be 1001 (binary 9); but it quite instructive to follow the summing process through the circuit.

In this particular example, the LSBs for numbers A and B are 1 and 0. Thus $S_0 = 1$ and $C_0 = 0$. Now the inputs to the first full adder are $A = 1$, $B = 1$ and $C_i = 0$. From the truth table in Fig. 16-2, S_1 is then 0 and C_1 is 1.

The next full adder then sees A and B inputs of 0 and 1, respectively, as well as a C_i of logic 1 from the previous full adder. Its sum output at S_2 is thus 0 and its carry-out bit is 1. The final adder must sum 0 and 0 from the MSB position of input numbers A and B, and a carry bit of 1 from the previous stage. The result is a S_3 output of 1 and a C_3 output of 0.

Overall, the sum outputs generate the number 1001; and if the C_3 output is considered the MSB of the final answer, the result is the five-bit number 01001, which is equal to decimal 9.

This 4-bit adder is capable of properly summing any two numbers whose sum is 30 or less. Try the foregoing exercise with $A = B = 1111$.

The 4-bit adder illustrated in Fig. 16-3 is called a 4-bit adder with ripple carry. The carry operations, in effect, ripple down the line of cascaded adders from the LSB to the MSB position. The overall operating speed is thus limited by the amount of time required to complete the ripple-carrying operation. That time interval is equal to the solution time of each adder multiplied by the number of adders in the system. Since the adding can take place in about 40 ns in each case, the carry settling time of the 4-bit adder is on the order of 160 ns.

Having a maximum solution time of about 160 ns limits the adding speed to a frequency of 6.3 MHz, which is below the 10-MHz clocking rate specified for most computer systems. Adder ICs thus include a special mechanism for speeding up the carrying process. The technique, known as *look-ahead carry*, actually anticipates the carry bits for each stage before the actual summing operation takes place. The difference between ripple and look-ahead carry adders is much the same as the difference between ripple (asynchronous) counters and synchronous counters. Recall that a synchronous counter anticipates the next count, thereby setting all outputs to their new state at exactly the same time. The same sort of operation takes place in an adder circuit featuring look-ahead carry.

16-1.1 The 7483 4-Bit Adder IC

Figure 16-4 shows the pin connections for the 7483 4-bit binary adder. With only two exceptions, the internal structure of this IC is identical to the 4-bit adder illustrated in Fig. 16-3.

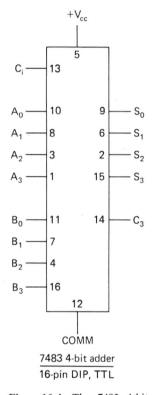

Figure 16-4   The 7483 4-bit full adder.

One essential difference is that the LSB adder in the 7483 IC device is a full adder instead of a half adder. This means the 7483 has a carry input (C_i) to the LSB adding stage. This input should be permanently connected to a logic-0 source if the device is to be used for adding nothing more than 4-bit positive numbers. This carry input, however, plays vital roles when cascading adder ICs and performing addition operations with signed numbers (subtraction operations).

Another difference between the 7483 adder and the block diagram in Fig. 16-3 is that the former includes the look-ahead carry feature to speed up summing operations.

If it is assumed for the moment that the C_i terminal is fixed at logic 0, the 7483 sums any two 4-bit binary numbers applied to the A and B inputs. The sum of the two numbers appears at outputs S_0 through S_3, while the carry-out bit from the MSB stage appears at C_3. The user does not have access to internally connected carry bits C_0 through C_2.

Two or more of these adder ICs can be cascaded to permit the summing of 8-, 12- and 16-bit numbers. Actually, the cascading can be carried out indefinitely, but there is rarely any reason to carry the process any further.

Cascading 7483's to achieve larger word sizes is a simple matter of connecting the C_3 carry-out terminal of one stage to the carry-in (C_i) terminal of the next. The circuit handling the four lower-order bits should have its C_i terminal fixed at logic 0. The C_3 carry-out terminal of the last stage can be used as either the MSB of the sum or an overflow detector.

16-1.2 A Simple 4-Bit Adder Demonstration Circuit

The circuit in Fig. 16-5 demonstrates the summing features of the 7483 4-bit adder IC. The two numbers to be summed are entered via SPDT toggle switches. The sum of the two numbers appears in an active-high format at the S outputs of the adder IC; but since the LED display responds only to active-low inputs, a set of inverters must be placed between the IC's summing outputs and the cathodes of the LEDs.

The C_3 carry-out connection from the adder circuit can be considered either an error indicator or a fifth bit in the sum display. If the user sets all eight switches to logic 1 (15 plus 15), the five-bit output will show 11110, where the left-hand bit is the C_3 output and the right-hand bit is S_0. That particular 5-bit binary number is equivalent to decimal 30.

If, however, the C_3 output is considered an error or overflow indicator, it lights up whenever the sum exceeds decimal 15 (1111 with the left-hand bit being S_3). Under these circumstances, setting $A = B = 1111$ yields a sum of 1110 and lights the C_3 "error" lamp. The 4-bit answer is shown as decimal 14, which, of course, is not the correct sum of 15 plus 15. The fact that the C_3 lamp is lighted indicates the displayed sum is incorrect.

The size of the answer thus depends on whether the C_3 output is used as a fifth bit in the sum or an error indicator. If C_3 is used as the fifth bit in the sum, the circuit in Fig. 16-5 is capable of summing any two 4-bit numbers whose sum does not exceed decimal 30. If the C_3 output is used as an error indicator, the circuit can sum any two 4-bit numbers whose sum does not exceed decimal 15.

16-1.3 An Adder with an Accumulating Register

The 4-bit adder described in connection with the circuit in Fig. 16-5 can sum only two numbers. If an arithmetic task calls for summing any more than two numbers, the circuit can't handle it. The sum of the first two numbers will indeed appear at the display output, but there is no way to enter a third number without destroying the results of summing the first two.

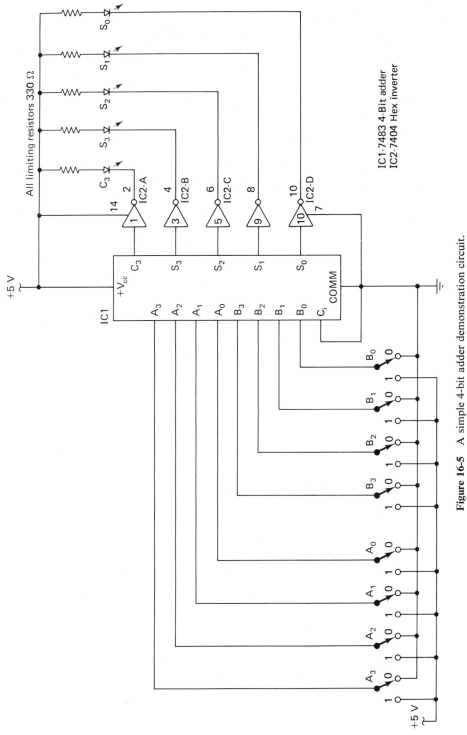

Figure 16-5 A simple 4-bit adder demonstration circuit.

The 4-bit adder in Fig. 16-6, however, is capable of adding together any number of 4-bit numbers, at least as long as the sum never exceeds the numeric capacity of the display assembly. The key to adding a string of two or more numbers is to accumulate the results in a register and then sum the contents of the register with the next number. The user enters the numbers one at a time, and their sums accumulate in the register and appear as an accumulating sum at the output display assembly.

Note in the circuit in Fig. 16-6 that the user has access to only one set of input switches. In this particular instance, these switches represent the "A"-number input to the adder. The summing outputs of the adder (terminals S_0 through S_3 of IC2) go to the parallel inputs of a parallel-in/parallel-out shift register. This shift register acts as the accumulating register, and its outputs (Q_A through Q_D of IC3) go both to the display assembly and to the "B"-number inputs of the adder circuit. Whatever number is stored in the register is thus automatically summed with whatever number the user enters via the "A"-number input switches.

To see how this accumulating-register adder works, begin by depressing the CLR push button. Depressing this button pulls the CLR input of the shift register, IC3, down to logic 0; as a result, all four register outputs are cleared to 0. All LEDs go to their off state, and the "B"-number inputs of the adder see all 0's. That operation simply clears the system to 0, performing the same function as a CLR key on an electronic calculator.

Next, the user sets the first number of the problem at the A switches. The adder circuit, IC2, immediately adds this number to the 0's from the output of the register; but the register itself cannot respond to this new figure until it is clocked.

As far as the user is concerned, the next step is to momentarily depress the PLUS key. Depressing this key initiates an 11-ms switch-debouncing interval that is set by the values of R_3 and C_3, which are the components of a 555-type monostable multivibrator. The OUT terminal of the timer, IC1, is an 11-ms positive pulse that is inverted by IC4-E and applied to the CLK input terminal of the shift register. The trailing edge of that CLK pulse is positive going; and since this particular shift register is clocked on positive-going edges, it follows that the sum outputs of the adder will appear at the Q outputs of the register about 11 ms after the user depresses the PLUS button.

To this point, then, the display should be showing the binary equivalent of the number the user just entered. But what is equally important is the fact that the "B"-number inputs are equal to that number the user entered and are no longer all 0's. Therefore, when the user sets the next number on the A switches and depresses the PLUS key, the shift register will respond to the sum of the previously determined sum and the new number.

The sum accumulates in the shift register and is added to any new number entered from switches A. In this fashion, the user can sum any number of 4-bit numbers, provided, of course, the accumulated sum doesn't exceed the bit capacity of the display.

It is rather easy to expand the circuit in Fig. 16-6 to handle larger sums. It is a matter of cascading adders, shift registers and the display assembly.

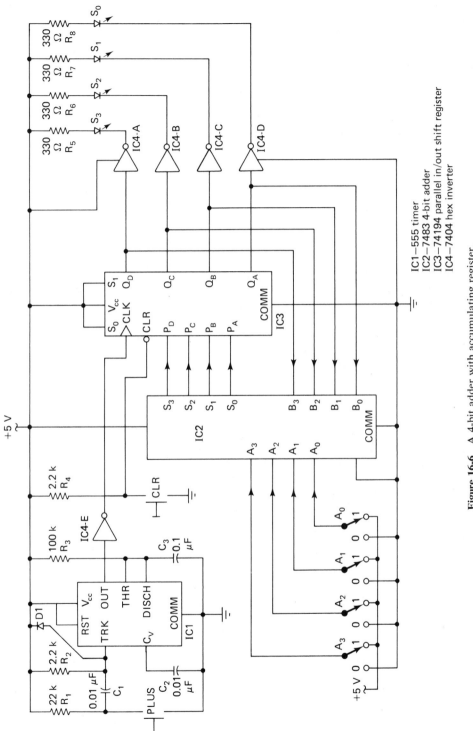

Figure 16-6 A 4-bit adder with accumulating register.

IC1—555 timer
IC2—7483 4-bit adder
IC3—74194 parallel in/out shift register
IC4—7404 hex inverter

In the context of electronic calculators, the "A" numbers arrive from the keyboard and the summing operation is initiated by depressing a "$+$" key. The adder itself is buried deep within the calculator chip, but the shift register serves as both an accumulating register and display register.

16-2 Adding Signed Numbers

Any arithmetic operation that is commonly called "subtraction" can be carried out by means of adding signed numbers. Instead of devising a whole new digital circuit for doing subtraction, it is far more economical to convert any so-called subtraction operation to an addition format and carry out those operations with the same adder circuit described in Sec. 16-1.

Suppose, for instance, it is necessary to work out the following arithmetic problem: $5 - 2 = ?$ To use an adder for working this particular problem, the problem can be restated as $+5 + (-2) = ?$ *Any subtraction operation can be reduced to one of summing signed numbers.*

In order to add signed numbers in a digital format, however, it is necessary to work out a scheme for indicating positive and negative numbers. By convention, positive and negative binary numbers are distinguished by means of a leading *sign bit*. This means adding an extra bit to the number, and that sign bit is usually added just to the left of the MSB in the number. The sign bit is logic 0 for positive numbers and logic 1 for negative numbers.

Thus any 4-bit number can be expressed as positive or negative by having a fifth bit at the MSB end. If that fifth bit is a 1, the 4-bit numeric value (or *mantissa*) is negative. The mantissa represents a positive number if the sign bit is 0.

So signed binary numbers must carry a sign bit to indicate whether they are negative or positive; but there is more to the scheme.

Negative numbers must be expressed in a complement form. It is not sufficient to merely attach a sign bit of logic 1 to a mantissa. A digital negative number must both carry a sign bit of 1 and be expressed in a complemented form.

16-2.1 Two Different Complement Forms

There are two commonly used complement forms of binary numbers: 1's complements and 2's complements. A 1's complement number is simply a binary number that has each of its bits complemented or inverted. The 1's complement of 0110, for example, is 1001. And as another example, the 1's complement of 0000 is 1111.

The 2's complement of a binary number can be determined by first working out the 1's complement and then adding binary 1 to the result. As an example, determine the 2's complement form of 1001. The 1's complement is 0110, and adding a binary 1 to the LSB position results in the binary number 0111. Thus the 2's complement of 1001 is 0111.

What is the 2's complement of 0000? First find the 1's complement, 1111; and then add 1 to the LSB of the result: $1111 + 1 = 10000$.

Now test a vital axiom of complement-form binary numbers by complementing a number that has been previously complemented. The 1's complement of 0011, for example, is 1100. Complementing that result again yields 0011, the original number.

Try double-complementing a number in 2's complement form. Suppose the original binary number is 0011 again. The 2's complement of this number is 1101. Complementing that number again yields 0011, again, the original number.

It is thus possible to "uncomplement" a number by complementing it. If the number is complemented originally by the 1's complement technique, it can be "uncomplemented" only by that same technique. Similarly, a number in 2's complement form can be properly "uncomplemented" only by the 2's complement method.

There is no way to tell whether a complemented number is expressed in 1's or 2's complement form simply by looking at that number. The 1's or 2's complement form must be clearly specified or at least understood from some preceding operations.

16-2.2 Signed and Complemented Binary Numbers

Negative binary numbers must be expressed with a sign bit of 1 and in one of the two complement forms. And reading that statement in reverse: Any binary number with a sign bit of 1 and expressed in complement form represents a negative binary number.

For the sake of clarity, the sign bit is often set off from the mantissa by a dot ($\cdot$) symbol, e.g., $1 \cdot 1100$. The sign bit in this particular example is 1, and the mantissa is 1100. A different binary number such as $0 \cdot 0011$ has a sign bit of logic 0, indicating the mantissa represents a positive binary number.

Now consider a few simple exercises: First express the decimal number -4 in 1's complement form. The number could be written $-0 \cdot 0100$, but adders do not recognize the minus sign. The proper complement form, then, is $1 \cdot 1011$. Note that the sign bit was complemented, too. That's a convenient feature, because now the symbols $1 \cdot 1011$ indicate (1) the number is in complement form and (2) it represents a negative number.

What is the 2's complement of -4? First express the number as -0.0100, then find its 1's complement, $1 \cdot 1011$ and finally add 1 to the LSB to get the 2's complement number $1 \cdot 1100$—a negative and complemented number representing decimal -4.

16-2.3 Adding Signed Numbers in 1's Complement Form

Signed and complemented binary numbers are summed in the same way as all-positive numbers. When summing signed numbers, however, it is also important to sum the sign bits; and what is done with any carry from the sign-bit position represents the difference between adding signed numbers in 1's and 2's complement form.

When summing numbers in 1's complement form, the carry bit from the sign-bit position is carried back to the LSB position of the mantissa. This process of adding any carry-out bit from the sign-bit position is called *end-around carry*. See the two examples in Fig. 16-7.

The problem in Fig. 16-7(a) involves summing two positive numbers; therefore, the sign bits in both instances are 0 and there is no need to consider complement forms. The end-around carry in this instance has a value of 0; therefore, it does not affect the result of the first summing operation.

In Fig. 16-7(b) the basic situation is one involving the addition of a positive and a negative number. The negative number (-2 in this instance) must be converted to a 1's complement form before it is added to $+8$. The steps shown in Fig. 16-7(b) show that the 1's complement of -2 is 1   1101. Adding these two numbers,

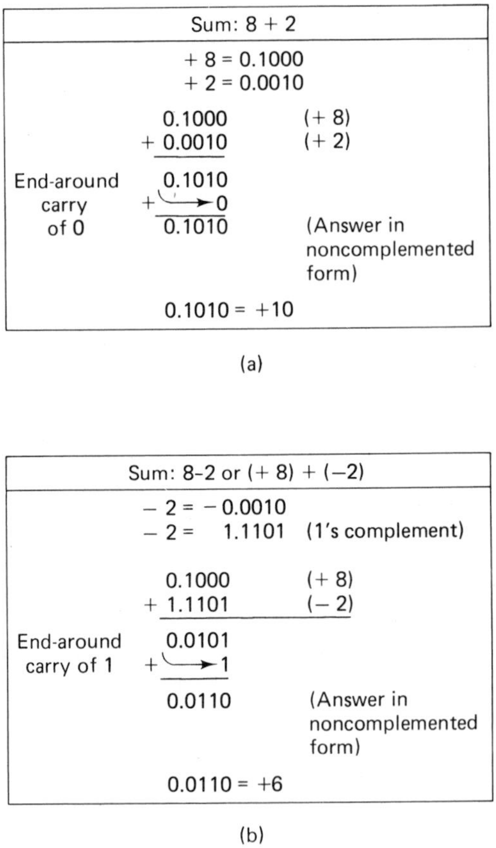

Sum: 8 + 2
$+8 = 0.1000$
$+2 = 0.0010$
0.1000   (+ 8)
+ 0.0010   (+ 2)
End-around carry of 0   0.1010
+ ⟶ 0
0.1010   (Answer in noncomplemented form)
0.1010 = +10

(a)

Sum: 8–2 or (+ 8) + (−2)
$-2 = -0.0010$
$-2 = $ 1.1101   (1's complement)
0.1000   (+ 8)
+ 1.1101   (− 2)
End-around carry of 1   0.0101
+ ⟶ 1
0.0110   (Answer in noncomplemented form)
0.0110 = +6

(b)

Figure 16-7 Examples of summing signed numbers in 1's complement form. (a) Adding two positive numbers. (b) Adding numbers with mixed signs, where the sum is positive.

```
┌─────────────────────────────────────────────┐
│         Sum: − 8  + 2 or (− 8) + (+ 2)        │
├─────────────────────────────────────────────┤
│        − 8 = − 0.1000                         │
│        − 8 =    1.0111   (1's complement)     │
│                                               │
│          0.0010         (+ 2)                 │
│        + 1.0111         (− 8)                 │
│          ─────                                │
│          1.1001                               │
│        +  ╰───→ 0                             │
│          ─────                                │
│          1.1001         (Answer in 1's        │
│                          complement           │
│                          form)                │
│                                               │
│        1.1001 = − 0.0110                       │
│                          (1's complement      │
│                          of answer)           │
│               = − 6     (True form of         │
│                          answer)              │
└─────────────────────────────────────────────┘
```

(c)

```
┌─────────────────────────────────────────────┐
│         Sum: − 8 − 2 or (−8) + (−2)           │
├─────────────────────────────────────────────┤
│        − 8 = 1.0111     (1's complement)      │
│        − 2 = 1.1101     (1's complement)      │
│                                               │
│          1.0111         (− 8)                 │
│        + 1.1101         (− 2)                 │
│          ─────                                │
│          1.0100                               │
│        +  ╰───→ 1                             │
│          ─────                                │
│          1.0101         (Answer in 1's        │
│                          complement           │
│                          form)                │
│                                               │
│        1.0101 = − 0.1010                       │
│                          (1's complement       │
│                          of answer)           │
│               = − 10    (True form of         │
│                          answer)              │
└─────────────────────────────────────────────┘
```

(d)

Figure 16-7 (*cont.*) (c) Adding numbers with mixed signs, where the sum is negative. (d) Adding two negative numbers.

including the sign bits, results in an end-around carry value of 1. Adding that value to the LSB of the mantissa yields the answer 0   0110. And since the sign bit is 0, we know that answer represents a positive number in a noncomplemented form. In this particular case, it is +6.

The example in Fig. 16-7(c) involves summing a larger negative number with a smaller positive number. The negative number must be converted to 1's complement form before it can be properly summed with the positive number. The 1's complement form of -8 is $1 \cdot 0111$; and when that is added to $0 \cdot 0010$, the result is $1 \cdot 1001$. The sign bit is 1, indicating that the answer is a negative number and that the mantissa is in a 1's complement form. Evaluating the result is thus a matter of "uncomplementing" it. As shown in the example, $1 \cdot 1001 = -0 \cdot 0110$, or decimal -6.

Finally, there is often a need to sum two negative numbers. In Fig. 16-7(d) both numbers must be complemented before they can be summed in an adder circuit. The sum of these two particular complemented numbers is $1 \cdot 0101$ after doing the end-around carry of 1.

That result is a negative number in complemented form, but it can be translated into a more meaningful decimal number by complementing it to $-0 \cdot 1010$, or -10.

16-2.4 Adding Signed Numbers in 2's Complement Form

The procedure for summing signed numbers in 1's complement form includes an end-around carry—adding any bit carried out of the sign-bit position to the LSB of the mantissa. This end-around carry is not necessary when summing signed numbers expressed in 2's complement form. In fact, the carry-out bit from the sign-bit sum is ignored altogether.

As in the case of summing signed numbers in 1's complement form, any sum from a 2's complement summing operation must be complemented again if the resulting sign bit is equal to 1. A number with a sign bit of 1 (whether resulting from 1's or 2's complement summing) is a negative and complemented number. And in the case of 2's complement summing, any result having a sign bit of 1 must be complemented by the 2's complement procedure to determine the true, noncomplemented value.

See the examples of 2's complement summing in Fig. 16-8 and then compare the results and procedures with the 1's complement examples in Fig. 16-7.

16-2.5 Implementing Addition of Signed Numbers

Figure 16-9 shows a rather simple and straightforward technique for generating the 1's complement of any 4-bit binary number. The scheme takes advantage of the fact that an EXCLUSIVE-OR gate can be programmed so that it either inverts or non-inverts incoming logic levels. (See Sec. 15-1.)

The input sign bit in this instance controls the operation of the quad EXCLUSIVE-OR function. If the sign bit (SB_{in}) is logic 0, the four bits of the A word emerge non-inverted, thus representing a positive binary number. If the sign bit is set to logic 1, however, the outputs of the EXCLUSIVE-OR function equal the 1's complement of the A inputs.

Thus the data emerging from the circuit in Fig. 16-9 is either noncomplemented or in a 1's complement form, depending on the value of the input sign bit.

Now the accumulating adder circuit in Fig. 16-6 can be extended to handle signed numbers in either 1's or 2's complement form. The circuit, in other words, can be modified to perform subtraction as well as summing operations.

Figure 16-10(a) shows how the EXCLUSIVE-OR, 1's complement generator can be inserted between the *A*-data inputs and the corresponding inputs of the 7483 4-bit adder IC. The scheme in this case works according to 1's complement addition, with the end-around carrying operation taking place between the carry-out connection of a sign-bit adder and the carry-in terminal of the adder.

The 4-bit data appearing at the *A* inputs of the adder circuit in Fig. 16-10(a) represents either a noncomplemented or 1's complement of the data from the input switches. The end-around carry bit to C_i completes the 1's complement adding format.

The circuit in Fig. 16-10(b) is a 2's complement adder. Note that the same 1's-

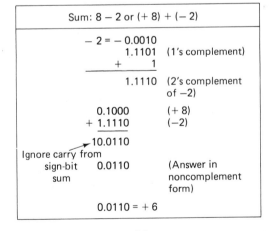

Figure 16-8 Examples of summing signed numbers in 2's complement form. (a) Adding two positive numbers. (b) Adding numbers with mixed signs, where the sum is positive.

Sum: $-8 + 2$ or $(-8) + (+2)$
$-8 = -0.1000$

$$
\begin{array}{ll}
1.0111 & \text{(1's complement)} \\
+1 & \\
\hline
1.1000 & \text{(2's complement of -8)} \\
\end{array}
$$

$$
\begin{array}{ll}
1.1000 & (-8) \\
+\,0.0010 & (+2) \\
\hline
1.1010 & \text{(Answer in 2's complement form)} \\
\end{array}
$$

$$
\begin{array}{ll}
-\,0.0101 & \text{(1's complement answer)} \\
+1 & \\
\hline
-\,0.0110 & \text{(Answer in noncomplemented form)} \\
\end{array}
$$

$$-0.0110 = -6$$

(c)

Sum: $-8 - 2$ or $(-8) + (-2)$

$$
\begin{array}{ll}
-8 = 1.1000 & \text{(2's complement)} \\
-2 = 1.1110 & \text{(2's complement)} \\
\end{array}
$$

$$
\begin{array}{ll}
1.1000 & (-8) \\
+\,1.1110 & (-2) \\
\hline
\end{array}
$$

$$\text{Ignore} \longrightarrow 11.0110$$

$$
\begin{array}{ll}
1.0110 & \text{(Answer in 2's complement form)} \\
\end{array}
$$

$$
\begin{array}{ll}
-\,0.1001 & \text{(1's complement of answer)} \\
+1 & \\
\hline
-\,0.1010 & \text{(Answer in noncomplement form)} \\
\end{array}
$$

$$-0.1010 = -10$$

(d)

Figure 16-8 (*cont.*) (c) Adding numbers with mixed signs, where the sum is negative. (d) Adding two negative numbers.

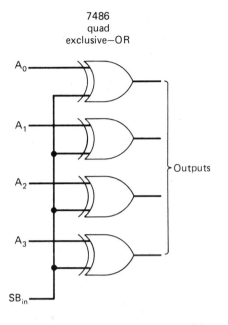

7486
quad
exclusive—OR

A_0

A_1

Outputs

A_2

A_3

SB_{in}

SB = 1, outputs are 1's complement of A
SB = 0, outputs = A

Figure 16-9   Using a quad EXCLUSIVE-OR function as a complement/noncomplement generator (1's complement format).

complement generator scheme is used between the *A* input switches and the *A* inputs of the adder circuit. Instead of using an end-around carry, however, a 1 is added to the *A* inputs whenever the sign-bit input (SB_{in}) is at logic one. Adding the 1 to generate the 2's complement is a simple matter of entering a logic 1 at the carry-input connection (C_i) of the adder circuit.

Either circuit in Fig. 16-10 can be interfaced directly with the accumulating adder in Fig. 16-6. The 2's complement scheme, however, has the advantage of not requiring a sign-bit summing circuit to generate the end-around carry bit.

16-3   Introduction to the 74181 ALU

The powerful TTL IC family includes a 74181 arithmetic logic unit (ALU). Until the advent of more efficient microprocessor and LSI calculator ICs, an ALU was at the heart of virtually every electronic calculating device.

If one does not have a thorough understanding of computer-oriented logic and arithmetic operations, it is rather difficult to appreciate the real power of an ALU device. Since the more sophisticated operations are beyond the scope of this text, we will have to be content with investigating some of the more elementary operations the 74181 can perform.

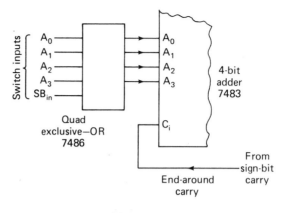

(a)

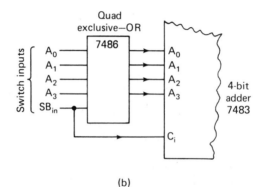

(b)

Figure 16-10 Complement generator as used with the accumulating adder circuit in Fig. 16-6. (a) 1's complement format. (b) 2's complement format.

Figure 16-11 shows the pin designations for the 4-bit 74181 ALU. The circuit accepts two sets of 4-bit words or numbers, A_0 through A_3 and B_0 through B_3. These inputs are treated much as the inputs to a 4-bit adder IC.

The system can be set up for 16 different kinds of operations, depending on the 4-bit code applied to the select inputs, S_0 through S_3. And it so happens the range of possible operations on words A and B can be tripled by applying various combinations of 1's and 0's to the MODE (M) and CARRY IN $(\overline{C_{in}})$ inputs.

The primary outputs of the circuit are the F terminals, F_0 through F_1. If 4-bit words A and B are to be summed, for example, the mantissa of the result would appear at the F outputs.

The $A = B$ output is simply a special EQUALITY function output. The three carry-oriented outputs $(P, \overline{C_n + 4}$ and $G)$ are used primarily for cascading ALUs and interfacing them with look-ahead carry functions.

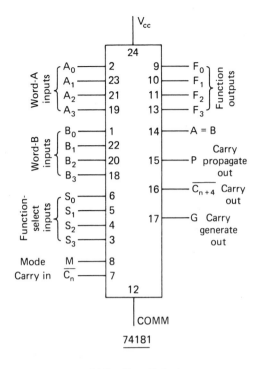

Figure 16-11   The 74181 ALU IC.

The function table for the 74181 in Fig. 16-12 summarizes about half the logic and arithmetic operations possible with this ALU device. Note that the operations being performed at any given moment depend on the 4-bit binary word applied to the S inputs and the status of the M and $\overline{C_n}$ inputs.

To get a general impression of the power of an ALU, suppose input M is fixed at logic 1. This condition applies to the first column of logic functions in Fig. 16-12. If the SELECT inputs are set to 0000, the table shows that $F = \overline{A}$. In other words, the F outputs are an inverted version of the A inputs—F is the 1's complement of A. If, however, the SELECT inputs are set to 1111 as in the last line of the truth table, $F = A$, the F outputs are identical to the A inputs.

The same two operations can be performed on the B inputs by selecting operation 0101 and 1010, respectively.

Also note that $F = 0$ (the F outputs are all 0's) at operation 0011, but the F outputs change to 1111 when the select inputs are at 1100.

An operation that results in an expression such as $F = AB$ requires some explanation. This is, indeed, the familiar AND operation, but it applies here to two

Select inputs	M = 1	M = 0		
S_3 S_2 S_1 S_0	$\overline{C}_n$ = X	$\overline{C}_n$ = 1		$\overline{C}_n$ = 0
0 0 0 0	F = $\overline{A}$	F = A		F = A plus 1
0 0 0 1	F = $\overline{A + B}$	F = A + B		F = (A + B) plus 1
0 0 1 0	F = $\overline{A}B$	F = A + $\overline{B}$		F = (A + $\overline{B}$) plus 1
0 0 1 1	F = 0	F = minus 1 (2's complement)		F = zero
0 1 0 0	F = $\overline{AB}$	F = A plus A$\overline{B}$		F = A plus A$\overline{B}$ plus 1
0 1 0 1	F = $\overline{B}$	F = (A + B) plus A$\overline{B}$		F = (A + B) plus A$\overline{B}$ plus 1
0 1 1 0	F = A $\oplus$ B	F = A minus B minus 1		F = A minus B
0 1 1 1	F = A$\overline{B}$	F = A$\overline{B}$ minus 1		F = A$\overline{B}$
1 0 0 0	F = $\overline{A}$ + B	F = A plus AB		F = A plus AB plus 1
1 0 0 1	F = $\overline{A \oplus B}$	F = A plus B		F = A plus B plus 1
1 0 1 0	F = B	F = (A + $\overline{B}$) plus AB		F = (A + $\overline{B}$) plus AB plus 1
1 0 1 1	F = AB	F = AB minus 1		F = AB
1 1 0 0	F = 1	F = A plus A		F = A plus A plus 1
1 1 0 1	F = A + $\overline{B}$	F = (A + B) plus A		F = (A + B) plus A plus 1
1 1 1 0	F = A + B	F = (A + $\overline{B}$) plus A		F = (A + $\overline{B}$) plus A plus 1
1 1 1 1	F = A	F = A minus 1		F = $\overline{A}$

Figure 16-12   Active-high function table for the 74181 ALU.

different 4-bit words instead of to a pair of 1-bit logic levels. In the context of 4-bit words, the expression $F = AB$ means that the corresponding elements of each word are AND-ed together: $F_0 = A_0 \cdot B_0$, $F_1 = A_1 \cdot B_1$, $F_2 = A_2 \cdot B_2$ and $F_3 = A_3 \cdot B_3$. The same general idea applies to all the other logic functions expressed in the $M = 1$ column of Fig. 16-12.

The $M = 1$ column, however, is merely the logic-function column. Note the sort of operations possible when $M = 0$. Most of the operations in the two $M = 0$ columns include an arithmetic function, but some are purely logic functions and many are combinations of logic and arithmetic functions. To prevent any confusion between the logic and arithmetic operations, the arithmetic operations are spelled out in plain text. The word *plus*, for instance, means the arithmetic operation of addition, while the $+$ operator stands for the logic OR operation.

There are two different columns of arithmetic/logic operations possible while $M = 0$. In one case, $\overline{C}_n = 1$; in the other case, $\overline{C}_n = 0$.

It should be pointed out again that the value or purpose of some of these functions is rather obscure to anyone who has not completed an in-depth study of binary arithmetic. The remaining discussion in this section thus points out some of the functions that can be considered apparent and useful in the light of the logic and arithmetic presented in this book.

In the column in which $M = 0$ and $C_n = 1$, the first three lines are rather obvious in their meaning. They are basic logic operations where $F = A$, $F = A + B$ and $F = A + \overline{B}$.

The fourth line, where the S inputs are set at binary 3, 0011, the F output generates the 2's complement form of -1. And the line where the S inputs equal 1001 causes the F outputs to be equal to the arithmetic sum of A and B.

Most of the other functions in the $M = 0/C_n = 1$ column do not have obvious applications as far as this chapter is concerned. One of these is worthy of special note, however: Note the line where the S inputs are 1100 and $F = A$ plus A.

Any binary number added to itself yields a sum that looks like the original number that has been shifted one bit toward the MSB position. As an example, suppose $A = 0111$. A plus A in this case equals 1110; the 1 in the LSB position of the original number has shifted to the left one position, leaving a 0 behind. The leading 0 in the original number is lost in the process.

The operation just described should look familiar. It looks very much like a serial-shift register. So if the result of the first sum is then summed with itself, the number shifts yet one more position to the left. This ALU can thus perform the function of a serial-shift register.

The third column in Fig. 16-12, where $M = 0$ and $C_n = 0$, also includes some interesting features. What is the value of the first line where the S inputs are 0000 and $F = A$ plus 1? Adding 1 to a number effectively increments it one count. Performing the same operation on the result increments the number yet another count. Here we have the effect of a binary up counter. The A minus 1 function in the last line of the $M = 0$, $C_n = 1$ column, incidentally, can be used for binary down counting.

A whole new table of functions can be generated if the A and B inputs are put into an active-low format. The revised table can be logically derived by inverting the A's and B's appearing in the table shown in Fig. 16-12. Students interested in pursuing the operation of ALU devices should consult the manufacturers' TTL data books.

Exercises

1. What is the main difference between the operational capabilities of the half-adder and full-adder circuits?

2. Add the following pairs of positive binary numbers:

   (a) 1101   (b) 1000   (c) 1111   (d) 1111
        0010   1010   0001   1111

3. What is the operational advantage of an adder circuit having an accumulating register as compared to an adder not having an accumulating register? (Compare the circuits in Figs. 16-5 and 16-6.)

4. Express the following numbers in binary 1's complement form and then in binary 2's complement form: (a) -2; (b) -9; (c) $+12$; (d) -1.

5. Assume the following binary numbers are in 1's complement form. Express them in signed, decimal form. (a) $1 \cdot 1000$; (b) $1 \cdot 1011$; (c) $1 \cdot 1010$; (d) $1 \cdot 1111$.

6. Assume the binary numbers in Ex. 5 are expressed in 2's complement form. Express them in signed, decimal form.

7. Add the following pairs of binary numbers; assume the complemented numbers are in the 1's complement format.

   (a) 0·1101   (b) 1·1000   (c) 0·1111   (d) 1·1001
        1·1101   1·1001   1·1111   1·1101

Express your results in both binary and decimal form.

8. Add the numbers in Ex. 7; assume the complemented numbers are in the 2's complement format. Express your results in both binary and decimal form.

ANSWERS TO SELECTED EXERCISES

Chapter 3

2. $(W + X)(Y + Z) = WY + WZ + XY + XZ.$
6. (a) 3-input NOR, $D = \overline{A + B + C}$;   (b) 3-input AND, $D = ABC$;
   (c) 3-input OR, $D = A + B + C$;   (d) 3-input OR, $D = A + B + C$.
7. See Fig. A-3.7.

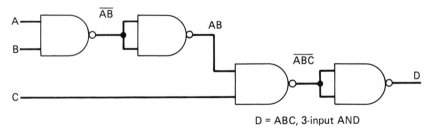

D = ABC, 3-input AND

(a)

Figure A-3.7

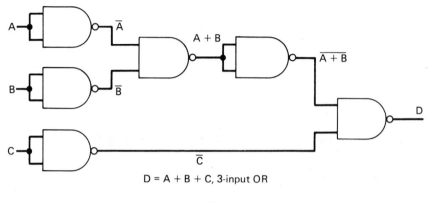

$D = A + B + C$, 3-input OR

(b)

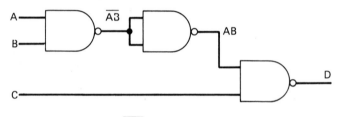

$D = \overline{ABC}$, 3-input NAND

(c)

Figure A-3.7 (*cont.*)

Chapter 4

2. (a) ST; (b) 0; (c) $A + \overline{B + C}$ or $A + \overline{B}\overline{C}$; (d) $S(T + \overline{U}) + T\overline{U}$;
   (e) $\overline{X}YZ\overline{W}$; (f) A.
3. (a) $A\overline{B}C\overline{D}$; (b) $\overline{A}B + \overline{C}$; (c) $E = \overline{A} + D, F = \overline{BD}, G = \overline{C} + D$;
   (d) $\overline{D}(AE + B\overline{E}) + CDE$.
4. See Fig. A-4.4.

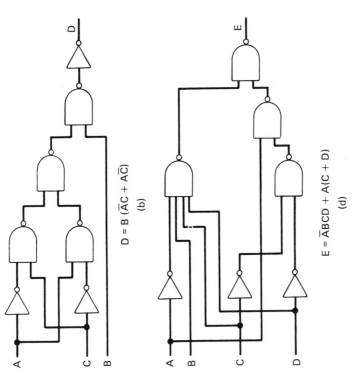

D = B̄ (ĀC + AC̄)
(b)

E = ĀBCD + A(C + D)
(d)

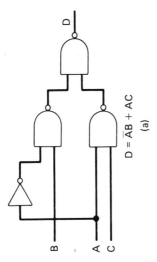

D = ĀB + AC
(a)

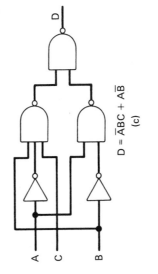

D = ĀBC + AB̄
(c)

Figure A-4.4

415

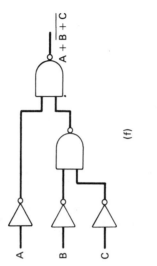

(f)

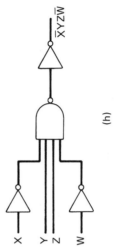

(h)

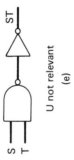

U not relevant

(e)

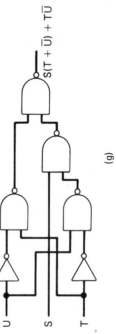

(g)

Figure A-4.4 *(cont.)*

Chapter 5

1. A switch-debouncing circuit eliminates the brief, rapid contact-bounce interval that always occurs whenever the contact components of a mechanical switch strike one another.
2. Noise margin is generally defined as the difference between the device's maximum-voltage definition of logic 0 and its minimum-voltage definition of logic 1, $V_{IH} - V_{IL}$. According to the data in Table 5-1, the noise margin for a typical TTL gate is 1.2 V. The noise margin for CMOS is about 2 V.
3. Connecting the discrete-component load from the TTL output to V_{cc} lets the IC's output drive the load with as much as 16 mA. A load connected to ground is limited to 400 μA of current from the IC.
4. Sink current flows into the output terminal of a digital device whenever the load is connected to the most positive voltage supply and the device is generating a logic-0 output. Source current flows out of the output terminal of the IC whenever it is loaded to common and the IC is generating a logic-1 output. As far as input sourcing and sinking are concerned, sink current flows into the device whenever it is seeing a logic-1 input; and source current flows out of the input is seeing a logic 0.
5. The LED in this instance ought to be connected from the output of the TTL device to $+V_{cc}$, thereby taking advantage of the higher-available amount of sink current. The LED thus lights when the output of the IC is at logic 0.
6. The V_{OH} minimum specification for TTL is below the V_{IH} minimum for CMOS.
7. The logic-0 output drive current for CMOS is barely adequate for operating the I_{IL} of TTL gates.

Chapter 6

1. See Fig. A-6.1. The invalid input transition is a simultaneous switching from $A = B = 1$ to $A = B = 0$.

A	B	C	D
0	0	C_{t-1}	D_{t-1}
0	1	0	1
1	0	1	0
1	1	0	0

Figure A-6.1

2. The memory mode for a simple R-S flip-flop only occurs when $S = R = 0$. The gated R-S flip-flop is in its memory mode anytime $G = 0$, regardless of the logic inputs at R and S.
3. $R = S = G = 1$ is an invalid input because it forces the Q and Q outputs to logic 1 at the same time. This condition destroys the complemented feature of the circuit's outputs; it makes the outputs logically inconsistent with the notation that implies complemented outputs under all normal operating conditions.
4. The Q output of a level-switched D flip-flop follows the D input as long as the gate input is at logic 1. The Q output of an edge-triggered D flip-flop responds to the D input only during the prescribed transition interval at the CLOCK input.

5. Recall from Ch. 5 that TTL circuits must be loaded from the output to $+V_{cc}$. This output configuration inverts the logic relationship between the circuit's actual output and the response of the lamps. The lamps must be reversed to "upright" the logic response.
6. $J_1 = J_2 = K_1 = K_2 = 1, \overline{J_3} = \overline{K_3} = 0.$

Chapter 7

1. The circuit in Fig. 7-1(b) quarantees a single output pulse in spite of any switching or power supply noise that might occur during the timing interval.
2. Setting input C to logic 0 stops the oscillator.
3. (a) See Fig. 7-6(a);   (b) See Fig. 7-6(b);   (c) See Fig. 7-12(a).
4. The output of the circuit in Fig. 7-8(a) must be manually reset to 0. The output of the circuit in Fig. 7-8(b) is automatically reset to 0 at the end of the timing interval for monostable IC1-B.
5. Increasing the value of R_2 increases the delay interval.
6. Exactly two-thirds or about 67 percent.
7. Setting the CLR input of either section would stop the oscillator. Driving one of the CLR inputs with a waveform from an astable multivibrator would create the effect of a tone-burst generator.

Chapter 8

1. The modulus of a counter is a number that expresses the number of different output states. The modulus of a single flip-flop is 2. The modulus of a BCD counter is 10. The modulus of a 5-bit binary counter is 32. Eight flip-flops are required for building a modulo-256 counter.
2. LED indicators having their anodes connected to $+V_{cc}$ effectively invert the logic level presented to their cathodes. By connecting the LEDs to the Q outputs of each flip-flop, they light whenever the Q output equals logic 1, and they go out whenever the corresponding Q output equals logic 0.
3. Using the clearing feature both stops the count and resets the outputs to 0. The enable feature stops the count, too, but it does not reset the outputs to 0; the outputs hold the state they had just prior to disabling the counter.
4. Up counting always occurs whenever the observed outputs and the clocking signals for each flip-flop are taken from the same point—the outputs and cascading clock signals are in phase. Down counting always occurs whenever the observed and clocking signals for each flop-flop are complements of one another—out of phase.
5. The largest decimal count possible with three cascaded BCD counters is 999.
6. This counter will automatically reset to 0 the instant it attempts to show binary 45 (101101). This is thus a modulo-45 counter.
7. The carry output of a counter is active when the count reaches the largest possible number (binary 15 for a 4-bit binary counter, for instance). The borrow output is active whenever the counter is showing a 0 count.
8. Synchronous counters can be operated at higher frequencies than ripple counters can.

Chapter 9

1. Apply the trigger input to CLK B; connect the B and D outputs to R_{01} and R_{02} inputs.
2. Connect the circuit as described in the answer to question 1 and then connect the D output also to CLK A. Output A is then the symmetrical divide-by-10 output.
3. Invert the outputs by means of external logic inverters.
4. It is used for dividing 120-Hz, full-wave rectified utility line frequency to 10 Hz for digital clock applications. Used as a modulo-6 counter, it serves as the 10's counter for minutes and seconds in a digital clock application.
5. The D output is symmetrical; the counter's decimal sequence skips decimal 6 and 7.
6. Both functions stop the counting action of the IC. The CLEAR function, however, resets the outputs to 0 and the LOAD function loads the outputs with any valid binary number appearing at the P inputs.
7. Fully synchronous, cascaded counters are distinguished by (1) a common CLOCK bus and (2) a ripple-carry type connection from one stage to the next.
8. The circuit in Fig. 9-14(a) uses a TTL circuit that must be interfaced with LEDs in an active-low logic format. The circuit in Fig. 9-15(a), however, uses a CMOS circuit that can drive light loads to ground (active-high format). Compare the drive characteristics of TTL and CMOS circuits as described in Ch. 5.

Chapter 10

1. $0 = 0000$ $1 = 0001$ $2 = 0010$ $3 = 0011$ $4 = 0100$
   $5 = 0101$ $6 = 0110$ $7 = 0111$ $8 = 1000$ $9 = 1001$
2. See Fig. A-10.2.
3. The current-limiting resistors are necessary for limiting the LED current and letting the LEDs clamp their maximum forward voltage at about 1.7 V.
4. (a) 1100;   (b) 1111000;   (c) 111000;   (d) 10011010010.
5. (a) 9, 1001;   (b) 26, 0010 0110;   (c) 42, 0100 0010;
   (d) 29, 0010 1001.
6. (a) 10011;   (b) 11011;   (c) 1001001100.
7. (a) 11;   (b) 64;   (c) 7122.
8. (a) 01010, 10;   (b) 10010, 18;   (c) 1011111010, 752;
   (d) 110111011101, 3549.

Chapter 11

1. An event counter accumulates a count over some unspecified period of time, ultimately showing the total number of events that occur. A frequency counter accumulates the number of events that occur within a fixed and precise interval of time, usually latching and updating the reading at regular intervals.
2. The 2-phase latch-and-clear operation is absolutely essential for frequency counters that feature automatic updating. The latch-and-clear operation provides a clear

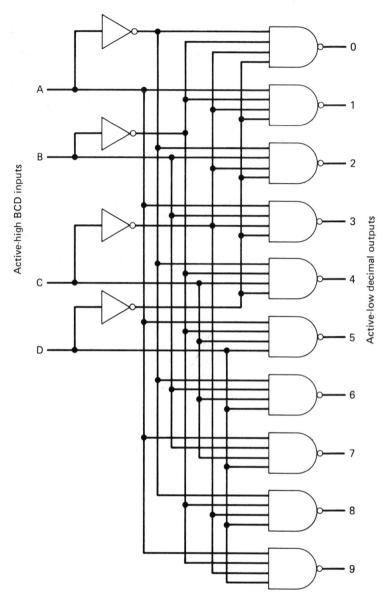

Figure A-10.2

output display, even though the counter is running between successive latch-and-clear operations.

3. A digital clock accumulates a count of events, namely, the number of 1-s or 1-min clock intervals that occur within a 12-hour period. It is thus more like an event counter than a frequency counter.

4. A digital tachometer *is* a frequency counter that has been scaled to read out revolutions per minute.

5. An analog voltage level can be applied to a VCO, and the output of the VCD can be applied to a frequency counter. The result would be a digital number proportional to the analog input voltage level.

6. A thermocouple generates a voltage proportional to the difference in temperature between its two ends; it is thus used with a VCO/frequency-counter combination. A thermistor is a resistance-changing sensor; therefore, it must be used in an RC oscillator that operates a pulse-duration counter.

Chapter 12

3. $I = (A + B)(C + D)(E + F)(G + H)$.
4. See Fig. A-12.4.

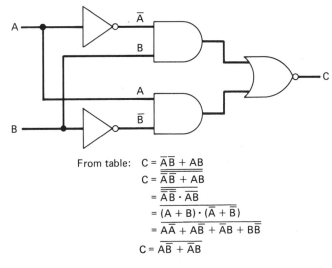

From table: $C = \overline{A}\overline{B} + AB$
$$C = \overline{\overline{\overline{A}\overline{B} + AB}}$$
$$= \overline{\overline{\overline{A}\overline{B}} \cdot \overline{AB}}$$
$$= \overline{(A + B) \cdot (\overline{A} + \overline{B})}$$
$$= \overline{A\overline{A} + A\overline{B} + \overline{A}B + B\overline{B}}$$
$$C = \overline{A\overline{B} + \overline{A}B}$$

Figure A-12.4

5. Four.

6. See Fig. A-12.6.

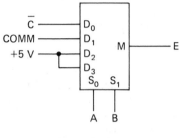

Figure A-12.6

7. The 7442 BCD-to-decimal decoder *is* a 1:10 demultiplexer used as a 4-line to 10-line decoder.

8. The circuit in Fig. 12-16(b) can be called a quad 4:1 demultiplexer.

9. The circuits in Figs. 12-10 and 12-16(b) are compatible in all respects *except* one: The circuit in Fig. 12-10 has an 8-word multiplexing capacity, whereas the demultiplexer in Fig. 12-16(b) has only a 4-word capacity.

10. The maximum display current for a multiplexed display is always equal to the maximum current rating of a single display unit. The maximum display current in a non-multiplexed display is equal to the product of the maximum current rating of one unit times the number of display units.

Chapter 13

5. The register must be clocked at least eight times to ensure loading it with all 0's, regardless of the bit pattern originally stored in it.

7. Serial shift data into the register in one direction and retrieve it by shifting in the opposite direction.

8. Serial-shift registers can be cascaded by connecting the serial output of the first unit to the serial input of the second and paralleling all other control connections such as CLK, CLR, etc.

9. Connecting the output of a serial-shift register to its own serial input allows data to circulate continuously. Propagating a single "1" bit through the system this way makes up a device commonly known as a ring counter.

10. See Fig. A-13.10.

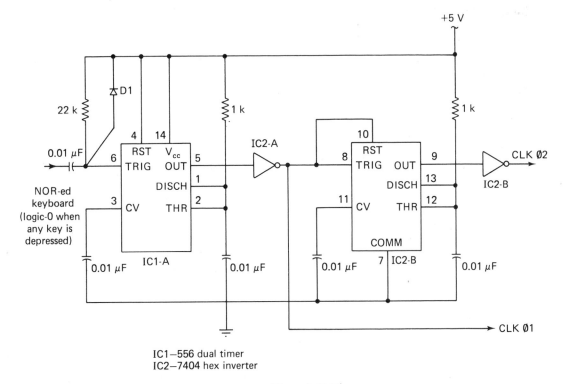

IC1—556 dual timer
IC2—7404 hex inverter

Figure A-13.10

Chapter 14

1. (a) Three address inputs; eight data inputs and outputs; two 7489 ICs.
   (b) Three address inputs; thirty-two data inputs and outputs; four 7489 ICs.
   (c) Eight address inputs; eight data inputs and outputs; thirty-two 7489 ICs.
   (d) Ten address inputs; eight data inputs and outputs; one hundred twenty-eight 7489 ICs.

2. The ROM can read only, but the RAM can both read and write.

3. Disabling the memory during addressing avoids the possibility of writing data into unwanted locations.

4. The data stored in a ROM remains intact, regardless of the status of the power supply. All stored RAM data is lost, however, the moment the power supply is interrupted. Also, the access time for ROM devices tends to be shorter than that of RAM devices of a comparable bit size.

Chapter 15

3. (a) $A > B$;   (b) $A < B$;   (c) $A < B$;   (d) $A > B$.
4. (a) 1;   (b) 1;   (c) 1;   (d) 0.

Chapter 16

1. A half-adder is incapable of accepting a carry-in bit from a previous adder stage. A full-adder circuit can accept a carry-in bit.
2. (a) 1111;   (b) 10010;   (c) 10000;   (d) 11110.
3. An adder circuit with an accumulating register can accept any number of digits to be summed in a serial fashion. There is no limit to the number of numbers that can be summed (as long as the result does not overflow the system's bit capacity).
4. 1's complement form: (a) 1·1101;   (b) 1·0110;   (c) 0·1100;
   (d) 1·1110.
   2's complement form: (a) 1·1110;   (b) 1·0111;   (c) 0·1100;
   (d) 1·1111.
5. (a) -7;   (b) -4;   (c) -6;   (d) 0.
6. (a) -8;   (b) -5;   (c) -7;   (d) -1.
7. (a) 0·1011, $+11$;   (b) 1·0010, -13;   (c) 0·1111, $+15$;
   (d) 1·0111, -8.
8. (a) 0·1010, $+10$;   (b) 1·0001, -15;   (c) 0·1110, $+14$;
   (d) 1·0110, -10.

INDEX